The near-Earth and interplanetary plasma

Volume 2

Plasma flow, plasma waves and oscillations

The near-Earth and interplanetary plasma

Volume 2
Plasma flow, plasma waves and oscillations

YA. L. AL'PERT

Formerly Deputy Director, Institute of Terrestrial Magnetism, Ionosphere and Propagation of Radio Waves (IZMIRAN) of the Academy of Sciences of the USSR

CAMBRIDGE UNIVERSITY PRESS

CAMBRIDGE
LONDON NEW YORK NEW ROCHELLE
MELBOURNE SYDNEY

Published by the Press Syndicate of the University of Cambridge
The Pitt Building, Trumpington Street, Cambridge CB2 1RP
32 East 57th Street, New York, NY 10022, USA
296 Beaconsfield Parade, Middle Park, Melbourne 3206, Australia

© Cambridge University Press 1983

First published 1983

Printed in Great Britain at the University Press, Cambridge

Library of Congress catalogue card number: 82–12879

British Library Cataloguing in Publication Data

Al'pert, Ya L.
The near-Earth and interplanetary plasma.
Vol. 2: Plasma flow, plasma waves and oscillations
1. Plasma (Ionised gases) 2. Solar system
I. Title
523.2   QC717.6

ISBN 0 521 24601 6

Contents

Preface ix

PART 1. FLOW AROUND BODIES MOVING IN A PLASMA

9 A brief description of some theoretical problems and experiments 3

10 Some remarks about conditions at the boundaries of bodies moving in a plasma 10
   10.1 Reflection of particles 11
   10.2 Production of particles 12
   10.3 Potential of a body 13
   10.4 Electron temperature around a body 16

11 Disturbances of the plasma near fast-moving bodies ($V_b \gg v_i$) 19
   11.1 Neutral approximation 19
   11.2 Effect of an external static magnetic field 26
   11.3 Effect of an electric field 29

12 Disturbances of the plasma near quasi-stationary bodies ($V_b \lesssim v_i$) 62
   12.1 A small stationary body ($\rho_b \ll D_e, V_b = 0$) 62
   12.2 A large stationary body ($\rho_b \gg D_e, V_b = 0$) 66
   12.3 Slowly moving bodies ($V_b \simeq, <,$ or $\ll v_i$; $\rho_b \ll D_e$ or $\rho_b \gg D_e$) 69

13 Scattering of radio waves in the wake of a fast-moving body 75

14 Some remarks about the excitation of waves and plasma instability near a fast-moving body 87
   14.1 The relationship between the perturbation of the electron density $\delta N_e(\mathbf{r})$ in the wake and ion-acoustic waves 88
   14.2 The interaction between incident electromagnetic waves and the wake 89
   14.3 Emission from the wake and plasma instability 91

PART 2. WAVES AND OSCILLATIONS IN THE NEAR-EARTH AND INTERPLANETARY PLASMA

15 Introductory remarks — 95
 15.1 A general outline of the results of various experiments — 96
 15.2 Classification of the observed wave processes — 98
 15.3 Generation mechanisms for waves of different types — 99

16 Results of studies of ELF waves — 103
 16.1 Hydromagnetic whistlers — 103
 16.2 Hydromagnetic waves (pulsations of the magnetic field) in the magnetosphere — 108
 16.3 Ion-cyclotron whistlers and waves — 110
 16.4 Hiss and chorus ELF radiation. Cutoff and intensification of emission as $n^2 \to 0$ — 116

17 Results of studies of VLF waves — 123
 17.1 Ion-acoustic waves; radiation at the proton gyrofrequency and its harmonics — 123
 17.2 Noise-like waves generated in the frequency range $\Omega_H < \omega < \omega_L$. Plasmaspheric hiss (PH) — 133
 17.3 Waves generated near the lower-hybrid frequency ω_L — 138
 17.4 Auroral hiss (AH). Saucer-shaped and V-shaped emissions — 142
 17.5 Trapping of VLF waves in the ionosphere and in the magnetosphere. Non-ducted waves — 149
 17.6 Broad-band emission of VLF waves. 'Lion's roar' LF waves — 154

18 Results of studies of LF waves — 159
 18.1 Whistling atmospherics — 159
 18.2 Emissions generated at the boundary of the plasmasphere by means of radio waves (artificially stimulated emissions, ASE). Intensification of radio waves in the whistler mode — 162
 18.3 Hiss and chorus low-frequency waves — 169

19 Results of studies of HF waves ($\omega \gtrsim \omega_H$, $\omega \gtrsim \omega_0$) — 178
 19.1 Resonances in the outer ionosphere (plasmasphere) — 179
 19.2 Waves in the magnetosphere, in the interplanetary medium, and in the solar wind — 187

20 Energy densities of various types of waves — 216

References — 224

Author index — 243

Subject index — 249

Contents of Volume 1
Preface
Introduction
1 General remarks
2 General properties of the near-Earth and interplanetary plasma
3 Fundamental equations
4 Refractive index for a cold magnetoplasma
5 Refractive indexes and attenuation factors in a warm plasma
6 Growth rates for the different oscillation branches
7 Nonlinear effects in a plasma
8 Group velocity, trajectories, and trapping of electromagnetic waves in a magnetoplasma

References

Author index

Subject index

Preface

This book is the second and final volume of a survey of the near-Earth and interplanetary plasma, of the plasma waves that are propagated within it, and of the plasma flow phenomena that occur around a solid body, such as an artificial satellite or a space probe, travelling through it.

Volume 1 began by summarizing the general properties of the plasma in various regions at different distances from the Earth, but its main subject was the theory of plasma waves. Starting from the fundamental equations and leading up to the theory of electromagnetic waves in a cold magnetoplasma, it went on to explain how the behaviour predicted for these waves is modified when the thermal motion of the charged particles is taken into account. Then it dealt with the theory of electrostatic waves, whose very existence depends on these kinetic effects. The various instabilities that occur when energetic particles are present in the plasma, and which can lead to amplification of electromagnetic or electrostatic waves, were also described. An account was given next of various nonlinear effects arising when the ionosphere is illuminated by a powerful beam of artificial radio waves from a transmitter on the ground; here, the review of the theory was supplemented by a presentation of the main experimental results. Finally, returning to linear phenomena, the first volume concluded with the theory of the group velocity, trajectories, and trapping of electromagnetic waves in a magnetoplasma.

This second volume is in two parts, distinct from one another and related in different ways to the contents of the first. Part 1 is concerned with plasma flow around moving bodies, Part 2 with natural plasma waves and oscillations.

Part 1 opens with an account of the phenomena that occur at the surface of a body moving through a plasma, then goes on to show how such effects govern the resulting disturbance in the ambient plasma, and what form this disturbance takes, particularly in the wake of the body. The approximations applicable in the special cases of fast-moving and slow-moving bodies are examined in detail. Wave theory from Volume 1 is invoked in studies of the scattering of electromagnetic waves from the wake of a fast-moving body, of the connection between ion acoustic waves and the perturbation of electron density in the wake, and of other related topics.

Part 2, which reviews the waves and oscillations present in the near-Earth and interplanetary plasma, is, of course, more intimately related to Volume 1. After an introduction where the main experimental findings are outlined, the observed wave processes are classified, and the various wave generation mechanisms are listed, detailed results concerning natural plasma waves are set out in order of ascending frequency, beginning with ELF and passing through LF to HF waves. Finally the energy densities of the different types of waves are discussed, and are presented in tabular form.

This entire review covers the first two decades of the space era, up to the time of writing, by which time the exploratory phase in the investigation of the near-Earth and interplanetary plasma was almost over, the survey phase was advancing rapidly, thanks to the efforts of space physicists worldwide during the 1976–9 International Magnetospheric Study, while the phase of detailed investigation of cause-and-effect relationships, in full progress at the time of publication, had already begun. At the end of Volume 2, the reader will find an extensive bibliography of the relevant work published before as well as during this historic period.

Part 1

Flow around bodies moving in a plasma

9

A brief description of some theoretical problems and experiments

Theoretical studies of effects in the vicinity of a body moving in a plasma generally call for a self-consistent solution of the system of kinetic equations (3.7) and of the Poisson equation (3.8) (see Volume 1) with the boundary conditions (10.1). This system of equations, as mentioned previously, becomes simplified if it is assumed that the electrons have a Maxwell–Boltzmann distribution (see formula (3.12)). Then the Poisson equation (3.8) can be replaced by (3.13). However, even this simplified system of equations is still very complicated. Thus only certain kinds of problems, mainly when the equations are linearized, are susceptible to complete analysis.

The linearization usually involves placing limitations on the potential of the disturbed region of the plasma and on the potential of the body (requiring these to be sufficiently small) or else on the distance from the body (requiring it to be sufficiently great). A number of problems, on the other hand, are solved either without these limitations or only imposing one of them. In this case the equations are already nonlinear. Another important factor is the placing of conditions on the linear size of the body ρ_b, namely that it be commensurable with the Debye length D_e and the Larmor radii ρ_{He} and ρ_{Hi}. The set of problems of this type, considered up to now for various velocities of motion of the body ($V_b \gg v_i$, $V_b \simeq v_i$ and $V_b \ll v_i$), in many respects provide, as we will see below, a quite orderly picture of the interactions between a body and a plasma and the effects arising in the vicinity of the body.

The following articles and monographs should be of use in getting acquainted with the theory explaining the effects in question, as well as with the methods and results of certain numerical calculations. They should also to some extent give the reader an idea of how this field of plasma physics has developed since the first satellites were launched. The works are: Jastrow & Pearse, 1957; Kraus & Watson, 1958; Al'pert, 1960; Gurevich, 1960; Pitaevskii, 1961; Davis & Harris, 1961; Chopra, 1961;

Al'pert et al., 1963, 1965; Brundin, 1963; Al'pert 1965; Maslennikov & Sigov, 1965, 1967; Singer (Editor), 1965; Brundin (Editor), 1967; Liu & Jew, 1967, 1969; Pan & Vaglio-Laurin, 1967; Taylor, 1967; Kiel et al., 1968; Call, 1969; Gurevich et al., 1969; Kasha, 1969; Liu, 1967, 1969; Maslennikov & Sigov, 1969; Vaglio-Laurin & Miller, 1971; Samir & Jew, 1972; Grard (Editor), 1973; Martin, 1974; Samir et al., 1975; plus the works of other investigators to be cited below, as the results of their studies are used. Quite a complete consideration and consistent exposition of this entire range of topics can be found in the monograph by Al'pert et al., 1963, and in the articles by Gurevich et al., 1969a, Liu, 1967, 1969 and Liu & Jew 1967, as well as in the proceedings of various symposia: Brundin (Editor), 1967; Grard (Editor), 1973; Singer (Editor), 1965.

However, a serious shortcoming of the present state of the theory is the absence of any studies of nonsteady-state problems, when in equations (3.7) and (3.8) the time dependence of the distribution function has to be taken into account, i.e., the term $\partial f/\partial t$. In spite of the fact that the solution of (3.7) and (3.8) is explicitly connected with the dispersion equation describing the spectra of the wave processes in a plasma, so far the excitation of waves and the nature of the plasma instability in the vicinity of a body have been little studied.

The following classes of problems have been given a fairly comprehensive theoretical analysis.

I. *Neutral approximation*, when the trajectories of ions are assumed to be rectilinear, as is the case when neutral atoms and molecules impinge upon a body and are reflected from it. Of course, then the body potential ϕ_b, the electric field of the plasma $\mathbf{E} = -\operatorname{grad}\phi$, and the external magnetic field $\mathbf{H}_0$ do not influence the motion of particles. Despite such a major simplification, it turns out that, in an *intermediate zone* of distances, and under some conditions close to the body as well, the results obtained in this approximation adequately describe the experimentally observed disturbance of the charged-particle density behind the body, provided that the potential of the body is not too high.

II. *Influence of an external magnetic field*, when the charged particles gyrate around $\mathbf{H}_0$, but the influence of the electric field of the plasma and the body potential are neglected. The corresponding theoretical results also show a good fit with the experimental results in a certain distance range.

III. *Far zone of the body at distances* $r \gg \rho_v V_b/v_i$, taking into account the electric field, the external magnetic field, and also the nonisothermal nature of the plasma; in these cases only linearized problems have been solved. Although this distance range has been little studied experimentally, the theoretical findings are apparently in good qualitative agreement with those few experimental data which have been obtained.

IV. *Near zone of the body*, taking into account the effect of the electric field for the case of a weakly charged body of small size ($\rho_b \ll D_e$), as well as large uncharged and charged bodies ($\rho_b \gg D_e$). In these cases nonlinear problems have been solved. The results of these studies also show quite a good fit with experimental results.

Experiments were conducted both using laboratory installations and directly in the near-Earth plasma (on satellites and rockets). However, since many of these experiments were carried out under conditions that do not correspond to those of the theoretical problems, frequently it is only possible to compare the findings of theory and experiment qualitatively. On the other hand, some data also show a good quantitative fit. The trouble with analysing laboratory experiments is that the particle fluxes impinging on the model (body) are insufficiently uniform over its cross-section and have a large spread of velocities, so that the degree of nonisothermality of the fluxes (the ratio T_e/T_i) is unknown, as are some of their other important properties. Moreover, the experiments using satellites are further complicated by the fact that the shapes and electrical properties of these bodies are very complex. They are generally equipped with auxiliary devices that cause the appearance of additional fluxes of charged particles impinging on the measuring instrument. Since the measurements are usually carried out close to the surface of the body, these additional effects can be quite pronounced. Moreover, the nature of the plasma disturbance is affected considerably by the fact that the near-Earth plasma is a multicomponent, nonisothermal medium, which is often hard to allow for.

The experimental results characterizing the plasma disturbance in the vicinity of a moving body were obtained directly in the ionosphere mainly with the aid of the following satellites:

1. *Explorer* 8, in the altitude range $z = 425$ to 2300 km: Bourdeau *et al.*, 1961; Bourdeau, 1962; Bourdeau & Donley, 1964.

2. *Ariel* 1, $z \simeq 400$ to 1200 km: Bowen *et al.*, 1964; Samir & Willmore, 1965, 1966; Henderson & Samir, 1967.

3. *Explorer* 31, $z \simeq 500$ to 3000 km: Hoffman, 1967; Samir & Wrenn, 1969, 1972; Samir, 1970, 1972; Samir *et al.*, 1973; Troy *et al.*, 1975.

4. *Gemini/Agena*, a two-body system, $z \simeq 300$ to 400 km: Medved, 1969; Troy *et al.*, 1970.

Numerous experiments have been devoted to the laboratory study of supersonic plasma flow around bodies of various shapes. The models of the bodies have included spheres, round and square disks, and cylinders, oriented in various ways relative to the direction of the oncoming plasma flow. Plasma flow around a half-plane has also been investigated. Many experiments have been carried out in isotropic plasmas. The experimental results were described in the following works: Meckel, 1961; Barrett, 1964; Hall *et al.*, 1964, 1965; Clayden & Hurdle, 1964; Kasha *et al.*, 1965; Kasha & Johnston, 1967; Osborne & Kasha, 1967; Skvortsov & Nosachev, 1968; Gurevich *et al.*, 1969b; Lederman *et al.*, 1969; Hester & Sonin, 1970a, b; Bogashchenko *et al.*, 1970, 1971; Astrelin *et al.*, 1971, 1972; Fournier, 1971; Fournier & Pigache, 1972; Schmitt, 1972; Oran *et al.*, 1974; Stone *et al.*, 1974; Oran *et al.*, 1975.

In Barrett, 1964, Clayden & Hurdle, 1964, Kasha & Johnston, 1967, Osborne & Kasha, 1967, Bogashchenko *et al.*, 1970, 1971, Gurevich *et al.*, 1969b and Astrelin *et al.*, 1972, measurements in a magnetized plasma are described. The experiments in the last three were subjected to theoretical calculations designed especially for analysis purposes. This enabled a quite accurate comparison of theory with experiment (see below).

In the various laboratory experiments the plasma parameters, the velocities of the on coming streams, and the body models were varied mainly within the following limits:

$$\left.\begin{array}{l} \rho_b/D_e \simeq 1 \text{ to } 10, \quad \rho_{Hi}/\rho_b \simeq 0.5 \text{ to } 1.5, \quad r/\rho_b \simeq 1 \text{ to } 10 \text{ and} > 10, \\ V_b/v_s \simeq 1 \text{ to } 10 \text{ and} > 10, \quad T_e/T_i \simeq 1 \text{ to } 5 \text{ and} \gg 1, \\ T_e = 1000 \text{ to } 3000 \text{ K}, \\ T_i \simeq 500 \text{ to } 1000 \text{ K}, \quad \phi_b = 0 \text{ to} - 5 V, \quad N \simeq 10^5 \text{ to } 10^8 \text{ cm}^{-3}, \end{array}\right\} \quad (9.1)$$

where D_e, ρ_{Hi} and v_s are, respectively, the Debye length of the electrons, the Larmor radius of the ions, and the nonisothermal sound velocity (see formulas (1.3), (1.4) and (1.9) of Volume 1); $\mathbf{V}_0 = -\mathbf{V}_b$ is the velocity of the oncoming plasma stream, ρ_b is the effective linear dimension of the body, r is the distance from its centre, T_e and T_i are the electron and ion temperatures, ϕ_b is the body potential, and N is the charged-particle density.

In a number of cases the values in (9.1) correspond to the conditions observed in the near-Earth plasma (see Tables 2.1 and 2.2 of Volume 1) or are close to the values of the corresponding parameters used in the theoretical calculations. On the other hand, it is very difficult in practice to compare closely the results of laboratory experiments with the results of measurements in the ionosphere or with theoretical results. This is because, for a suitable quantitative comparison, it is important, first, to ensure a close correspondence between the experimental conditions and the conditions of the theoretical problems, and, second, as mentioned previously, to take into account a number of vital conditions which are realized in the experiments but which are either unknown or else have not as yet been evaluated theoretically. Nevertheless, the principal features typifying the state of a plasma in the vicinity of a body moving in it, which are observed under both laboratory and natural conditions, agree in many respects with one another and also with the anticipated theoretical effects. These will be described in detail later in Part 1 of this volume. Briefly, these features are as follows.

I. The wake of the body, i.e., the plasma region behind the body relative to the direction of its velocity $\mathbf{V}_b$, is highly rarefied in the *near zone*, namely at distances $r \ll$ or $< \rho_b V_b / v_i$. On the axis of the body ($\theta = 0$), close to its surface ($r/\rho_b = 1$), the electron density N_e ($\theta = 0$) and ion density $N_i(\theta = 0)$ are considerably lower than the undisturbed particle densities N_{e0} and N_{i0}; here r is the distance from the centre of the body, ρ_b is its mean linear dimension (the 'radius' of the section of the body), and θ is the angle between the vectors $\mathbf{r}$ and $\mathbf{V}_0$, $= -\mathbf{V}_b$ where $\mathbf{V}_b$ is the velocity of the body.

Under different conditions for $r/\rho_b \simeq 1$, the following ratios N_e/N_{e0} and N_i/N_{i0} are observed:

$$N_e(\theta = 0)/N_{e0}, N_i(\theta = 0)/N_{i0} \simeq (0.1-0.2) \text{ to } (10^{-2}-10^{-4}),$$

where $N_e(\theta) \gg N_i(\theta)$. The degree of rarefaction of each of the charged-particle densities depends considerably on the mean mass of the ions M^+, decreasing with a drop in M^+. The degree of rarefaction of the ions is quite a bit lower than the rarefaction of the neutral particles. Figures 9.1 and 9.2 give some experimental data illustrating the indicated properties of a body wake, obtained with Explorer 31. These effects have been well examined theoretically.

II. In the *intermediate zone* of the body's wake, at distances $r \lesssim \rho_b V_b/v_s$, focusing of charged particles appears. The influence of the electric field of the plasma manifests itself here. The degree and the nature of the focusing, i.e., the dependences of N_e and N_i on θ, vary with an increase in the distance r from the body, being a function of the cross-section relative to the vector $\mathbf{V}_0$, of the radius ρ_b, of the potential ϕ_b, and of the velocity V_b of the body, as well as of the various plasma parameters.

In the vicinity of, for example, the axis of a sphere at distances $r \lesssim \rho_b V_b/v_i$ a relatively narrow region of particle focusing is observed, with a maximum

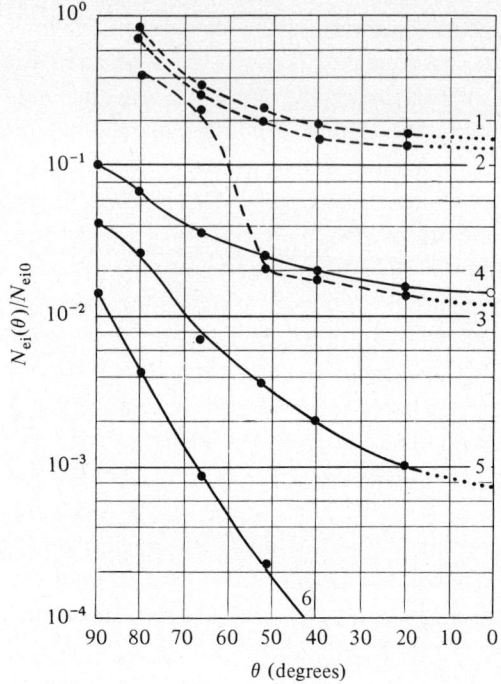

Fig. 9.1. Angular dependences of the ratios $N_e(\theta)/N_{e0}$ (dashed lines) and $N_i(\theta)/N_{i0}$ (solid lines) in the wake of a body close to its surface ($r/\rho_b \simeq 1$), obtained under different conditions in experiments on Explorer 31 (Samir et al., 1975). The different curves correspond to the following experimental conditions (the asterisks denote dimensionless normalized quantities):

1. $\rho_b^* = \rho_b/D_e \simeq 7.8$, $\phi_b^* = e\phi_b/kT_e = -5.5$, $a_{bs} = V_b/2^{1/2}v_s \simeq 3.6$;
2. $\rho_b^* = 9.5$, $\phi_b^* = -3.8$, $a_{bs} \simeq 3.7$;
3. $\rho_b^* \simeq 43$, $\phi_b^* \simeq -5.1$, $a_{bs} \simeq 5.6$;
4. $\rho_b^* \simeq 20$, $T_e/T_i \simeq 1.1$, $\phi_b^* \simeq 4.6$;
5. $\rho_b^* \simeq 19$, $T_e/T_i \simeq 1.06$, $\phi_b^* \simeq -3.2$, $a_{bs} \simeq 3.9$;
6. $\rho_b^* \simeq 21$, $T_e/T_i \simeq 1.23$, $\phi_b^* \simeq -3.6$, $a_{bs} \simeq 5.8$.

The results of the measurements 1, 2, 3 were obtained, respectively, on the same orbits as the data 4, 5, 6, but in different altitude ranges.

at $\theta \simeq 0$ and a rarefaction region on each side of it. In the focusing region the charged-particle density may be considerably higher than that in the undisturbed plasma.

In the *far zone* of the wake of a sphere, $r \gg \rho_b V_b/v_i$, on both sides of its axis focusing zones appear, with maxima at certain angle values θ_M. The values of θ_M, as well as the degree and the nature of the functions

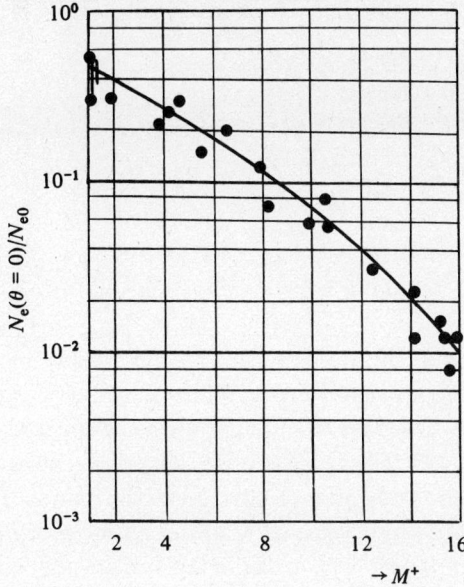

Fig. 9.2. The ratio $N_e(\theta = 0)/N_{e0}$ as a function of the mean mass of the ions M^+ in the wake of a body close to its surface (see Samir, 1970).

$N(\theta)$, depend on the velocity of the body and on the various plasma parameters. The effect of an external static magnetic field $\mathbf{H}_0$ will be considerable in this case. The far zone of a body's wake has been well studied theoretically and in a number of laboratory experiments.

III. In some experiments in the near-Earth plasma, angular dependences of the electron temperature $T_e(\theta)$ and of the potential $\phi_b(\theta)$ have apparently been observed, using a sensor that formed part of the surface of the satellite body but was insulated from the rest of it (see §10.3). In the wake of a body, on its axis ($\theta = 0$) close to its surface $r/\rho_b \simeq 1$, the functions $T_e(\theta)$ and $\phi_b(\theta)$ go through maxima (see Samir & Wrenn, 1972; Troy et al., 1975; Medved, 1969). A dependence of T_e on the distance from the body was also detected (see Troy et al., 1970). There are indications that the temperature T_e is also governed by the Earth's magnetic field $\mathbf{H}_0$ (Clark et al., 1973). These effects have not, however, as yet received a sufficiently clear theoretical explanation.

10

Some remarks about conditions at the boundaries of bodies moving in a plasma

The phenomena arising in a plasma in the vicinity of a moving body depend in a number of cases on the shape of the body and on the physical properties of its surface, as well as on its structure and on the material of which it is made. This is because, *firstly*, particles that experience *reflection* from the body surface have their distribution function changed in its immediate vicinity. *Secondly*, this is due to the *production* of new particles via evaporation from the body surface and the destruction of this surface by, for instance, the oncoming streams. *Thirdly*, the potential acquired by a body moving in a plasma also depends on the nature of the interaction and on the properties of its surface, which in turn influence considerably the structure and character of the disturbance of particles in its vicinity.

However, the role of the boundary conditions at the body surface is not always so great. It may, for instance, be a determining factor when considering the stability and excitation of plasma oscillations. On the other hand, the boundary conditions have little effect on the scattering of radio waves in the body wake, a process in which the entire disturbed region of the plasma behind the body participates. Thus the effective scattering cross-section is mainly determined by the part of the wake quite far away from the body.

It is convenient to portray both of the above groups of phenomena, reflection and production of particles, with the aid of functions of the general form:

$$A_e(\mathbf{r}_b, \mathbf{v}_i, \mathbf{v}_r)\delta(b), \quad A_i(\mathbf{r}_b, \mathbf{v}_i, \mathbf{v}_r)\delta(b) \tag{10.1}$$

which are to be added, together with the collision integrals Y_e and Y_i, to the right-hand sides of equations (3.7). In (10.1) the vector $\mathbf{r}_b$ characterises a point on the surface of the body, while $\mathbf{v}_i$ and $\mathbf{v}_r$ are the particle velocities before and after colliding with the body. Depending on the physical formulation of the problem, the functions A_e and A_i assume different forms and they may, in particular, include several terms describing processes of various kinds. The delta function $\delta(b)$ in (10.1) signifies that the functions A_e and A_i differ from zero only at the surface of the body, while their physical meaning can be explained as follows. The dimensions of the quantity $A\delta(b)$ must be the same as those of the derivative $\partial f/\partial t$. But this means that the integral

$$\int A d^3 \mathbf{v} = J \qquad (10.2)$$

defines the *variation of the particle flux* per unit time ($J\,\mathrm{cm}^2\,\mathrm{s}^{-1}$), due to the presence of the body surface.

The *reflection* and *production* of particles, as well as the establishing of a *body potential* and its influence on the nature of the plasma disturbance, have in many respects been little studied, either experimentally or theoretically, since such studies are fraught with diverse, sometimes fundamental, difficulties. Thus let us confine ourselves here to just some brief remarks about these topics.

10.1. Reflection of particles

The term 'reflection' includes the following phenomena, which may be quite different physically.

a. Specular reflection, when the incidence angle of the particle equals its reflection angle and $|\mathbf{v}_i| = |\mathbf{v}_r|$. Such particle reflection is possible only from an absolutely smooth dielectric surface. For instance, in the case of a spherical surface, the boundary conditions for ions can be written as

$$A_i \delta(b) = \frac{\mathbf{r}\cdot\mathbf{V}_b}{r} f_i\left(\mathbf{r}, \mathbf{v} - \frac{2\mathbf{r}(\mathbf{r}\cdot\mathbf{v})}{r^2}\right) \delta(r - \rho_b) \qquad (10.3)$$

if the scalar product $\mathbf{r}\cdot\mathbf{V} > 0$, and as

$$A_i \delta(b) = \frac{\mathbf{r}\cdot\mathbf{V}_b}{r} f_i(\mathbf{r}, \mathbf{v}) \delta(r - \rho_b) \qquad (10.4)$$

if $\mathbf{r}\cdot\mathbf{v} < 0$, where $\mathbf{V}_b$ is the velocity of a sphere of radius ρ_b and $V_b \gg v_i$.

b. Elastic diffuse reflection, when all the different velocity directions of the reflected particles are equally probable, while the moduli of their velocities remain unchanged. In this case *scattering* of the particles is usually referred to.

c. Inelastic reflection (partial accommodation), when velocity directions of reflected particles $\mathbf{v}_r$ are equally probable, but the moduli become smaller because the particles give up part of their energy to the surface when they collide with the body. For total accommodation of the particles, i.e., complete *absorption* of them by the body surface, we have

$$A_i \delta(b) = 0 \text{ for } \mathbf{r} \cdot \mathbf{v} > 0$$

$$A_i \delta(b) = \frac{\mathbf{r}}{r} f_i \delta(r - \rho_b) \text{ for } \mathbf{r} \cdot \mathbf{v} < \mathbf{0}. \qquad (10.5)$$

10.2. Production of particles

The plasma surrounding the body must certainly be filled continually with particles leaving the body surface as a result of *evaporation* and *erosion*, caused by bombardment of the body with streams of particles or with meteoric material. The particles are produced by electron and ion photoemission and other processes. Neutral atoms or molecules leaving the surface become ionized very gradually. According to various estimates, in the media of interest to us the *ionization time* is

$$\tau_i \simeq 10^7 \text{ s}, \qquad (10.6)$$

while the velocity of particle outflow from the body is

$$v_s = (2kT_0/M_s)^{1/2} \simeq 10^2 \text{ m s}^{-1} \qquad (10.7)$$

(the subscript s refers to the species of particle). Thus, the evaporating particles first gradually recede from the body and only at great distances do they acquire the thermal speed of the surrounding medium, after which they then diffuse rapidly. The time needed for a particle to recede from the body surface, say to a distance $r_s \simeq 1$ m, will be

$$\tau_s = \frac{r_s}{v_s} \simeq 10^{-2} \text{ s}. \qquad (10.8)$$

Consequently, in this region the ratio of the density of 'produced' charged particles N_s to the density of 'produced' neutral particles n_s will be negligibly small:

$$\frac{N_s}{n_s} \simeq \frac{\tau_s}{\tau_i} \simeq 10^{-9}. \qquad (10.9)$$

Moreover, satellite measurements of streams of produced particles indicate that around the body itself the number of produced neutral particles n_s may be quite large. Therefore, in plasma regions where $n_s \gg n_0$, i.e., where the density of the natural neutral particles is negligible, produced particles may well play a major role in the processes taking place around the body. Apparently, this circumstance should not in some cases be overlooked when analysing experimental data. For instance, in one series of experiments (McKeown, 1961) it was shown that in the altitude range $z \simeq 216$ to $810\,\mathrm{km}$ the particle flux lost by a gold plate positioned normally to the oncoming stream (vector $\mathbf{V}_0$) varied over the limits

$$J_s \simeq 10^7 \text{ to } 10^{10} \text{ atoms cm}^{-2}\,\mathrm{s}^{-1}, \tag{10.10}$$

which under the conditions of the experiments corresponded to an evaporation rate of the order of 5×10^{-6} gold atoms per particle in the oncoming stream. For other metals (aluminum, zinc, iron, magnesium, lithium), for $T_s \simeq 10^2$ to 10^3, the following value was observed:

$$J_s = \overline{(nv_s)} \simeq 10^{10} \text{ to } 10^{14} \text{ atoms cm}^{-2}\,\mathrm{s}^{-1}. \tag{10.11}$$

The sublimation rates in vacuo for polymers, nylon, sulfides, and vinyl chloride are about 3×10^{-9} times the weight of the substance per second. Consequently, using the data in (10.10) and (10.11), we find that for $v_s \simeq 10^2\,\mathrm{m\,s}^{-1}$ it can be assumed that in different cases

$$n_s \simeq \overline{(nv_s)}/v_s \simeq 10^3 \text{ to } 10^{10}\,\mathrm{cm}^{-3}. \tag{10.12}$$

10.3. Potential of a body

The potential ϕ_b acquired by a body moving in the near-Earth plasma is a very important factor. In a number of experiments a knowledge of ϕ_b is decisive, determining the accuracy of the interpretation of the measurement results. A precise theoretical calculation of the potential ϕ_b is scarcely possible, however, in view of the complexity of both the geometry and the electrical structure of the body surface, and the lack of initial data on the interaction of the body material with the streams impinging on it and the radiation incident upon it.

Let us consider briefly how an approximate value can be obtained. The potential ϕ_b is governed by a balance between the fluxes of electrons and ions onto the surface. For an insulating (i.e. nonconducting, or dielectric) body, these fluxes must balance locally, and in general ϕ_b is nonuniform over the surface, whereas for a conducting body they balance globally and ϕ_b is uniform. If a small part of a conducting surface, such as the sensor of a measuring instrument, is insulated from the rest, then it acquires approximately the potential that the surface of an insulating body would have at the same point.

At each point b on the surface of an insulating body, the potential ϕ_b is determined by the ratio of the numbers of electrons and ions absorbed by the surface per unit time. Thus, since $v_e \gg v_i$, the body will be negatively charged. Actually, let us assume that electrons and singly charged ions impinge upon the body. Assume, too, that $N_e \simeq N_i$ and $T_e \simeq T_i$. At an arbitrary point in the plasma the ratio of their fluxes will be

$$J_e/J_i \simeq v_e/v_i \gg 1. \tag{10.13}$$

Consequently, when the body surface is bombarded by these particles, it will become charged until at any given point b the electron and ion fluxes are equal. This will be possible only if the electron flux impinging upon the body decreases, i.e., if the surface becomes negatively charged and repels electrons.

Let us first take the case of a body at rest. Then the density of the electron flux at any point b on its surface can be written as

$$J_{eb} = J_{e0} \exp(-e\phi_b/kT), \tag{10.14}$$

where $J_{e0} = Nv_e/2\pi^{1/2}$ corresponds to an electron stream undisturbed by a body, i.e., when $\phi_b = 0$. The dependence of the ion flux on the potential is more complex. However, in the limiting case of low potentials ($|\phi_b| \ll kT/e$) it can be assumed that

$$J_{ib} \simeq J_{i0} \simeq N_0 v_i / 2\pi^{1/2}, \tag{10.15}$$

i.e., that J_{ib} is equal to the undisturbed ion flux J_{i0}. Now, if the coefficients of reflection of ions and electrons from the body are, respectively, ρ_i and ρ_e, then ϕ_b can be found from the equation

$$J_{ib}(1 - \rho_i) = J_{eb}(1 - \rho_e). \tag{10.16}$$

Using (10.14) and (10.15), we get

$$\phi_b = \frac{kT}{e} \ln\left[\frac{v_e(1-\rho_e)}{v_i(1-\rho_i)}\right]. \tag{10.17}$$

Hence, for a completely absorbing, quasi-stationary body ($v_i \ll V_b$), when $\rho_i, \rho_e \ll 1$ in the media of interest to us (see Tables 2.1 and 2.2 of Volume 1), it follows that

$$\phi_b \simeq -1 \text{ to } -2 \text{ volts}. \tag{10.18}$$

In order to determine the potential of the forward surface of a rapidly moving body ($V_b \gg v_i$), the ion flux (10.15) has to be substituted into (10.16) in the form

$$J_i = N_0 V_0 \cos\theta_0, \tag{10.19}$$

where θ_0 is the angle of incidence of the particles upon the body. For $\cos\theta_0 \simeq 1$. we have

$$\phi_b = \frac{KT}{e} \ln\left[\frac{v_e(1-\rho_e)}{V_0(1-\rho_i)}\right] \simeq -0.5 \text{ to } -1 \text{ volts} \tag{10.20}$$

10.3. Potential of a body

(Al'pert, 1965). The value of ϕ_b behind the body is difficult to calculate, since for this region no simple accurate formulas for the particle fluxes exist, but clearly it should be more negative here than it is elsewhere, due to the decrease of the ion flux compared with the electron flux. Even for a dielectric body with a uniform surface, the potential will vary considerably from point to point, due to variations, in particular, of the reflection coefficients as well as due to the effects of different emission processes.

On the other hand, for a conducting body, such as one of bare metal, the surface potential is the same everywhere. It is approximately equal to the potential that an insulating body of the same shape would take up on its forward surface.

In some experiments the potential acquired by a body as a plasma stream flows around it has been determined. The most complete data are apparently the experimental results obtained with Explorer 31 (see Samir & Wrenn, 1969; Samir, 1970; Troy et al., 1975) and Gemini/Agena (Troy et al., 1970). These experiments established that the potential of a body in the ionosphere varies under different conditions approximately within the following limits:

$$\phi_0 \simeq -[(0.4-0.5) \text{ to } (1-1.2)] \text{ volts.} \qquad (10.21)$$

It is easy to show that these values of ϕ_b agree well with theoretical estimates (10.18) and (10.20). In the experiments on Explorer 31 it was found that $\phi_b = \phi_b(\theta)$, i.e., the body potential depends on the position of the measurement point relative to the direction of motion V_b. Fig. 10.1 shows the corresponding experimental results. Despite the considerable spread of individual ϕ_b values and the smallness of their range of variation, it can nevertheless be concluded that $|\phi_b|$ has a maximum in the wake of

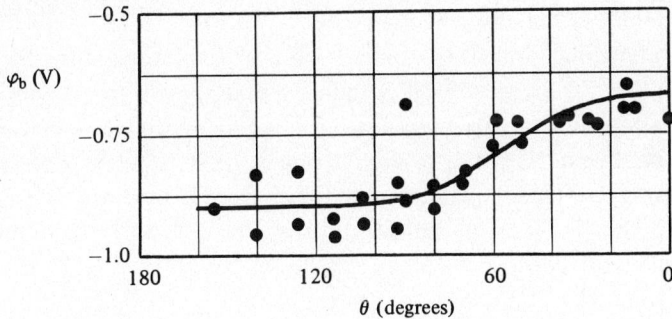

Fig. 10.1. Potential ϕ_b (in volts) measured aboard Explorer 31 for various instrument orientations (angle θ) relative to the plasma flow velocity vector V_0 (Troy et al., 1975).

the body, behind the oncoming flux at $\theta = 0$. This, in particular, agrees with the remarks made above.

Note, too, that in some earlier experiments (Sharp et al., 1963; see also Whipple & Troy, 1965) potentials $\phi_b \simeq -12$ to -14 volts were observed, values considerably higher than the experimental results (10.21) or the theoretical estimates (10.18) and (10.20). With regard to this, the following should be mentioned. A body surface on which measurements are carried out often has a complex geometry and electrical structure (sharp protrusions, conducting regions alternating with dielectric regions). This means that the potential distribution over the surface $\phi_b(\theta)$ may be quite complicated. It is particularly so if the satellite is equipped with instruments, such as Langmuir probes, that are exposed to the plasma and are biased to substantial potentials with respect to the body. For similar reasons, the uniform potential of a body of which the entire surface is conducting may depend on its orientation with respect to the plasma flow velocity vector $\mathbf{V}_0$. This may well explain the above-mentioned discrepancies between the different data. Of course, studies of the body potential ϕ_b constitute a very important activity. It was mentioned above, for example, that without a knowledge of ϕ_b it is often impossible to interpret correctly either the experimental results or the theoretical explanations.

10.4. Electron temperature around a body

It will be helpful here to present some measured electron temperatures T_e in the immediate vicinity of bodies moving in the ionosphere. Certain factors, still requiring theoretical analysis, indicate that the behavior of T_e is related to the potential $\phi_b(\theta)$ of the body. This ensues, in particular, from the data to be presented here.

The above-mentioned experiments aboard Explorer 31 (see Samir & Wrenn, 1972; Troy et al., 1975) indicate that the electron temperature close to the surface of a body has an angular dependence $T_e(\theta)$ with a maximum behind the body at $\theta = 0$ (see Fig. 10.2). In the experiments on Gemini/Agena, the following ratio of the electron temperature in the wake close to the body to the temperature of the undisturbed plasma was obtained:

$$\frac{T_e(\theta = 0)}{T_{e0}} \simeq (1.9\text{--}2.9).$$

These data are in good agreement with other experimental results

10.4. Electron temperature around a body

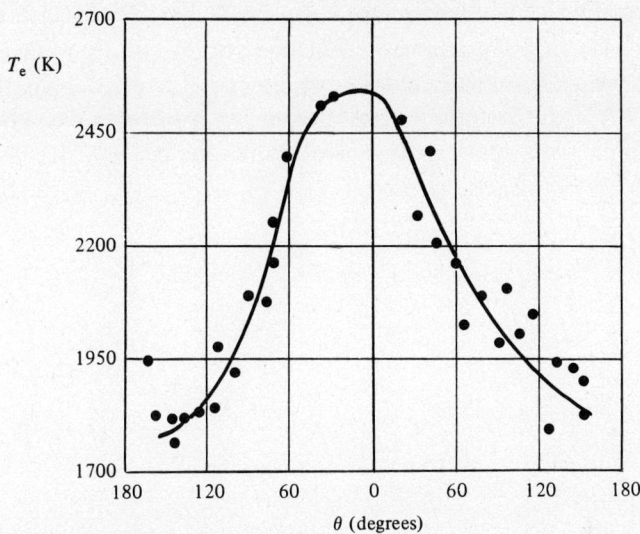

Fig. 10.2. Angular dependence of the electron temperature $T_e(\theta)$ according to data from Explorer 31 (Samir & Wrenn, 1972).

indicating ratios

$$\frac{T_e(\theta = 0)}{T_{e0}} \simeq (1.3\text{–}2.3). \tag{10.22}$$

In the Gemini/Agena experiments a 10–12% reduction of the temperature T_e in the wake of Gemini was also recorded, as the second test body Agena receded. Here the potential $-\phi_b$ of the body was also seen to drop (see Troy et al., 1970). On the other hand, in the Ariel 1 experiments no sizable variation of temperature T_e in the wake of the body was detected (Henderson & Samir, 1967). However, in these experiments, in contrast to those described in Samir & Wrenn, 1972, the measurements were carried out at a distance $r \simeq 5\rho_b$ from the body. In Troy et al., 1975, it was also decided that the angular dependence $T_e(\theta)$ is apparently not determined by the orientation of the body wake relative to the direction of the Earth's magnetic field $\mathbf{H}_0$. The opposite conclusion was drawn by Clark et al., 1973 on the basis of measurements aboard the magnetically stabilized satellite ESRO 1a. In these experiments it was noted that the electron temperature $T_{e\parallel}$ parallel to the magnetic field $\mathbf{H}_0$ was greater than the perpendicular temperature $T_{e\perp}$. But the variations of the temperature T_e on ESRO 1a did not take place right at the surface of the body either, being measured by instruments mounted on booms. Naturally, therefore, only further quite comprehensive temperature measurements, together with an adequate theoretical study of the corresponding topics, can tell us whether, in particular, the experimental results described above are contradictory. The appropriate investigations involve the solution of some complex and subtle problems regarding the near-Earth plasma.

11

Disturbances of the plasma near fast-moving bodies ($V_b \gg v_i$)

11.1. Neutral approximation

Let us first consider the supersonic flow of a stream of neutral particles around a body. In this case the particles behind the body are, as it were, 'swept out,' and an extended rarefied wake forms. The particle trajectories are rectilinear. They are not influenced by the electric or magnetic fields, so the problem reduces to solution of the kinetic equation

$$A\frac{\partial f_n}{\partial r} = A\delta(s) \tag{11.1}$$

determining the distribution function $f_n(\mathbf{r}, \mathbf{v})$, and thus the disturbed distribution of the particle density near the body.

In (11.1), the variable $\mathbf{r}$ is the radius vector drawn from the body centre to the observation point, and θ is the angle between r and the z-axis. The positive z-axis is anti-parallel to the velocity vector $\mathbf{V}_b$ of the body with respect to the plasma; in other words, it is parallel to the plasma flow velocity $\mathbf{V}_0$ relative to the body. Thus, behind the body, along the axis, the angle $\theta = 0$, while ahead of it $\theta = \pi$; the magnitude of z is reckoned from the body centre.

A basic feature of the disturbance in the given case is a concentration of the particles ahead of the body and a rarefaction behind it. The concentration occurs in a small volume, and it is significant only up to distances from the body of the order of its linear size. On the other hand, the rarefied wake of the body is quite extensive and becomes smoothed out along the z-axis only at distances of the order of the particle mean free path. Close to the body, the distribution of the particle density depends considerably on the body shape.

Neglecting collisions, we can describe the concentration of neutral particles $N_n(\rho, z)$ ahead of a sphere with the aid of the following formulas:

$$\left.\begin{array}{c}\dfrac{N_n(\rho,z)}{N_{n0}} = 1 + \left(\dfrac{\rho_b}{\rho}\right)^2 \dfrac{\sin^2\theta_0 \cos^2\theta_0}{1 - \dfrac{\rho_b}{\rho}\sin^3\theta_0}, \\[2mm] \dfrac{z}{\rho_0}\cos\theta_0 + \dfrac{\rho}{\rho_b}\sin\theta_0 = \tfrac{1}{2}\left(1 + \dfrac{\rho}{\rho_b \sin\theta_0}\right),\end{array}\right\} \quad (11.2)$$

where N_{n0} is the undisturbed density and z, ρ and θ_0 are the polar coordinates of the observation point, z being reckoned in (11.2) from the sphere centre along the direction of the velocity of the body $\mathbf{V}_b$, while $\theta_0 = \pi - \theta$. In the derivation of (11.2) it was assumed that specular reflection of particles from the sphere surface takes place. The particle density in the rarefaction region of the sphere, and in general behind any body of circular section, will be

$$\dfrac{N_n(\rho, z)}{N_{n0}} = 1 - 2\exp(a_b\rho/z)^2 \int_0^{a_b\rho_b/z} u\exp(-u^2) I_0(\rho a_b u/z) du, \quad (11.2a)$$

where $a_b = V_b/v_n$, and $I_0(x) = (1/\pi)\int_0^\pi \exp(-x\cos\theta)d\theta$ is a Bessel function of zero order with imaginary argument. On the z-axis, where $\rho_0 = 0$ and $I_0 = 1$, we have

$$N_n(0, z) = N_{n0}\exp(-a_0\rho_b/z)^2 \quad (11.2b)$$

(Gurevich, 1960). Consequently, for high values of z, when $a_b\rho_b/z \ll 1$, the disturbance of the neutral-particle density in the body wake is

$$\delta N_n(z) = [N_n(z) - N_{n0}]/N_{n0} = -(V_b/v_n)^2(\rho_b/z)^2. \quad (11.2c)$$

Figs. 11.1, 11.2, and 11.3 show the curves for $N_n(\theta, r)/N_{n0}$ as a function of z/ρ_b for bodies of various shapes having the same cross-section s_b (here ρ_b is the radius of the body). The curves of equal values of $N_n(\theta, r)/N_{n0}$ plotted in Fig. 11.3 for a sphere were obtained by solving equations (11.2). These curves, like those in Fig. 11.1 and 11.2, were calculated for $a_b = V_b/v_n = 8$. Inspection of these figures indicates that in the near zone of the body wake, where $r \ll$ or $< \rho_b V_b/v_n$, for angles $\theta \simeq \rho_b/r \lesssim 1/a_b$ the particle density drops quite markedly, going to zero. The angular variation of the disturbed density takes place very rapidly at short distances behind the body. This is evident from Fig. 11.1, which portrays the functions $N_n(\theta, r = \text{const.})/N_{n0}$ at distances $r/\rho_b = 4, 2, 1$ for a sphere and for an ellipsoid. With an increase in distance, namely if

$$r/\rho_b \gtrsim V_b/v_n \quad (11.3)$$

11.1. Neutral approximation

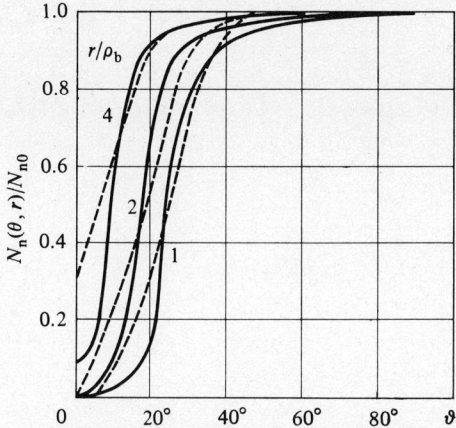

Fig. 11.1. Angular dependences of the normalized neutral particle density $N_n(\theta,r)/N_{n0}$ in the vicinity of a rapidly moving ($V_b/v_n = 8$) sphere (solid curves), and ellipsoid (dashed curves) at distances from body centre $r/\rho_b = 4, 2, 1$ (Gurevich, 1960).

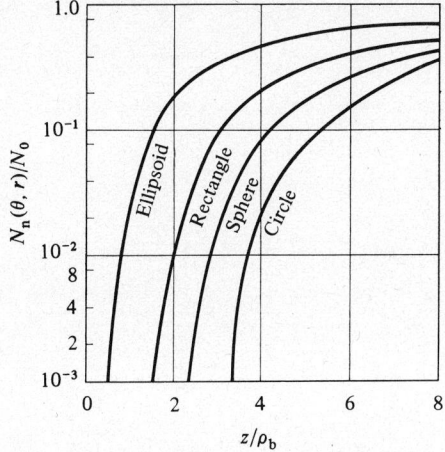

Fig. 11.2. Normalized neutral particle density $N_n(\theta,r)/N_{n0}$ as a function of z/ρ_b along the axis ($\theta = 0$) behind rapidly moving bodies of various shapes ($V_b/v_n = 8$) (Gurevich, 1964).

(this region is known as the *intermediate zone*), $N_n(\theta, r)$ already depends very little on the shape of the body, being affected only by the body cross-section s_b in the plane perpendicular to the vector $\mathbf{V}_b$. If the condition (11.3) is satisfied, then the disturbance of the particle density along the z-axis is described quite well by the formula

$$\delta N = [N_n(z,0) - N_0]/N_{n0} = (s_b/\pi)(a_b/z)^2 \exp[-(a_b\rho_b/z)^2], \quad (11.4)$$

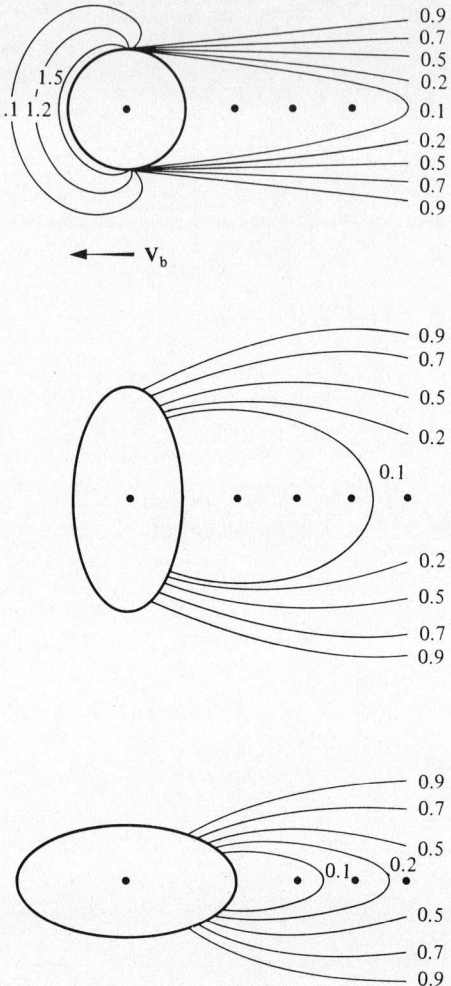

Fig. 11.3. Curves of equal values of $N_n(\theta, r)/N_{n0}$ behind a rapidly moving sphere (Gurevich, 1960) or ellipsoid (Sawchuck, 1963) ($V_b/v_n = 8$).

where $\rho_b \simeq (s_b/\pi)^{1/2}$ is effective radius of the largest cross-section of the body. In the *far zone* of the body wake, where $r/\rho_b \gg V_b/v_n$, the disturbance is

$$\delta N_n \simeq (s_b/\pi)(a_b/z)^2, \qquad (11.2d)$$

i.e., it decreases as z^{-2}. It is easy to show that the formula (11.2d) is the same as the formula (11.2c) when $s_b = \pi \rho_b^2$.

The neutral approximation is helpful in determining the ion density in

11.1. Neutral approximation

a certain intermediate zone of distances from the body, namely for

$$\rho_b V_b/v_i \geqslant r \leqslant \rho_{Hi} V_b/v_i. \qquad (11.3a)$$

Closer than this zone, the influence of the electric field gradually begins to make itself felt. At distances $r > \rho_{Hi} V_b/v_i$ an external magnetic field $\mathbf{H}_0$ already has considerable effect on the motion of particles, and the structure of the disturbance becomes quasi-periodic (see below). In the far zone, where $r \gg \rho_b V_b/v_i$, the electric field has an effect and the angular dependences $N_i(\theta, r)$ become complicated. Naturally, the condition (11.3a) will be meaningful only if $\rho_{Hi} >$ or $\gg \rho_b$, which is the case in the near-Earth plasma but which is not always true in laboratory experiments. In a nonisothermal plasma for $T_e \gg T_i$, the influence of the electric field is enhanced and thus the applicability of the neutral approximation is more limited.

Figs. 11.4 to 11.6 present some experimental results illustrating the applicability of the neutral-approximation theory to determine the angular dependence of the charged-particle density. Fig. 11.4 shows the ratio of the electron flux J_e, measured at a probe located at a distance of $5\rho_b$ from the centre of Ariel 1, to the undisturbed electron flux J_{e0}, where $J_e/J_{e0} \sim N_e/N_{e0}$ (Bowen et al., 1964; Samir & Willmore, 1965; Henderson & Samir, 1967). The shape of this satellite was close to spherical, but it had a complex surface structure, which apparently is the primary reason for the large scatter of the experimental results. Fig. 11.4 gives these results in a somewhat smoothed form (broken and solid curves). The broken curve in this figure gives the $N_n(\theta, r)/N_{n0}$ relation calculated in the neutral approximation for a mean value $a_b \simeq V_b/v_i$ corresponding to the conditions

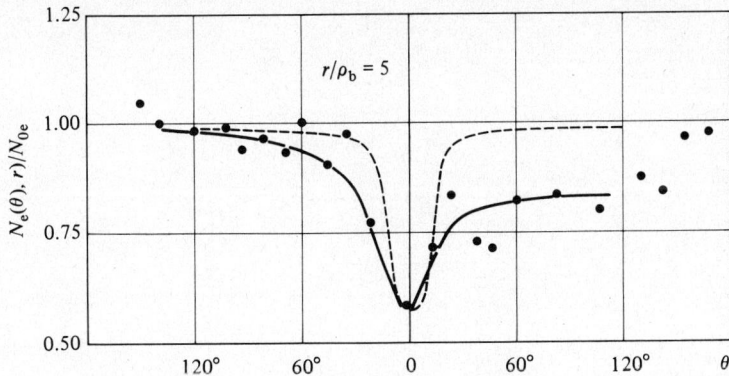

Fig. 11.4. Angular dependences of $N_e(\theta, r)/N_{n0}$ obtained in experiments on Ariel 1 (dots and solid line). The broken line indicates the theoretical $N_n(\theta, r)/N_{n0}$ relation (Gurevich et al., 1969a).

of the experiments (Gurevich et al., 1969a). In these experiments, because of the considerable variation in the ion composition and temperature with altitude, a_b was found to change greatly, thereby complicating considerably the theoretical analysis of the experimental results.

Taking these things into account, the fit between the experimental results and the calculations can be said to be satisfactory. It is important to note that the minimum values of $N_e(\theta, r)$ are almost exactly equal to the anticipated values of the the neutral-particle density for $\theta = 0$. These data imply the following significant conclusion: at a distance $r = 5\rho_b$ there apparently already exist in the plasma conditions for which $N_e(\theta, r) \simeq N_i(\theta, r)$.

Fig. 11.5 shows the results of laboratory measurements of the angular

Theory

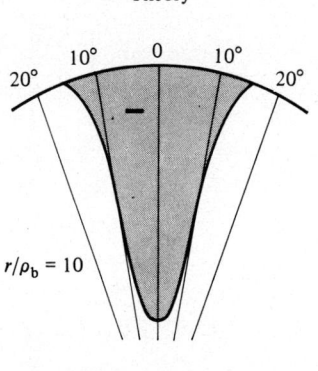

Experiment

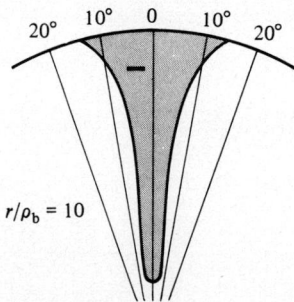

Fig. 11.5. Angular dependences of the perturbations of ion density $\delta N_i(\theta, r) = [N_i(\theta, r) - N_{i0}]/N_{i0}$ (experiment under laboratory conditions; $\phi_b = 0$, $V_b/v_s = (6-50)$, $\rho_b/D_e = 5$, $r/\rho_b = 10$, Clayden & Hurdle, 1964) and neutral-particle density $\delta N_n(\theta, r)$ (theory; $\phi_b = 0$, $V_b/v_i = 8$, $\rho_b/D_e \gg 1$, $r/\rho_b = 10$, Gurevich et al., 1969a) behind a body around which particles are flowing rapidly.

11.1. Neutral approximation

dependence $\delta N_i(\theta, r) = [N_i(\theta, r) N_{i0}]/N_{i0}$ for $\phi_0 = 0$ and the corresponding theoretical curve of $\delta N_n(\theta, r)$; in this case, too, the fit is quite good (Clayden and Hurdle, 1964). The theoretical $\delta N_n(\theta, r)$ relation was plotted in this figure from the data of Gurevich et al., 1969a.

It should be noted that below we will often use representations of the angular dependences of various quantities that are similar to the one in Fig. 11.5. In these figures the line in the form of an arc corresponds to the zero level $\delta N = 0$. Above the arc, values of $\delta N > 0(+)$ are plotted as a function of θ, while below it are values of $\delta N < 0(-)$, the angle θ being reckoned behind the body on either side of the z-axis, which is parallel to $\mathbf{V}_0$.

The results of a more accurate quantitative comparison of laboratory measurements of δN_i with theoretical calculations are presented in Fig. 11.6 (Bogashchenko et al., 1970). These data pertain to measurements in a magnetized plasma with an ion stream impinging upon a disk. The curves labelled 1 in Fig. 11.6 were plotted according to calculations in the neutral approximation, while curve 2 takes into account the effect of the

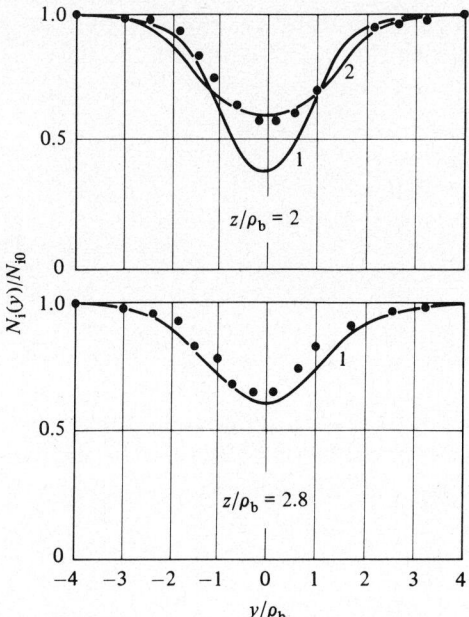

Fig. 11.6. Variation $N_i(y)/N_{i0}$ behind a sphere with rapid flow around it, in the direction normal to the direction $\mathbf{V}_0 \| z$ (i.e. $y \perp z$), at various distances from the axis ($z = \text{const}$); $V_0/v_i = 2$, $\rho_{Hi}/\rho_b = 5$ (dots = experiment; 1 and 2 = theoretical curves). (Bogashchenko et al., 1970).

electric field (see below); the dots correspond to measurements. For $z/\rho_b = 2$, the neutral approximation is seen to portray the experimental results poorly in the shadow region of the body $z/\rho_b \lesssim \pm 1$, while at a distance $z/\rho_b = 2.8$ the agreement is good for all y. According to these experimental data, the neutral approximation is useful up to $z/\rho_b \simeq 4$ to 5. At greater distances the effect of the magnetic field begins to manifest itself, but the influence of the electric field apparently need be taken into account only for $z/\rho_b \gtrsim 18$ to 20, i.e., when $z/\rho_b \gtrsim 9V_b/v_i$ to $10V_b/v_i$. Flow around a plate, half-plane, or cylinder, and similarity effects for plasma flow around bodies of various shapes, were studied in Astrelin et al., 1971, the experimental and theoretical results being compared, just as in Fig. 11.6.

11.2. Effect of an external static magnetic field

In the presence of an external magnetic field $\mathbf{H}_0$, the particles gyrate around the vector $\mathbf{H}_0$ and their trajectories become complicated. However, it is easy to see that $\mathbf{H}_0$ will influence the shape of the body wake only at distances $z \gtrsim (V_b/v_i)\rho_{Hi}$.

For a body with a circular cross-section moving along a magnetic field $(\mathbf{H}_0 \parallel \mathbf{V}_b)$, the ion density is described, without taking into account the effect of either the electric field or collisions, with the aid of the formula

$$N_i(\rho,z) = N_{i0}\left\{ 1 - 2\exp\left[-\left(\frac{\rho}{2\rho_{Hi}\sin(z\Omega_{Hi}/2V_b)}\right)^2 \right] \right.$$
$$\left. \times \int_{b_0}^{\infty} u\exp(-u^2) I_0(bu)\,du \right\} \tag{11.5}$$

where

$$b_0 = \frac{1}{2}\frac{\rho}{\rho_{Hi}}\sin\left(\frac{z\Omega_{Hi}}{2V_b}\right),\ b = \frac{\rho}{\rho_{Hi}}\sin\left(\frac{z\Omega_{Hi}}{2V_b}\right),\ \Omega_{Hi} = \frac{\mu_0 eH}{M},\ \rho_{Hi} = \frac{v_i}{\Omega_{Hi}}.$$

During the derivation of (11.5) it was assumed that the undisturbed plasma is quasi-neutral ($N_{e0} = N_{i0}$), isothermal ($T_e = T_i$), and consists of electrons together with ions of only one species. Obviously, at small distances from the body, when

$$z \ll \frac{1}{2}\frac{V_b}{\Omega_{Hi}},\ \sin\left(\frac{z\Omega_{Hi}}{2V_b}\right) \simeq \frac{z\Omega_{Hi}}{2V_b} \tag{11.6}$$

formula (11.5) becomes formula (11.2a), i.e., in the near zone of the body, as noted earlier, the effect of the magnetic field disappears.

On the z-axis, where $\rho = 0$ and $I_0 = 1$, formula (11.5) assumes the

11.2. Effect of an external static magnetic field

following simple form:

$$N_i(z,0) = N_{i0} \exp\left[-\left(\frac{\rho_b}{2\rho_{Hi}\sin(2\pi z/2\Lambda_z)}\right)^2\right] \quad (11.5a)$$

indicating that the ion density has a periodic structure along the z-axis, with a spatial period

$$\Lambda_z = 2\pi \frac{V_b}{v_i}\rho_{Hi} = 2\pi \frac{V_b}{\Omega_{Hi}} \quad (11.7)$$

(Gurevich, 1960). The case with $\mathbf{V}_b$ parallel to $\mathbf{H}_0$ is special, inasmuch as the density perturbation δN_i does not diminish with increasing distance in the absence of collisions: in this case the ions move helically along $\mathbf{H}_0$, and the rarefied region (the wake of the body) is not filled with particles; it has a cylindrical shape. However, under the influence of collisions, the oscillatory nature of the wake gradually becomes blurred, its structure is no longer strictly periodic, and the wake itself disappears at distances of the order of the mean free path of the particles.

From formula (11.5) it is evident that, if the conditions (11.6) are satisfied, we have

$$\delta N_i \simeq -\left(\frac{V_b}{v_i}\right)^2\left(\frac{\rho_b}{z}\right)^2 = -\frac{S_b}{\pi}\left(\frac{a_b}{z}\right)^2. \quad (11.8)$$

Therefore, along the axis of the body the perturbation of the ion density decreases as z^{-2}, even when no magnetic field $\mathbf{H}_0$ is present.

If the body moves perpendicularly to the magnetic field ($\mathbf{V}_b \perp \mathbf{H}_0$), then it is simpler to calculate the particle density in the wake of a rectangular plate. For a circular plate, when $\mathbf{V}_b \perp \mathbf{H}_0$ there is no axial symmetry and the formulas are very complicated. For a plate with a rectangular section, with sides ρ_x and ρ_y,

$$N_i(x,y,z) = N_{i0}\left\{1 - \frac{1}{4}\left[\left[\phi\left(\frac{V_b}{v_i}\frac{x-\rho_x}{z}\right) - \phi\left(\frac{V_b}{v_i}\frac{x+\rho_x}{z}\right)\right]\right.\right.$$
$$\left.\left.\times\left[\phi\left(\frac{y-\rho_y}{2\rho_{Hi}\sin(z\Omega_{Hi}/2V_b)}\right) - \phi\left(\frac{y+\rho_y}{2\rho_{Hi}\sin(z\Omega_{Hi}/2V_b)}\right)\right]\right]\right\}.$$
$$(11.9)$$

It follows from (11.9) that along the axis of the plate (in the direction $\mathbf{V}_b \| z$), where $x=0$ and $y=0$, we have

$$N_i(z=0) = N_{i0}\exp\left[1 - \phi\left(\frac{a_b\rho_x}{z}\right)\phi\left(\frac{\rho_y/2\rho_{Hi}}{\sin(z\Omega_{Hi}/2V_b)}\right)\right], \quad (11.9a)$$

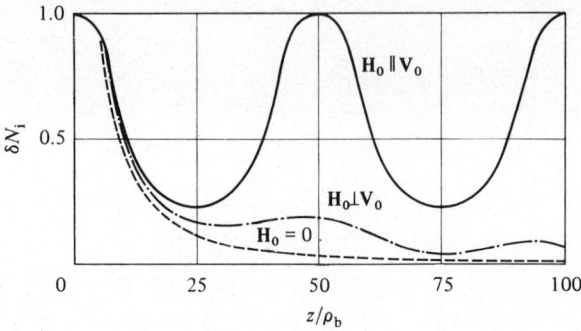

Fig. 11.7. Theoretical curves for the perturbation of ion density δN_i as a function of z/ρ_b, behind a sphere moving rapidly through a magnetoplasma (Gurevich, 1960).

where

$$\Phi(x) = (2/\pi^{1/2})\int_0^x \exp(-u^2)\,du \qquad (11.10)$$

is the error function. Obviously, for $\mathbf{V}_b \perp \mathbf{H}_0$ the perturbation of the ion density also varies periodically, but, in contrast to the case where $\mathbf{V}_b \parallel \mathbf{H}_0$, here the quantity δN_i decreases with the distance even in the absence of collisions, albeit more gradually than in the neutral approximation (as z^{-1}, on the average). For comparison, Fig. 11.7 shows the δN_i curves calculated using formulas (11.4), (11.5a), and (11.9).

A quasi-periodic structure of the body wake was detected by Barrett, 1964, and a quantitative comparison of the corresponding theoretical and experimental results was carried out in the previously cited work (Bogashchenko et al., 1970), taking into account the effect of collisions on the ion temperature. Fig. 11.8 presents the results of two series of such measurements of $J_i(0,z)/J_{i0} \simeq N_i(0,z)/N_{i0}$, for values $V_b/v_i = 2$ and 2.6 and $\rho_{Hi} = 0.4$ and 0.2 cm. In part (a) of the figure the fit between experiment (dots) and theory (solid curve) is quite good. In (b), however, which corresponds to measurements with a magnetic field about twice as strong, the periodicity of the wake agrees well with that predicted theoretically but the experimental values of the ion density are higher than the theoretical values. This state of affairs was verified in many cases during the course of these experiments, the discrepancy being greater for higher values of ρ_b/ρ_{Hi}. The authors of the work assume that, since in a magnetic field the effect of the electric field increases the ion density along the z-axis (Pitaevskii, 1961; Vas'kov, 1966), therefore beginning at certain values of $\mathbf{H}_0$ the influence of the electric field has to be taken into account. With

11.3. Effect of an electric field

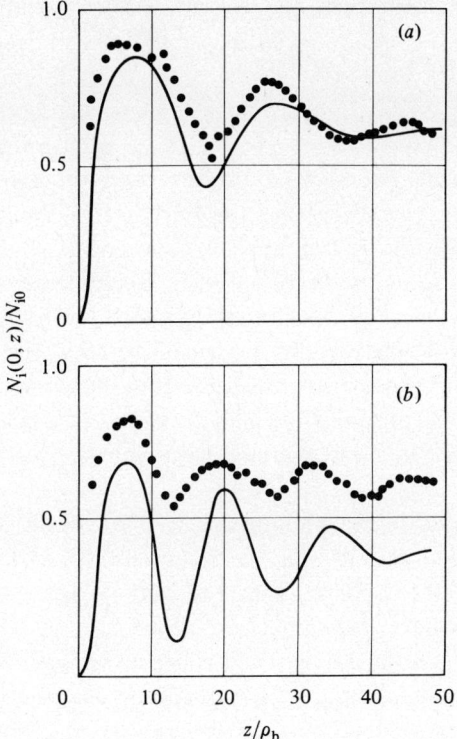

Fig. 11.8 Experimental (dots) and theoretical (solid curves) dependences of $N_i(0,z)/N_{i0}$ on z/ρ_b in a magnetoplasma (Bogashchenko *et al.*, 1970).

an increase in ρ_b/ρ_{Hi}, the rarefaction of the body wake is enhanced and the role of the electric field becomes greater.

11.3. Effect of an electric field

In the vicinity of a body moving in a plasma an electric field $E(\theta, r)$ is produced due to the differences between the trajectories of the ions and the electrons. This leads to a corresponding difference of the densities: $N_e(\theta, r) \neq N_i(\theta, r)$. Quite understandably, the body potential ϕ_b also produces an electric field, and it is precisely this field source that is, in a number of cases, a decisive factor with regard to the structure of the plasma disturbance. Naturally, the role of the electric field is important right up to the boundary of the region of Debye screening, and also in the far zone of the body, where the plasma is quasi-neutral: $N_e \simeq N_i$. At great distances, because the perturbation of the density δN is small, the relative effect of the field is enhanced and, as we saw above, the plasma perturbation has a complicated structure: the angular dependences $N_i(\theta, r)$ are more complex than close to the body. In addition, the role of the external magnetic field is greater far away from the body. On the other hand, in the immediate vicinity of the body, behind it in the region of high rarefaction the structure of the perturbation becomes simpler, and in a certain zone, as we saw earlier, the perturbation is in general described satisfactorily using the neutral approximation.

The relative size of the body ρ_b/D_e, in combination with the electric field, also plays a major role in determining the form of the plasma perturbation; with a decrease in the size of the body, the relative effect of its potential is enhanced and in the near zone effects will arise which are analogous to those observed for high potentials of a large body at great distances from it. The nonisothermal nature of the plasma ($T_e \neq T_i$) has a similar effect.

In the literature the terminology of gas dynamics is often invoked to describe the effects arising in a plasma around a rapidly moving body. For instance, a Mach cone is referred to, with regard to the wake behind a body. Moreover, even the simple analogy between the Mach cone, which is observed in a 'continuous gas' and is hydrodynamic in nature, and the wake of a body in a plasma is a purely formal one; this could be termed an analogy of the 'geometric' type. In contrast to a Mach cone, which has sharp boundaries, the structure of a body wake in a plasma is a kinetic in nature and depends considerably on the effect of the electric field. Moreover, a wake has quite blurred boundaries, a fact which is connected with the attenuation of ion-acoustic waves. In a number of cases the wake has a multi-lobed structure: regions of rarefaction and concentration of particles, etc. Consequently, by its very nature as well as by its structure, the wake of a body in a plasma differs considerably from a Mach cone. It should also be noted that in gas dynamics the appearance of shock waves ahead of a body is typical, whereas this does not take place in the phenomena being considered here. Thus it is inadvisable and physically incorrect to employ hydrodynamic terminology and to call the wake of a body moving rapidly in a plasma a Mach cone. It is the opinion of the author that such a misuse of terminology leads to confusion, preventing a physical understanding of the various phenomena.

Before going on to a more detailed description of the phenomena, let us list briefly their main general features.

1. In the immediate vicinity of the surface of a body, under the influence of the electric field the charged-particle density will be considerably higher

than that predicted in the neutral approximation. The angular dependences $N_{e,i}(\theta, r)$ and $N_n(\theta, r)$ are qualitatively the same, but the electron density $N_e(\theta, r)$ is much higher than the ion density $N_i(\theta, r)$.

2. Behind the body, focusing of charged particles takes place. The region of maximum rarefaction lies on both sides of the axis, on a conical surface with an opening angle $\theta_{max} = \sin^{-1}(v_i/V_b)$ or $\theta_{max} = \sin^{-1}(v_s/V_b)$ (where $v_s = (kT_e/M)^{1/2}$ is the velocity of nonisothermal sound).

3. The focusing in the vicinity of the body axis is in some cases so great that the density $N_i(\theta, r)$ in a certain angle range $\Delta\theta$ exceeds the density of the undisturbed plasma N_{i0}, i.e., $\delta N_i > 0$, meaning that particle concentration takes place.

4. These focusing effects become intensified if the negative potential of the body or the degree of nonisothermality of the plasma is increased, or if the linear dimensions of the body decrease. They depend on the relative velocity of the body $a_b = V_b/v_i$ and on its shape.

5. Further away from the body, at quite great distances from it, but slightly off the axis, two concentration regions may appear. In this case one or three rarefaction regions are observed.

6. The above effects are possible either in the presence of or in the absence of an external magnetic field $\mathbf{H}_0$. Under the influence of a magnetic field the structure of the disturbance in the far zone is smoothed out and becomes asymmetrical about the rotation axis, if the velocity vector $\mathbf{V}_b$ makes an angle $\theta \neq \pi/2$ with $\mathbf{H}_0$.

11.3.1. Near and intermediate zones. Large and small bodies ($\rho_b \gg D_e$ and $\rho_b \ll D_e$)

The theoretical formulas taking into account the effect of the electric field are very complicated and are generally integral formulas. Thus the results of solving the various problems become usable only with the aid of numerical calculations. In a number of cases general use is made of a numerical solution of the differential equations of particle motion together with the Poisson equation, and the results are presented in graphical form, rather than analytically. Only in certain cases is it possible to obtain sufficiently simple formulas. This enables a direct comparison to be made between the results of experiment and of theory, provided that the experimental conditions correspond to the limitations of the theory or, on the other hand, provided that the experiments can be so carried out as to satisfy the conditions complying with the theoretical formulas.

For an infinitely long circular cylinder with a radius $\rho_b \gg D_e$, moving through an isothermal plasma in a direction normal to its axis, in a region close to the cylinder surface, where $z \ll \rho_b V_b/v_i$, the ion density in the vicinity of the body can be described approximately by the following formula:

$$N_i(r,\theta) = N_0 A\left(\frac{T_e}{T_i}\right)\left\{\exp\left[-(V_b/s)\left(\pi-\theta-\sin^{-1}\frac{\rho_b}{r}\right)\right.\right.$$
$$\left.-\frac{1}{2}\left(\pi-\theta-\sin^{-1}\frac{\rho_b}{r}\right)^2\right] + \alpha\exp\left[-\frac{V_b}{v_s}\right.$$
$$\left.\left.\times\left(\pi-\theta+\sin^{-1}\frac{\rho_b}{r}\right)-\frac{1}{2}\left(\pi-\theta+\sin^{-1}\frac{\rho_b}{r}\right)^2\right]\right\} \quad (11.11)$$

(Gurevich et al., 1969a). In (11.11) the coefficient $A(T_e/T_i) \simeq 1$, while $\alpha = 1$ or 0, depending on whether the radius vector makes an angle

$$\theta > \frac{\pi}{2} - \sin^{-1}\left(\frac{\rho_b}{r}\right) \quad \text{or} \quad \theta < \frac{\pi}{2} - \sin^{-1}\left(\frac{\rho_b}{r}\right)$$

with the direction of $\mathbf{V}_b$, i.e., depending on whether the observation point is outside or inside the 'shadow' of the cylinder (the volume enclosed by the tangent to the cylinder parallel to the vector $\mathbf{V}_b$).

In the wake of a circular disk of radius $\rho_b \gg D_e$, on its axis ($\theta = 0$) we have

$$N_i(z,0) = N_{i0} A\left(\frac{T_e}{T_i}\right)\left(\frac{2\pi\rho_b}{z^2+\rho_b^2}\right)^{1/2}\left(\frac{T_e}{T_i}\right)\cdot\left[\frac{V_b}{v_s}+\frac{z}{\rho_b}+\tan^{-1}\left(\frac{\rho_b}{z}\right)\right]$$
$$\times \exp\left[-\left(\frac{V_b}{v_s}\right)\cdot\tan^{-1}\left(\frac{\rho_b}{z}\right)-\frac{1}{2}\left[\tan^{-1}\left(\frac{\rho_b}{z}\right)\right]^2\right], \quad (11.12)$$

where $A(T_e/T_i)$ varies over the same limits as in formula (11.11).

Fig. 11.9 shows the contours $N_i(\rho,z)/N_{i0} = \text{const.}$ for two ratios T_e/T_i, calculated in Gurevich et al., 1969a, according to a more general formula than (11.12), which will not be given here because of its clumsiness. The results of these calculations illustrate the effect of the nonisothermal nature of the plasma on the structure of the wake of a rapidly moving body.

Simple formulas are obtained for an isothermal plasma and a weakly charged body of small size $\rho_b \ll D_e$, i.e., strictly speaking, for a point charge $Q = \phi_b \rho_b$, with $Q < 0$, assuming that the conditions

$$b = \frac{e|Q|}{kT}\left(\frac{v_i}{V_b}\right)\frac{1}{D_e} = \frac{e|\phi|\rho_b}{kT D_e}\left(\frac{v_i}{V_b}\right) \ll 1, \quad \frac{e|\phi_b|}{kT} \simeq \frac{V_b}{v_i}. \quad (11.13)$$

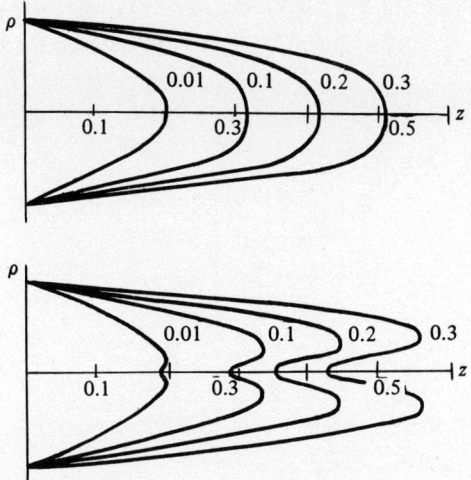

Fig. 11.9. Lines of equal values of the ratio $N_i(\rho,z)/N_{i0}$ in the wake of a rapidly moving circular disk. The ratio $\rho_b V_b/2^{1/2} v_s$ is plotted on the z-axis. The upper family of curves was calculated for $T_e/T_i = 1$, the lower family for $T_e/T_i = 4$ (Gurevich et al., 1969a).

are satisfied. In this case at distances

$$r \ll D_e(V_b/v_i) \tag{11.14}$$

the perturbation of the ion density δN_i is

$$\delta N_i(r,\theta) = \frac{e|\phi_b|}{kT} \frac{\rho_b}{r} F\left(\sin\theta, \frac{V_b}{v_i}\right), \tag{11.15}$$

where asymptotically behind the body

$$F\left(\sin\theta, \frac{V_b}{v_i}\right) = -2.2\left\{\mathrm{Im}\left[\left(\frac{V_b}{v_i}\sin\theta\right)^2 + 0.6 + 0.87i\right]^{-3/2}\right\} \tag{11.16}$$

(Dubovoi, 1972). In (11.16) Im denotes the imaginary part of the expression in square brackets. The field inside the region of Debye screening ($r \ll D_e$) is a Coulomb field, i.e.,

$$\phi(r) = -|Q|/4\pi\varepsilon_0 r = -|\phi_b|\rho_b/r \tag{11.17}$$

and, accordingly, the perturbation of the electron density is

$$\delta N_e(r) = -(e|\phi_b|/kT_e)(\rho_b/r). \tag{11.18}$$

At greater distances, i.e., for $r \lesssim D_e$, we have

$$\phi(r) = -(|\phi_b|\rho_b/D_e)\exp(-r/D_e), \quad \delta N_e(r) = -e|\phi(r)|/kT_e. \tag{11.19}$$

In the transition zone, where r/D_e increases gradually, becoming comparable to and then greater than V_b/v_i, the perturbation δN_i can be determined for $\rho_b \ll D_e$ using numerical integration (Dubovoi & Yaroslavtsev, 1976):

$$\delta N_i(\theta, r) = \frac{\rho_b}{r} \frac{e|\phi_b|}{kT} \operatorname{Im}\left\{ \frac{2}{\pi} \int_0^{\pi/2} W'\left(\frac{V_b}{v_i} \cos\chi \sin\theta\right) d\chi \right.$$

$$- \frac{1}{\pi^{1/2}} \int_0^{V_b/v_i} [W'(p)S(p)] dp \int_0^{\pi} \exp\left[-S(p)\frac{r}{D}\frac{v_i}{V_b}\right] \quad (11.20)$$

$$\left. \times |p\cos\theta + \sin\theta \cos\chi(1-p^2)^{1/2}| \right] d\chi \Big\},$$

where Im, as in (11.16), is the imaginary part of the expression in braces;

$$W(p) = \exp(-p^2)\left[1 + (2i/\pi^{1/2})\int_0^p \exp(t^2) dt\right] \quad (11.21)$$

is the Kramp function (see (5.26) and (5.28) of Volume 1); $W'(p)$ is its derivative; and

$$S(p) = [2 + ip\pi^{1/2} W(p)]^{1/2}. \quad (11.22)$$

In the far zone, where $\delta N_i \simeq \delta N_e \simeq -e|\phi(r)|/kT$, namely at distances $r/D_e \gtrsim 10 V_b/v_i$, the disturbance of the electron density is found using formula (11.29), to be given below. The $\delta N(\theta, r)$ curves obtained with the aid of (11.20) and (11.29) will be given below.

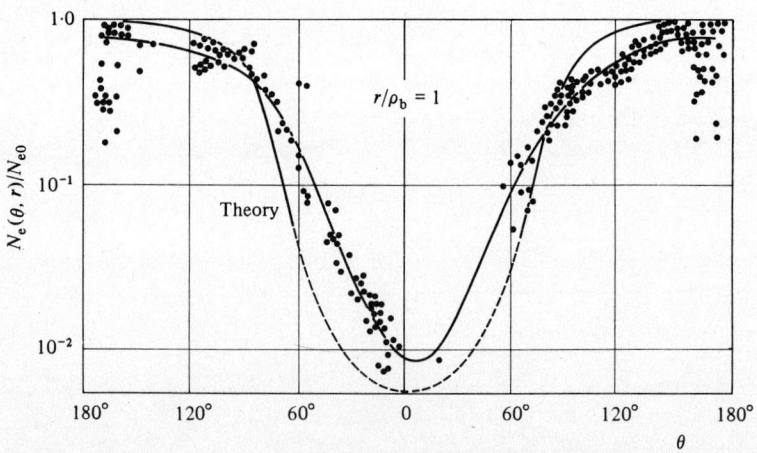

Fig. 11.10. Angular dependences of the ratio $N_e(\theta,r)/N_{e0}$. The experimental points were obtained with Ariel 1 (Bowen et al., 1964; Henderson & Samir, 1967; Samir & Willmore 1966); the theoretical curve was calculated using formula (11.11), the dashed part being the region of limited applicability of the formula (Gurevich et al., 1969a).

11.3. Effect of an electric field

Fig. 11.10 shows the $N_i(\theta,r)/N_{i0}$ curve calculated using formula (11.11), compared with the data of measurements of $N_e(\theta,r)$ at the surface of Ariel 1, which had an approximately spherical shape; these two curves are comparable because the dimensions of the wake are much greater than the Debye length D_e, so the plasma there should be quasi-neutral ($N_i \simeq N_e$). An average value of V_b/v_i satisfying the experimental conditions was used for the calculations. In general, the fit between theory and experiment is seen to be quite good, not only qualitatively but quantitatively as well. Note that the minimum value $N_e(\theta,r)/N_{e0} \simeq 10^{-2}$ for $\theta = 0$ obtained in these experiments is close to the theoretical value, being about three orders of magnitude higher than the corresponding minimum $N_n(0,r)/N_{n0} \simeq 10^{-5}$ to be expected under these conditions using the formulas of the neutral approximation. Here, as mentioned above, the electric field close to the body has a considerable effect.

Fig. 11.11 shows a similar comparison between theory and experiment, for measurements carried out aboard Explorer 31, which had the shape

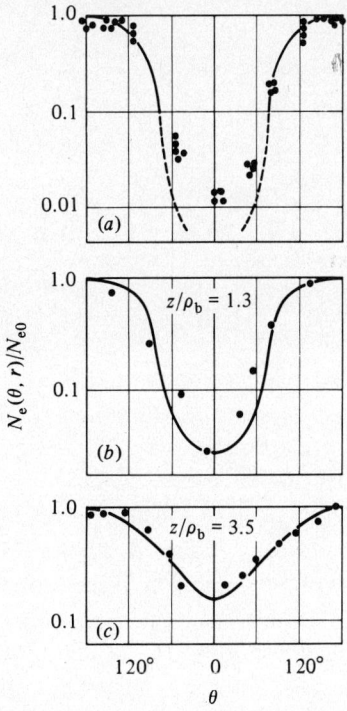

Fig. 11.11. Comparison of the measurement points for $N_e(\theta,r)/N_{e0}$ from Explorer 31 with the theoretical curves (Gurevich et al., 1969a).

of an octahedral parallelepiped. The data in Fig. 11.11a correspond to conditions when the electron density was close to the ion density, since one species of ion (O^+) predominated in the ionosphere, their relative content being $\simeq 99\%$ and $a_b = V_b/v_i(O^+) \simeq 5$. In this case there is good agreement between the experimental results and the calculations using formula (11.11), except for the range of angles $-60° < \theta < 60°$, where this formula is not too applicable. However, in another case protons (H^+) predominated in the makeup of the ionosphere, and $a_b = V_b/v_i(H^+) \simeq 1.2$, i.e., a value about $\frac{1}{4}$ of the measured values given in Fig. 11.11a.

When H^+ ions predominate, due to the decrease in a_b the effect of the electric field is weakened substantially and does not have to be taken into account when analysing experimental results for the boundary of maximum rarefaction, where the values of N_e and N_i become almost identical. This boundary is characterized by the angle ψ_0 at which it is viewed from the point where the corresponding probes are located. For the experiments described by Gurevich et al., 1969a, an angle $\psi_0 \simeq 45°$ was chosen; apparently this value of ψ_0 is common to many experiments. To sum up, when the relative proton content is greater than 30%, the following formula of the neutral approximation is quite satisfactory for the plasma density close to the body:

$$N(\theta) = \frac{N(O_1^+)}{2}\left\{1 + \phi\left[\frac{V_b}{v_i(O_1^+)}\cos\psi_0\cos\theta\right]\right\}$$
$$+ \frac{N(H_1^+)}{2}\left\{1 + \phi\left[\frac{V_b}{v_i(H_1^+)}\cos\psi_0\cos\theta\right]\right\}, \qquad (11.23)$$

where ϕ is the error function (11.10). Fig. 11.11b and c shows two series of experimental results (dots), compared with theoretical curves calculated using formula (11.23) for relative values of H^+ equal to 0.23 and 0.94, at distances from the body centre $z/\rho_b \simeq 1.3$ and 3.5. The fit is seen to be good even quite far away from the body.

Some results of measuring the angular dependences of the electron and ion densities on Explorer 31 close to the surface of the body have been analysed in detail theoretically (Samir & Jew, 1972; Samir et al., 1973, 1975). Figure 11.12 shows a comparison of the ratios $N_e(\theta)/N_{e0}$, averaged over intervals $\Delta\theta = 15°$ and $30°$ for eight series of measurements, with the corresponding theoretical relations. The $N_e(\theta)$ values were calculated in these studies on the basis of numerical integration of formulas derived in the theory of flow around a sphere in an isotropic plasma worked out by Liu and Jew (Liu, 1967, 1969; Liu & Jew, 1967). Due to their complexity, these formulas will not be given here. For the theoretical calcultions, Samir

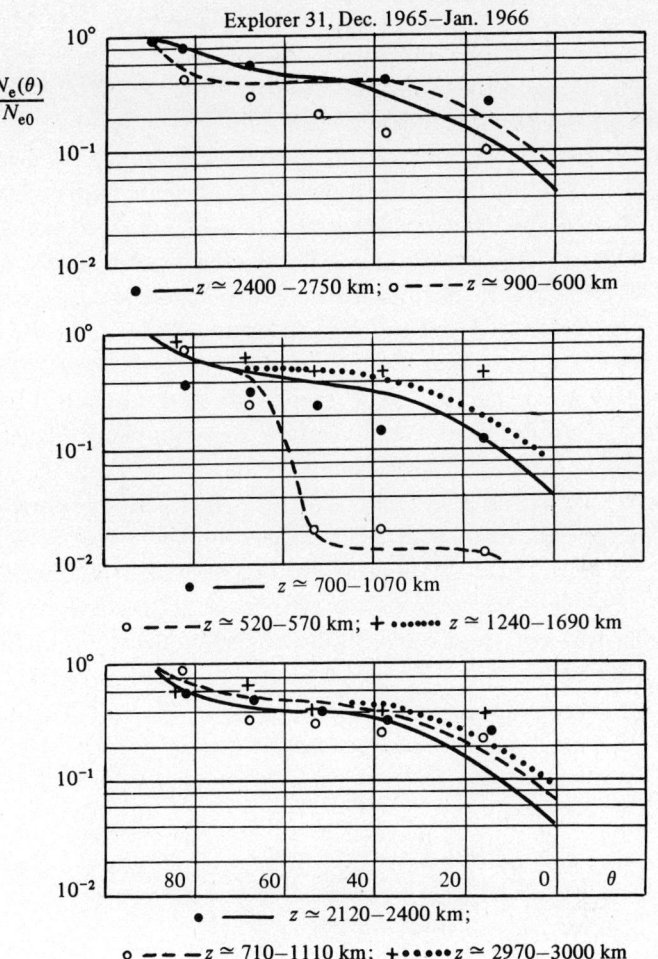

Fig. 11.12. Comparison of the ratios $N_e(\theta)/N_{e0}$ measured close to the surface of Explorer 31 (dots, circles, crosses) with theoretical calculations corresponding to the experimental conditions (curves). (Samir & Jew, 1972.)

et al., 1973 and 1975, selected the most reliable experimental data obtained in various altitude intervals from $z \simeq 520$ km to $z \simeq 3000$ km, where the other characteristic parameters varied as follows:

$$\left.\begin{array}{l} M^+ \simeq 1 \text{ to } 16, T_e \simeq 2 \times 10^3 \text{ to } 3 \times 10^3 \text{ K}, N_{e0} \simeq 10^3 \text{ to } 8 \times 10^3 \text{ cm}^{-3}, \\ V_b \simeq 6 \text{ to } 8 \text{ km s}^{-1}, a_b = V_b/v_i \simeq 0.85 \text{ to } 3.7, \\ \rho_b/D_e \simeq 2 \text{ to } 43, (e\phi_b/kT_e) \simeq -3.8 \text{ to } -5.7. \end{array}\right\}$$

(11.24)

M^+ is the average ionic mass, in units of the proton mass.

An analysis of the results (see Fig. 11.12) indicates that in general there is a satisfactory fit between the theoretical and experimental data, in spite of the fact that the values of some parameters used in the theory are not too close to the experimental conditions. In some cases there was good quantitative agreement, but the experimental values of $N_e(\theta)/N_{e0}$ generally were higher than the theoretical values. Their mean values exceeded the calculated values by a factor of less than 1.5 to 2.5. Note that the spread of individual values of $N_e(\theta)/N_{e0}$ relative to the mean values given in Fig. 11.12 is in some cases quite great, exceeding the observed discrepancy between the theoretical and experimental data (see Samir & Jew, 1972).

The results of measuring the angular dependence $N_i(\theta)$ were analysed by Samir *et al.*, 1973, 1975, with the aid of three kinds of theoretical formulas: the formulas of the neutral approximation suggested by Gurevich *et al.*, 1969a, the formulas derived by Liu, 1969, and the modified formula (11.23) suggested by Samir *et al.*, 1973, themselves. Formula (11.23) describes the experimental data adequately for low values of $a_b = V_b/v_i$. This agrees with the conclusions drawn by Gurevich *et al.*, 1969a.

The *focusing of particles* in the wake of a body gradually leads to conditions whereby the disturbed value of the ion density $N_i(\theta)$ begins in a number of cases to exceed considerably the undisturbed value N_{i0}. Focusing already appears in the near zone of the body, at distances $r < \rho_b V_b/v_i$. This effect manifests itself especially strongly in a nonisothermal

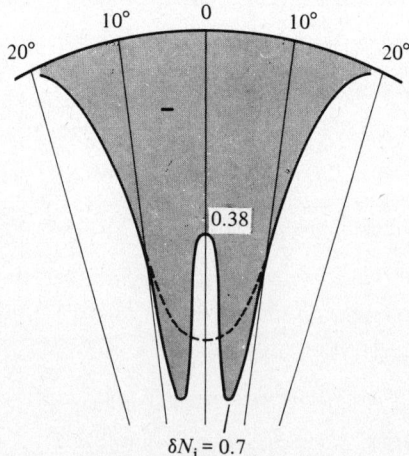

Fig. 11.13. Theoretical angular dependences of the perturbation of ion density δN_i in a nonisothermal plasma ($\phi_b = 0$, $V_b/v_i = 8$, $\rho_b/D_e \gg 1$, $r/\rho_b = 4.5$); solid curve, $T_e/T_i = 4$; dashed curve, $T_e/T_i = 1$ (Gurevich *et al.*, 1969a; Alpert *et al.*, 1963).

11.3. Effect of an electric field

plasma, when the body surface is negatively charged and its potential $\phi_b < 0$. Fig. 11.10, above, already showed some theoretical results illustrating this effect. Here let us take a look at the corresponding theoretical data and results of laboratory experiments.

The theoretical dependence of the perturbed density $\delta N_i(\theta)$ for $T_e/T_i = 4$ and $r/\rho_b = 4.5$ is shown in Fig. 11.13. This relation was calculated taking into account the effect of the electric field. It is seen that for $\theta = 0$ the rarefaction of the ions is only about half as great as for $\theta = \pm 5°$; an analogous effect is not observed when $T_e = T_i$. The experimental $\delta N_i(\theta)$ relation depicted in Fig. 11.14a illustrates this effect; it shows a good fit with the theoretical curve in Fig. 11.12. According to estimates in these

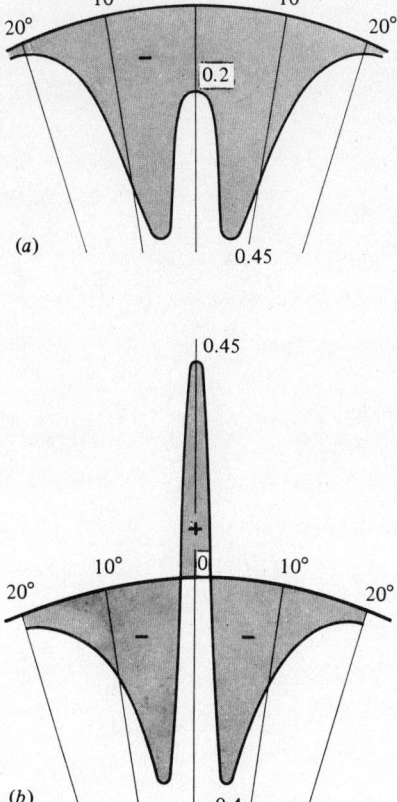

Fig. 11.14. Experimental angular dependences of the ion-density perturbation δN_i behind a sphere in a rapidly moving stream. Measurements under laboratory conditions: (a) $\phi_b \simeq 0, V_b/v_s \simeq 1.1, \rho_b/D_e \simeq 30, T_e/T_i \simeq 5, r/\rho_b = 7.3$; (b) $\phi_b \simeq -kT/e$, $V_b/v_s \simeq 1.1, \rho_b/D_e \simeq 10, T_e/T_i \simeq 5, r/\rho_b = 6.8$ (Skvortsov & Nosachev, 1968).

measurements, $T_e/T_i \simeq 5$. However, Fig. 11.14b, which shows the $\delta N_i(\theta)$ relation obtained in these same experiments but with a negative body potential $\phi_b < 0$, corresponds to a case of quite strong focusing. In a certain range of angles $\delta N_i > 0$, i.e., the disturbed value of the ion density N_i is even higher than N_{i0}. Values of $\delta N_i > 0$ were also detected in other experiments (Clayden & Hurdle, 1964; Hall et al., 1964), and this effect

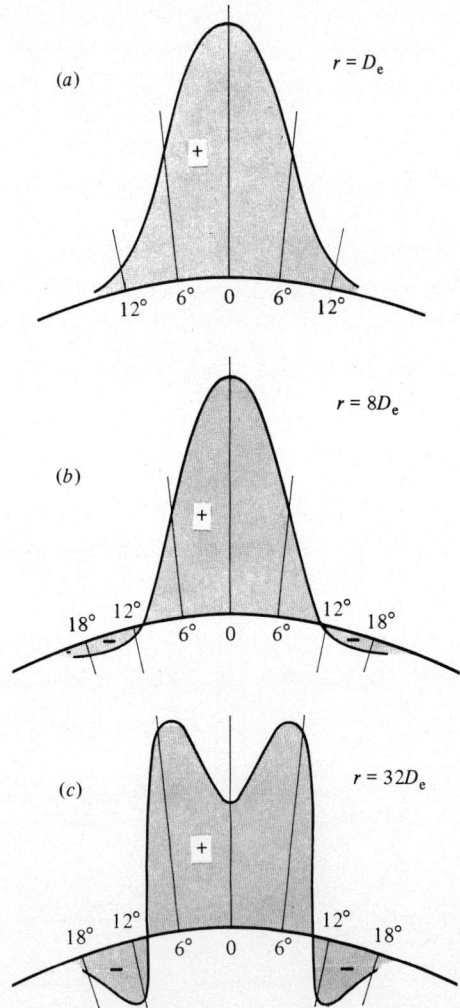

Fig. 11.15. Theoretical angular dependences of the ion-density perturbation behind a rapidly moving point charge at various distances from it ($\rho_b \ll D_e$, $V_b/v_s = 8$, $T_e/T_i = 1$). (Dubovoi, 1972).

was predicted theoretically even before this (Moskalenko, 1964a).

For very small bodies ($\rho_b \ll D_e$) the focusing effect is enhanced significantly. For example, at distances $r \simeq D_e$, as Fig. 11.15 shows, there is in general no rarefaction region behind the body. Only with an increase in the distance does a region gradually appear in which $\delta N_i < 0$, the values $|\delta N_i|$ becoming significant at great distances.

Detailed studies were made of the evolution of the angular dependence $\delta N_i(\theta, r)$ behind spheres of two sizes: $\rho_b/D_e = 14$ and $\rho_b/D_e = 1.8$ (Hester & Sonin, 1970a). The body potentials in these experiments amounted to fractions of a volt. Fig. 11.16a and b, shows some of the experimental results. At distances from the body $r/\rho_b \simeq V_b/v_s$, as may be seen from Fig. 11.15 as well, positive focusing of the ions was observed: $\delta N_i > 0$. The angular dependences $\delta N_i(\theta, r)$ thus gradually become more complicated for $r/\rho_b > V_b/v_s$ and they closely approximate the corresponding theoretical relations to be considered in the next section, for the far zone of the body $r/\rho_b \gg V_b/v_i$.

Fig. 11.17 shows some similar results of measurements of the ion-density distribution behind a small body ($\rho_b \ll D_e$)–small, at any rate, in two

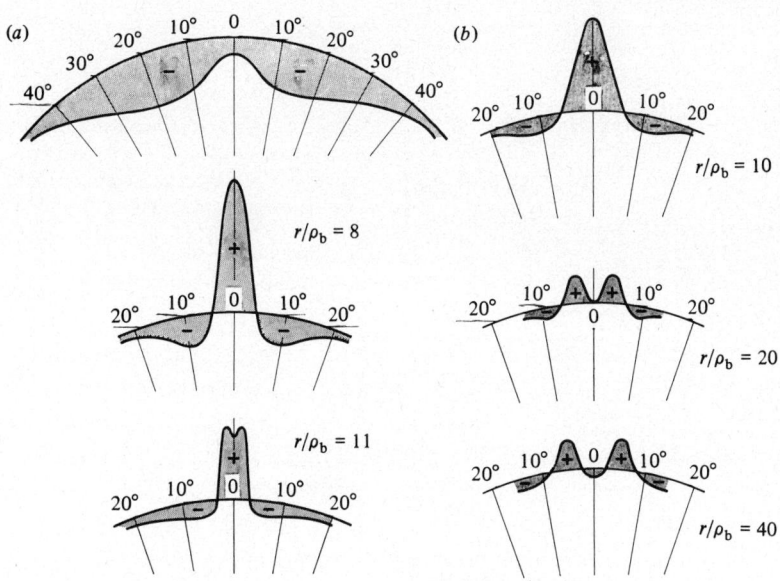

Fig. 11.16. Experimental angular dependences $\delta N_i(\theta, r)$ behind a body at various distances from it: (a) $e\phi_b/kT \simeq -3.5$, $V_b/v_s \simeq 10.5$, $\rho_b/D_e \simeq 14$, $T_e \gg T_i$; (b) $e\phi_b/kT \simeq -3.5$, $V_b/v_s \simeq 8$, $\rho_b/D_e \simeq 1.8$, $T_e \gg T_i$ (Hester & Sonin, 1970a).

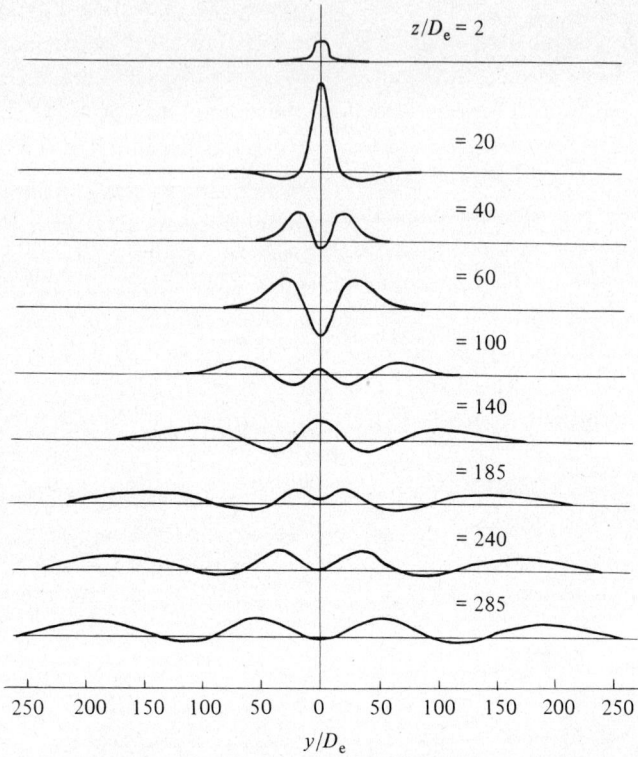

Fig. 11.17. Theoretical perturbation of the ion density $\delta N_i(x,z)$ behind a cylinder of radius $\rho_b \simeq 5 \times 10^{-3} D_e \ll D_e$, for $a_{bs} = V_b/v_s \simeq 8$ and $e\phi_b/kT = -3.5$, at various distances from it. The z-axis is parallel to $\mathbf{V}_0 = -\mathbf{V}_b$ (Hester & Sonin, 1970b).

dimensions out of the three – namely an infinitely long cylinder whose radius ρ_b was 5×10^{-3} times the Debye length D_e (Hester & Sonin, 1970b). The axis of the cylinder was perpendicular to the velocity of motion $\mathbf{V}_b$. The coordinates were such that the x-axis coincided with that of the cylinder, while the z-axis was parallel to $\mathbf{V}_0$. Curves were drawn showing how N_i varied in the third direction, as a function of y at fixed values of z/D_e. Such curves, which are presented in this figure for distances ranging from $z = 2D_e$ to $z = 285 D_e$ from the body, give us a good idea of the change in the nature of the plasma disturbance in the body wake. The fit with the predicted theoretical effects is good (see Fig. 11.15 and below).

In a series of laboratory experiments studies were made of the disturbance of a plasma in the wakes of large bodies ($\rho_b \gg D_e$) of different shapes, over a wide range of distances z/ρ_b (Oran et al., 1974, 1975). In

11.3. Effect of an electric field

these experiments the parameters varied within the following limits:

$$T_e \simeq 900 \text{ to } 1000 \text{ K}, \quad D_e \simeq 0.1 \text{ to } 0.3 \text{ cm}, \quad T_e/T_i \gg 1, \quad \phi_b \simeq -0.3 \text{ to } -4 \text{ V},$$
$$\rho_b/D_e \simeq 12 \text{ to } 25, \quad a_{bs} = V_b/v_s \simeq 6 \text{ to } 17, \quad r/\rho_b \simeq 10^{-1} \text{ to } 34. \quad (11.25)$$

An important circumstance was revealed during these measurements, namely that, even quite close to the body, for $r < \rho_b V_b/v_i$, the nature of the focusing of the ion density $N_i(y,z)$ depends considerably on the shape of the body. During the analysis of these data, it was kept in mind that the experimental conditions differed from those of other laboratory experiments, in that the potentials ϕ_b on the test bodies were considerably more negative than $-kT/e$. Figs. 11.18 and 11.19 show the transverse profiles $N(y,z/\rho_b)$ of the ion density in the wake of a sphere, cylinder, and Apollo model. Inspection of these figures indicates that the focusing effect is the most pronounced in the wake of a sphere. The ion density $N_i(0,z/\rho_b)$ behind a sphere exceeds the undisturbed density N_{i0} by about an order of magnitude at distances from the body $r \lesssim \rho_b V_b/v_i$. The $N(y,z/\rho_b)$ distributions behind test bodies with noncircular sections, bodies of square

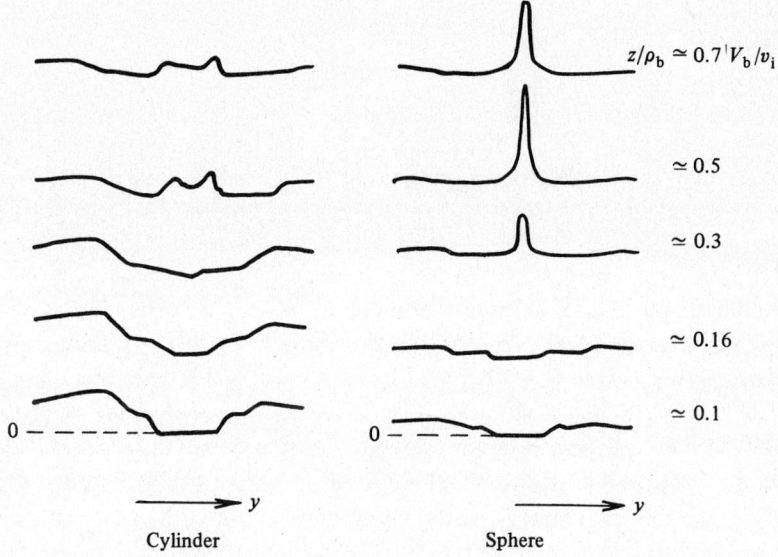

Fig. 11.18. Comparison of theoretical transverse profiles of the perturbation of ion density $N_i(y,z)$ behind a sphere and a cylinder ($z \| -\mathbf{V}_b$). The radii ρ_b of the sphere and the cylinder and the height h of the cylinder are, respectively: $\rho_b = 12 D_e$; $\rho_b = h/2 = 13 D_e$ ($T_e = 900$, 1000 K; $\phi_b = -2 \text{ V}$; $a_{bs} = V_b/v_s \simeq 12$, 13). (Oran et al., 1974.)

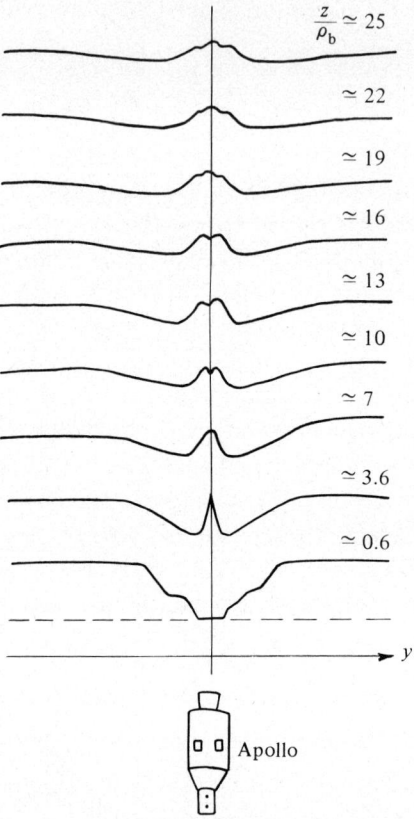

Fig. 11.19. Theoretical transverse profiles of the perturbed ion density $N_i(y,z)$ behind an Apollo model at various distances from it ($\rho_b/D_e \simeq 20$, $a_{bs} = V_b/v_s \simeq 6$, $e\phi_b/kT_e \simeq -10$) (Stone et al., 1974).

section in particular, exhibit a number of special features. The region $\delta N_i > 0$ is more blurred and it has additional maxima, which are not especially pronounced.

Of course, the distribution of the space potential of the plasma in the wake of a body is also complex. Similar results were obtained, apparently for the first time, for a large sphere with a radius $\rho_b = 50 D_e$. Fig. 11.20 shows the lines of equal potential $\phi(r)$ behind a body (the potentials are given in arbitrary units by the curves), obtained via a self-consistent solution of the kinetic equation and the Poisson equation (Liu & Jew, 1967). The potential of the plasma is a maximum near the z-axis, in the distance region $r/\rho_b \simeq 2.5$.

Fig. 11.21 shows the results of theoretical calculations of the potential

11.3. Effect of an electric field

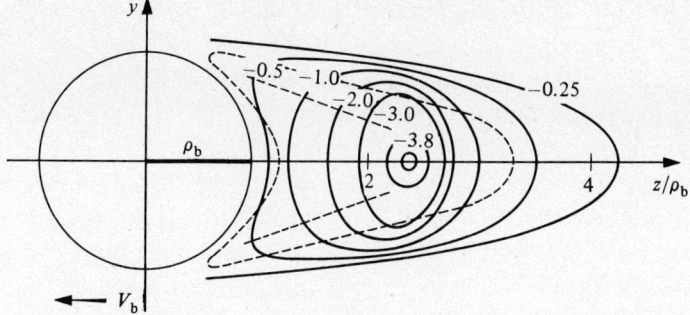

Fig. 11.20. Lines of equal potential behind a rapidly moving sphere (the potentials are marked on the curves, in arbitrary units), calculated taking into account the effect of the electric field (solid curves) and also in the quasi-neutral approximation (dashed curves) $\rho_b/D_e = 50$ (Liu & Jew, 1967).

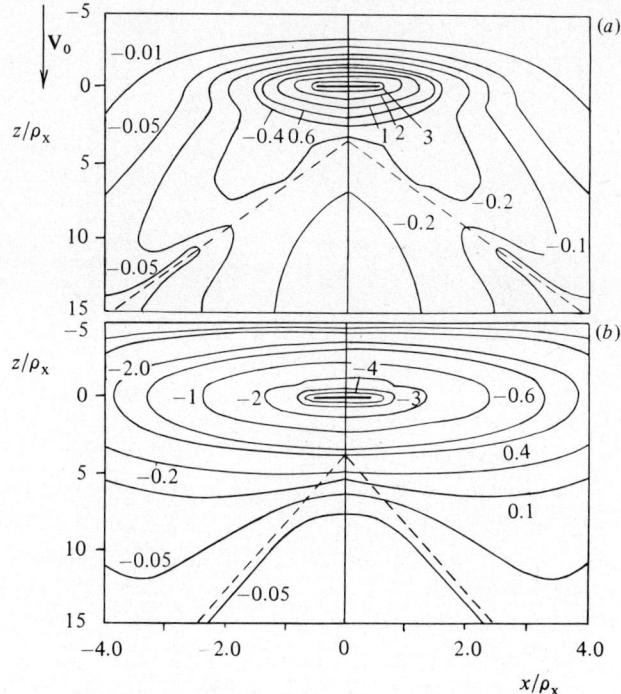

Fig. 11.21. Curves of equal potential $\phi(x\rho_x, z/\rho_x)$ in the vicinities of square plates: (a) $\rho_x = \rho_y \simeq 2D_e$; (b) $\rho_x = \rho_y = 0.1D_e$ ($e\phi_b/kT \simeq -5$, $V_0/v_s = 4.9$). (Martin, 1974.)

distribution in the vicinity of rather small square plates with side lengths $\rho_x = \rho_y = 2D_e$ and $0.1D_e$ (Martin, 1974). The plate potential $\phi_b = -5kT/e$, and $V_b/v_s = 4.9$. The figure gives the lines of equal potential $e\phi(x,z)/kT_e$. The potential distribution is seen to be more complicated than for a sphere (Fig. 11.20). Typically, a plate with a side length $\rho_x = 2D_e$ can already be considered to be a large body. As shown by Martin, 1974, along the dashed lines in Fig. 11.21a are situated the maxima of the ratio $N_i(x,z)/N_{i0}$, i.e., the maxima of the regions of focusing of the ion density; these correspond to distances from the body centre $z \gtrsim \frac{1}{2}\rho_x V_b/v_s$. This is typical, as will be seen below, for bodies with $\rho_b \gg D_e$. Such a $\phi(x,z)$ distribution is not observed for a plate with $\rho_x = 0.1D_e$ (Fig. 11.21b) or for a body with $\rho_b \ll D_e$.

Fig. 11.22 shows the calculated potential distributions in the wake of a point charge ($\rho_b \ll D_e$). The angular distribution of the potential in this case also exhibits some idiosyncrasies. For example, in the near zone of the wake in a nonisothermal plasma the potential is positive ($\phi(\theta) > 0$) in a certain angular range close to the $\mathbf{z} \| \mathbf{V}_0$ axis.

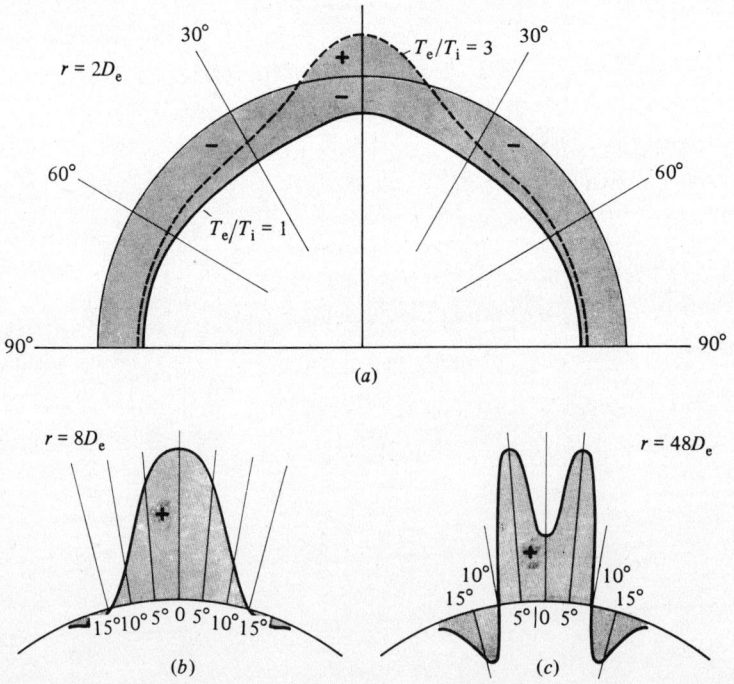

Fig. 11.22 Theoretical angular dependences of the potential $\phi(\theta)$ in the vicinity of a point charge at various distances from it ($\rho_b \ll D_e$, $V_b/v_i = 8$, $T_e/T_i = 1$). (Dubovoi, 1972.)

11.3. Effect of an electric field

11.3.2. The far zone ($r \gg \rho_b V_b/v_i$)

The far zone of the wake of a body cannot be defined with any great precision, since neither calculated not experimental results are available that could pinpoint it better. The far zone is generally thought of as the region at distances $r \gg \rho_b V_b/v_i$ or $r \gg \rho_b V_b/v_s$. However, if these conditions are satisfied, the predicted theoretical effects are observed in experiments with large bodies ($\rho_b \gg D_e$) even at distances $r \lesssim \rho_b V_b/v_i$ (see Fig. 11.16), while around small bodies ($\rho_b \ll D_e$) they are observed at distances $r = 2D_e$ to $10D_e$. Naturally, the division into different zones is quite arbitrary, and the zone boundaries will depend considerably on the relative size ρ_b/D_e and potential $e\phi_b/kT$ of the body, as well as on the degree of deviation from isothermality of the plasma T_e/T_i. The theoretical results presented in this section, however, were obtained with satisfaction of the aforementioned conditions.

Focusing of electrons and ions in the vicinity of the axial direction behind the body is the main special feature of the angular distribution of the density in the far zone of a large body (Panchenko & Pitaevskii, 1964; Bud'ko, 1969, 1970). Figs. 11.23 and 11.24 show the corresponding angular dependences $\delta N_i(\theta)$ for various values of V_b/v_i, for a sphere and a cylinder in the absence of a magnetic field and for a sphere with $\mathbf{H}_0 \neq 0$ and $V_b/v_i = 8$ and with various angles θ_0 between the magnetic field $\mathbf{H}_0$ and the velocity of the body $\mathbf{V}_b$ (Vas'kov, 1966). In a magnetic field $\mathbf{H}_0$ the degree of focusing of particles is reduced, and for angles $\theta \neq \pi/2$ the distribution is asymmetrical relative to the velocity vector $\mathbf{V}_b$. Figs. 11.23 and 11.24 show the results of the corresponding calculations, carried out for uncharged bodies ($\phi_b = 0$). Note here that for $(\phi_b| \ll (\rho_b/D_e)^{4/3} kT/e$ the effect of the potential is in general small and can be neglected.

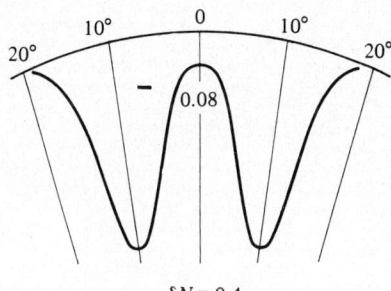

Fig. 11.23. Theoretical angular dependence of the perturbation of ion density $\delta N_i(\theta)$ behind a sphere ($\phi_b = 0$, $V_b/v_i = 8$, $\rho_b/D_e \gg 1$, $r/\rho_b \gg V_b/v_i$). (Panchenko & Pitaevskii, 1964.)

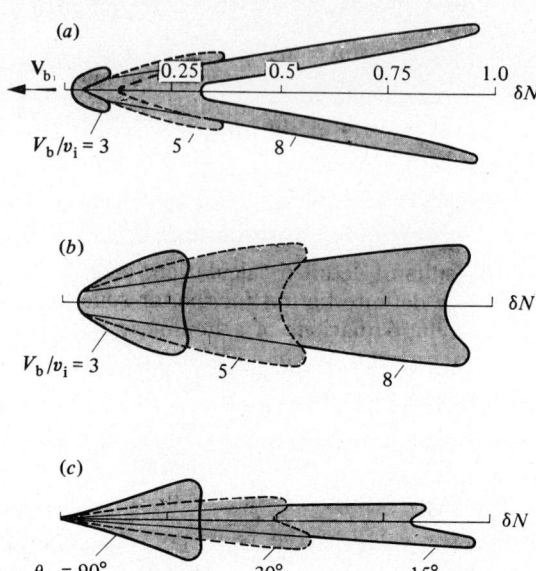

Fig. 11.24. Theoretical curves of equal values of the ion-density perturbation $\delta N_i(\theta)$ behind a rapidly moving body: (a) a sphere, $\mathbf{H}_0 \simeq 0$, (Panchenko, 1965); (b) a cylinder $\mathbf{H}_0 = 0$ (Panchenko, 1965; Vas'kov, 1966); (c) a sphere, $V_b/v_1 = 8$, $\mathbf{H}_0 \neq 0$. (Vas'kov, 1966.)

Analytically, for a sphere with $\rho_b \gg D_e$ the perturbation of the charged-particle density in an isothermal plasma is

$$\delta N(r) = -(\rho_b/r)^2 (V_b/v_i)^2 B_0(\theta, V_b/v_i) \qquad (11.26)$$

which implies that in the far zone $\delta N \sim r^{-2}$. In order to determine $B_0(\theta, V_b/v_i)$, we can use graphs of the universal function $F_0((V_b/v_i)\sin\theta)$ (Fig. 11.25), where

$$B_0\left(\theta, \frac{V_b}{v_i}\right) = F_0((V_b/v_i)\sin\theta)\cos\theta \qquad (11.27)$$

$$\left.\begin{aligned}F_0\left(\frac{V_b}{v_i}\sin\theta\right) &= -\frac{1}{4\pi}\int_L \frac{Q(s)\,ds}{(s^2 - a^2)^{3/2}} \\ &= -\frac{1}{4\pi}\int_L \frac{\pi^{1/2}W(s)\,ds}{[2 + i\pi^{1/2}W(s)]\{s^2 - [(V_b/v_i)\sin\theta]^2\}^{3/2}}\end{aligned}\right\}$$

$$(11.28)$$

(Bud'ko, 1970). In (11.28), $a = (V_b/v_i)\sin\theta$, the path of integration L contains the pole of the function $Q(s)$ along the s-axis ($-\infty$ to $+\infty$), and $W(s)$ is the Kramp function (see (11.21)). For numerical calculations the following

11.3. Effect of an electric field

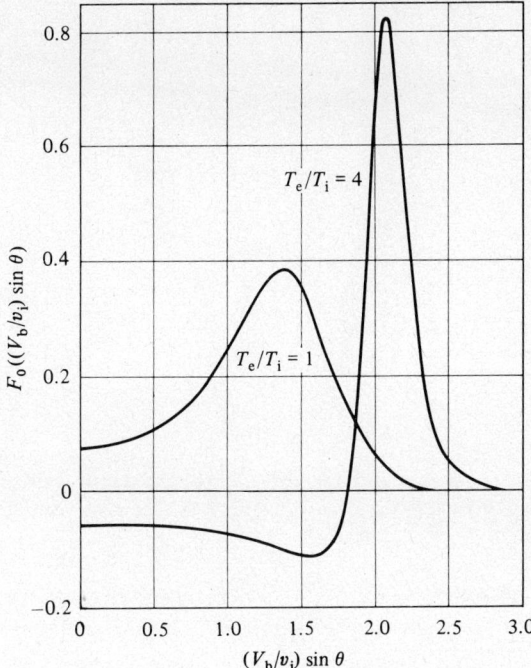

Fig. 11.25. Universal functions $F_0((V_b/v_i)\sin\theta)$ determining the perturbation of ion density $\delta N_i(\theta)$ in the far zone of the wake of a rapidly moving large uncharged sphere (Bud'ko, 1969b).

expansion is useful

$$F_0\left(\frac{V_b}{v_i}\sin\theta\right) = -\sum_{n=0}^{\infty} \text{Re}\left\{\frac{s_n}{(s_n^2-1)\{s_n^2-[(V_b/v_i)\sin\theta]^2\}^{3/2}}\right\}, \quad (11.29)$$

where s_n is the pole (11.28). Since the series (11.29) converges rapidly as n^{-2}, we need retain only a few of its terms. The F_0 curves given in Fig. 11.25 were constructed for the following three values of the poles:

$$s_0 = 1.42 - i0.64, \quad s_1 = 2.35 - i1.78, \quad s_2 = 2.98 - i2.47 \quad (11.30)$$

the use of which gives F_0 quite precisely.

For a nonisothermal plasma, the function $Q(s)$ in (11.28) is replaced by the function

$$Q\left(s, \frac{T_e}{T_i}\right) = \frac{\pi^{1/2} W(s)}{1 + (T_e/T_i)[1 + i\pi^{1/2} s W(s)]}. \quad (11.31)$$

Thus, we ultimately obtain a formula for F_0 analogous to (11.29), but in

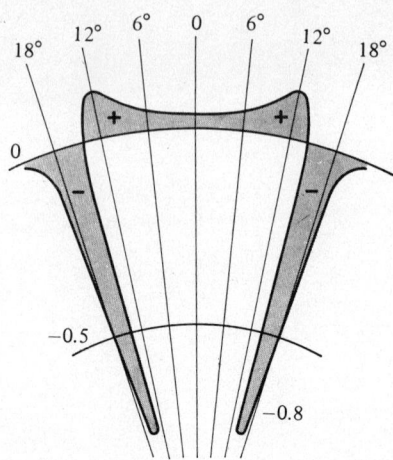

Fig. 11.26. Theoretical angular dependence of the ion-density perturbation $\delta N_i(\theta)$ behind a sphere ($\phi_b = 0$, $V_b/v_i = 8$, $\rho_b/D_e \gg 1$, $T_e/T_i = 4$). (Bud'ko, 1969b.)

which s_n is replaced by $s_n[1 + (T_e/T_i)]$ in the numerator and $(s_n^2 - 1)$ by $[2s_n^2 - 1 + (T_e/T_i)]$ in the denominator. It is important to note that, in a highly nonisothermal plasma, when $T_e \gg T_i$, we have

$$Q(s) = \frac{\pi^{1/2} W(s)}{1 + i\pi^{1/2}[(m/M)(T_e/T_i)]^{1/2} + (T_e/T_i)(1 + i\pi^{1/2} s W(s))} \quad (11.32)$$

and in the denominator of (11.32) the term proportional to $(m/M)^{1/2}$ has to be taken into account. In this case, instead of (11.28) we have

$$F_0\left(\frac{V_b}{v_i}\sin\theta\right) = -\operatorname{Re}\left\{\frac{s_0}{2\{s_0^2 - [(V_b/v_i)\sin\theta]^2\}^{3/2}}\right\} \quad (11.33)$$

where $\operatorname{Re}\{(s_0^2 - a^2)^{1/2}\} > 0$. Here $\operatorname{Im}\{(s_0^2 - a^2)^{1/2}\} < 0$; $a = (V_b/v_i)\sin\theta$.

Figure 11.26 shows the angular dependence of the disturbance $\delta N_i(\theta)$, calculated with the aid of the universal function F_0 (see (11.28) and (11.32)) for $T_e/T_i = 4$. An enhancement of the particle focusing in directions slightly off the axis of the wake is seen to be a typical effect of the nonisothermal nature of the plasma in the far zone. Consequently, in the given case, when $T_e/T_i \gtrsim 1.76$, regions of positive values $\delta N_i(\theta) > 0$ appear, even for $\phi_b = 0$. With a further increase in T_e/T_i, the perturbation gradually acquires a narrow-beam, lobed structure (Fig. 11.27), apparently foreshadowing the appearance in a collisionless plasma of a phenomenon similar to a shock wave. However, this subject requires further study. Let us just note here that, when T_e/T_i is reduced to less than unity, the effect of the electric field

11.3. Effect of an electric field

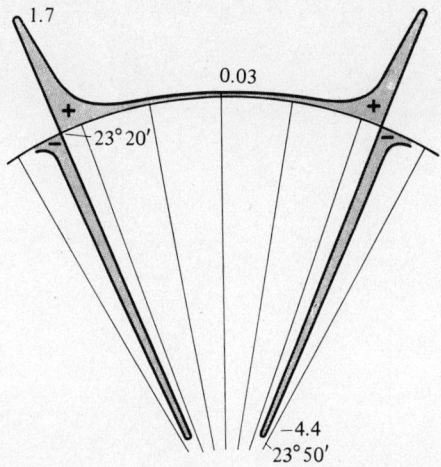

Fig. 11.27. Theoretical angular dependence of the ion-density perturbation $\delta N_i(\theta)$ behind a sphere ($\phi_b = 0$, $V_b/v_i = 10$, $T_e/T_i = 32$, $\rho_b/D_e \gg 1$). (Bud'ko, 1969b.)

weakens significantly. For $T_e/T_i \lesssim 0.23$, the rarefaction is a maximum on the axis behind the body, just as for neutral particles, while for $T_e/T_i \to 0$ the corresponding formulas for $\delta N_i(\theta)$ become the formulas of the neutral approximation.

Let us consider a cylindrical body of infinite length whose cross-section projected onto a plane perpendicular to V_b is $2\rho_b$ long. The wake of this body is described by the following formulas (Bud'ko, 1970):

$$\delta N_i(\theta, r) = -2\frac{\rho_b}{r}\left(\frac{V_b}{v_i}\right) B_\|\left(\frac{V_b}{v_i}\sin\theta\right), \tag{11.34}$$

$$B_\|\left(\frac{V_b}{v_i}\sin\theta\right) = \frac{1}{\pi^{1/2}}\operatorname{Re}\left\{\frac{W((V_b/v_i)\sin\theta)}{2 + i[(V_b/v_i)\sin\theta]\pi^{1/2}W[(V_b/v_i)\sin\theta]}\right\}$$

$$= \frac{1}{\pi^{1/2}}\frac{[1 + (T_e/T_i)]\exp(-a^2)}{\pi a^2(T_e/T_i)^2\exp(-2a^2) + \{[1 + (T_e/T_i)] - a\pi^{1/2}(T_e/T_i)\operatorname{Im}[W(a)]\}^2}, \tag{11.35}$$

where $a = (V_b/v_i)\sin\theta$, and the function W was defined above (see (11.21)) (Panchenko & Pitaevskii, 1964). Fig. 11.28 shows $B_\|((V_b/v_i)\sin\theta)$ curves for $T_e/T_i = 1$ and $T_e/T_i = 4$. Note that in this case a series expansion of (11.35) in terms of the poles, as was done above for a sphere (11.29), does not facilitate the calculations, in view of the slow convergence of the corresponding series.

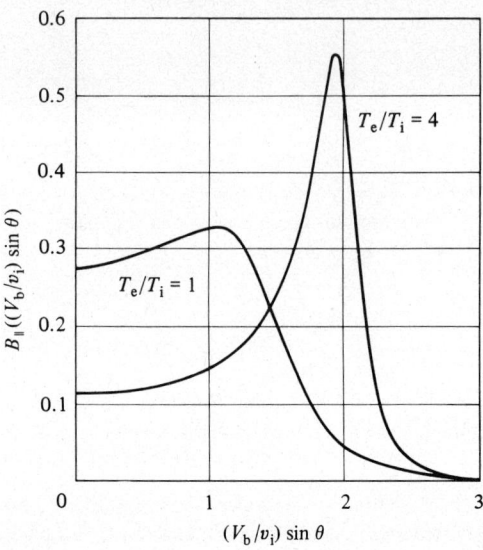

Fig. 11.28. Universal functions determining the ion-density perturbation $\delta N_i(\theta)$ in the far zone of the wake of a rapidly moving long uncharged cylinder ($\rho_b \gg D_e$). (Bud'ko, 1969b.)

In the far zone of a cylinder, $\delta N_i(\theta)$ diminishes more slowly than behind a sphere, namely as r^{-1}. In the wake of a cylinder, throughout the entire angular range the disturbance $\delta N_i < 0$, while $B_\| > 0$. This is connected with the fact that the filling in of the wake of an infinite cylinder with particles takes place only at its lateral surface. Around a cylinder of finite length in a nonisothermal plasma, a region of particle concentration inevitably appears, just as around a sphere. However, the corresponding calculations have not as yet been carried out. In this case the problem is two-dimensional, and it is complicated to take into account the boundary conditions in terms of a cylinder.

Taking into account the influence of the external magnetic field ($\mathbf{H}_0 \neq 0$), the disturbance of the electron density in an isothermal plasma behind a large body, whose section is normal to the vector $\mathbf{V}_b$ and is equal to $\pi \rho_b^2$ (where ρ_b is the effective radius of the body), δN_e being averaged over the direction normal to $\mathbf{H}_0$, is given by the following expression:

$$\delta N_e(r, \theta_0) = -\pi \left(\frac{V_b}{v_i}\right) \frac{\rho_b^2}{\rho_{Hi} r} B_H\left(\theta, \theta_0, \frac{V_b}{v_i}\right) \quad (11.36)$$

(Vas'kov, 1966). As in a cylinder wake, $\delta N_e(r, \theta)$ decreases in the given

11.3. Effect of an electric field

case with the distance as r^{-1}, while

$$B_{\text{H}}\left(\theta,\theta_0,\frac{V_{\text{b}}}{v_{\text{i}}}\right) = \frac{1}{\sin(\theta_0 - \theta)} F_{\text{H}}\left[\frac{V_{\text{b}}}{v_{\text{i}}}\frac{\sin\theta}{\sin(\theta_0 - \theta)}\right] \quad (11.37)$$

where θ_0 is the angle between the vectors $\mathbf{V}_{\text{b}}$ and $\mathbf{H}_0$. The form of the universal function F_{H} entering into (11.37) is the same as that of the universal function for a cylinder $B_{\parallel}$ in an isotropic plasma, shown graphically in Fig. 11.28. It is

$$F_{\text{H}}(a_{\text{H}}) = F_{\text{H}}\left(\frac{V_{\text{b}}}{v_{\text{i}}}\frac{\sin\theta}{\sin(\theta_0 - \theta)}\right) = \frac{1}{\pi^{1/2}}\text{Re}\left\{\frac{W(a_{\text{H}})}{2 + ia_{\text{H}}\pi^{1/2}W(a_{\text{H}})}\right\}. \quad (11.38)$$

Since in this case the disturbance only spans small angles $\pm\theta$, in (11.37) the factor $\sin\theta$ can be replaced by θ. It should also be kept in mind that formula (11.37) is valid only for $\sin(\theta_0 - \theta) \geq 0$. If $\sin(\theta_0 - \theta) < 0$, then $F_{\text{H}} = 0$.

In general, if δN_{e} is not averaged over the direction $\mathbf{Z}\parallel[\mathbf{V}_{\text{b}}\times\mathbf{H}_0]$, the perturbation of the electron density in the far zone ($r \gg \rho_{\text{b}}V_{\text{b}}/v_{\text{i}}, \rho_{\text{Hi}}V_{\text{b}}/v_{\text{i}}$) of the wake of a sphere of radius $\rho_{\text{b}} \gg D_{\text{e}}$ can be calculated using numerical integration. In this case

$$\delta N_{\text{e}}(\theta,\theta_0,r) = \pi\frac{\rho_{\text{b}}^2}{\rho_{\text{Hi}}r}\left(\frac{V_{\text{b}}}{v_{\text{i}}}\right)\frac{G(\theta,Z)}{\sin(\theta_0 - \theta)}, \text{ if } \sin(\theta_0 - \theta) > 0$$

$$\delta N_{\text{e}}(\theta,\theta_0,r) = 0, \text{ if } \sin(\theta_0 - \theta) < 0 \quad (11.39)$$

where Z is the length of a normal dropped from the point (r,θ) to the $(\mathbf{V}_{\text{b}},\mathbf{H}_0)$ plane, passing through the body centre. Here the function

$$G(\theta,\theta_0) = -\frac{1}{2\pi^{3/2}}\int_{-\infty}^{+\infty} L(s^2/2)\left[\exp i\left(s\frac{|Z|}{\rho_{\text{Hi}}}\right)\right]$$

$$\times \text{Re}\left[\frac{W(a_{\text{H}})}{2 + (D_{\text{e}}/\rho_{\text{Hi}})^2 s^2 + i\pi a_{\text{H}}W(a_{\text{H}})L(s^2/2)}\right]ds, \quad (11.40)$$

where $a_{\text{H}} = (V_{\text{b}}/v_{\text{i}})[\sin\theta/\sin(\theta_0 - \theta)]$, while the functions W and L are defined in formulas (11.21) and (11.55) respectively (Dubovoi & Yaroslavtsev, 1976). The $\delta N_{\text{e}}(\theta,\theta_0)$ relations behind a body, calculated using formulas (11.39) and (11.40), are shown in Fig. 11.29 for an angle $\theta_0 = \theta + \pi/4$ between the direction of motion of the body $\mathbf{V}_{\text{b}}$ and the direction of the magnetic field $\mathbf{H}_0$ and for various ratios of the sphere radius ρ_{b} to the Larmor radius of the ions ρ_{Hi}. The asymmetrical nature of the relation $\delta N_{\text{e}}(\theta)$ relative to the axis $\mathbf{z}\parallel\mathbf{V}_0$ (direction $\theta = 0$) and the focusing of particles, which becomes significant with an increase in body

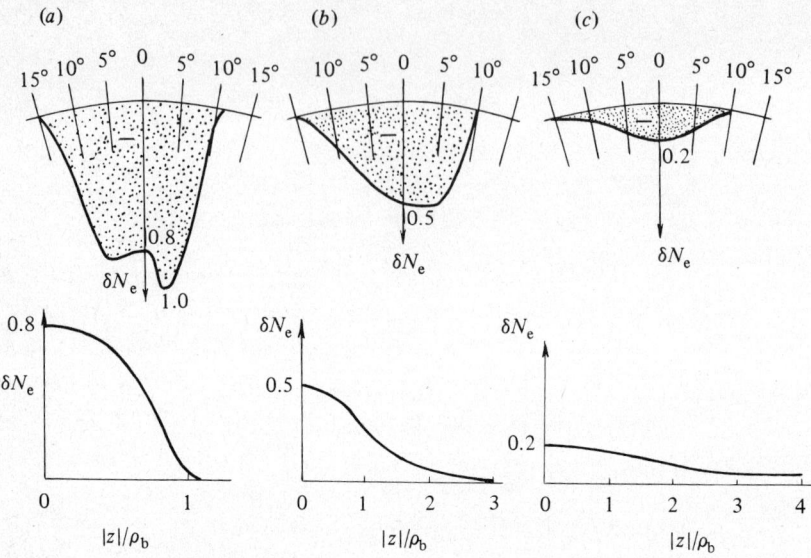

Fig. 11.29. Theoretical relations for the density perturbation $\delta N_e(\theta)$ and $\delta N_e(z/\rho_b)$ behind a sphere ($\rho_b \gg D_e$, $H_0 \neq 0$, $\theta_0 - \theta = \pi/4$, $r/D_e \gg V_b/v_i$, $V_b/v_i = 8$). (a) $\rho_b/\rho_{Hi} = 10$, (b) $\rho_b/\rho_{Hi} = 1$, (c) $\rho_b/\rho_{Hi} = 0.2$. In the lower part of the figure $\theta = 0$ (Dubovoi & Yaroslavtsev, 1976).

size, have been mentioned above (see Fig. 11.24). With an increase in z, which means that the observation point (r, θ) is further from the body, the degree of rarefaction of the particles δN_e will be lower; naturally, this lowering is especially rapid for quite large values of ρ_b/ρ_{Hi}. This is evident from the corresponding curves in the lower part of Fig. 11.29 for $\theta = 0$ and various ρ_b/ρ_{Hi}.

The wake of a long cylinder in the presence of a magnetic field is described by the same universal function, namely:

$$\delta N_e(\theta, r) = -2\left(\frac{V_b}{v_i}\right)\frac{\rho_b}{r}B_H\left(\theta, \theta_0, \frac{V_b}{v_i}\right), \quad (11.41)$$

where $2\rho_b$ is the length of the projection of the cylinder cross-section onto a plane perpendicular to $\mathbf{V}_b$.

In the far zone of the body, as mentioned repeatedly above, the plasma is quasi-neutral. Thus the plasma potential will be proportional to the electron-density perturbation; in fact

$$\phi \simeq (kT_e/e)\delta N_e \quad (11.42)$$

and, for instance, for a sphere, formula (11.26) can be used to obtain the following expressions for the radial and angular components of the electric

11.3. Effect of an electric field

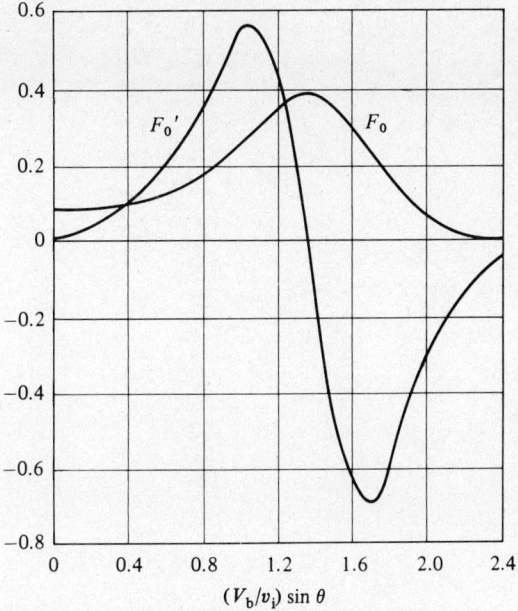

Fig. 11.30. The universal function $F_0((V_b/v_i)\sin\theta)$ and its derivative F_0', which determine the electric field in the far zone of a rapidly moving uncharged large sphere (Bud'ko, 1969b).

field in an isothermal plasma:

$$E_r = -\frac{2kT}{e}\left(\frac{V_b}{v_i}\right)^2\left(\frac{\rho_b^2}{r^3}\right)B_0\left(\theta,\frac{V_b}{v_i}\right), \quad E_\theta = -\frac{kT}{e}\left(\frac{V_b}{v_i}\right)^2\left(\frac{\rho_b}{r^3}\right)B_0'\left(\theta,\frac{V_b}{v_i}\right) \quad (11.43)$$

and the derivative of B_0 with respect to θ is

$$B_0'\left(\theta,\frac{V_b}{v_i}\right) = -F_0\left(\frac{V_b}{v_i}\sin\theta\right)\sin\theta + \left(\frac{V_b}{v_i}\right)^2 F_0'\left(\frac{V_b}{v_i}\sin\theta\right)\cos^2\theta. \quad (11.44)$$

Fig. 11.30 shows F_0 and F_0' as functions of $(V_b/v_i)\sin\theta$, and Fig. 11.31 gives the angular dependences of the field components E_r and E_θ and of the total field $(E_r^2 + E_\theta^2)^{1/2}$, constructed with the aid of Fig. 11.30 for $V_b/v_i = 8$. The values of functions B_0 and B_0' are indicated near the extreme points in Fig. 11.31. Inspection of this figure shows that in the far zone of a large body the angular distribution of the electric field is quite complex (it has several lobes).

The far zone of the wake of a small body ($\rho_b \ll D_e$), namely a point charge, has been investigated theoretically in some detail (Kraus & Watson,

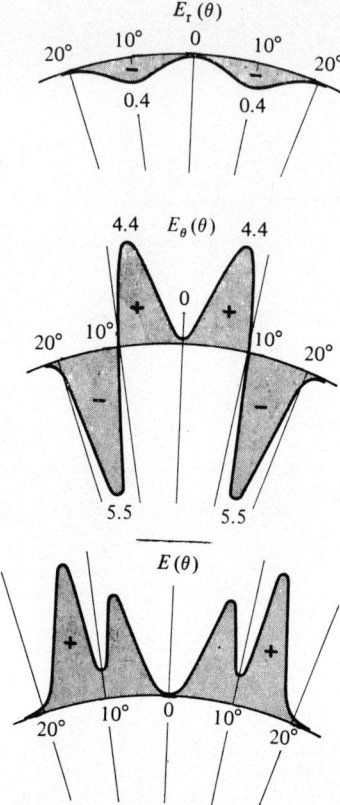

Fig. 11.31. Theoretical angular dependences of the electric-field components E_r, E_θ and $(E_r^2 + E_\theta^2)^{1/2}$ in the far zone of a rapidly moving, uncharged sphere ($\rho_b \gg D_e$) (Bud'ko, 1969b).

1958; Pitaevskii & Kresin, 1961; Bud'ko, 1966; Vas'kov, 1966). At large distances from the body, the body shape, and to some extent its size as well, play a minor role. Consequently, the results of these studies are of great interest, in that they can be used for comparison with various experimental results. The theory of the wake of a small body is also interesting for another reason: the corresponding calculations can in this case be carried out taking the charge of the body into account, albeit only for a charge that is sufficiently slight to satisfy the conditions (11.13).

In the absence of magnetic field, the perturbation of the electron density in the far zone of a small body is described by the following formula:

$$\delta N_e = \frac{1}{4\pi}\left(\frac{V_b}{v_i}\right)^2 b^2 \ln\left(\frac{1}{b}\right)\left(\frac{D_e}{r}\right)^2 \left[2\pi \frac{r_b}{r_1} B_1\left(\theta, \frac{V_b}{v_i}\right) + B_2\left(\theta, \frac{V_b}{v_i}\right)\right], \quad (11.45)$$

11.3. Effect of an electric field

where
$$b = (e|\phi_b|/kT)(\rho_b/D_e)(v_i/V_b) \ll 1 \quad (11.46)$$
(see (11.13), above) and
$$r_b = 4\left(\frac{V_b}{v_i}\right)\frac{D_e}{b\ln(1/b)}. \quad (11.47)$$

The function B_2 in formula (11.45) is equivalent to the universal function F_0 (see (11.28) and (11.29)), describing the perturbation of the electron density by a large body ($\rho_b \gg D_e$) in the far zone. The function B_1, on the other hand, can be expressed in terms of F_0 and its derivative F_0' (see (11.44)).

Fig. 11.32 shows the form of the functions B_1 and B_2 at various distances from the body. From (11.45) it follows that two zones of the body wake are characteristic: distances $r \ll r_b$ and $r \gg r_b$. In the first zone, i.e., at distances considerably less than the characteristic distance r_b (see (11.47)), the perturbation $\delta N_e(\theta)$ decreases as r^{-3}, and in the more remote zone as r^{-2}. With a variation in the distance, the angular structure of the disturbance gradually becomes deformed. In the zone $r \ll r_b$, where $\delta N_e(\theta)$

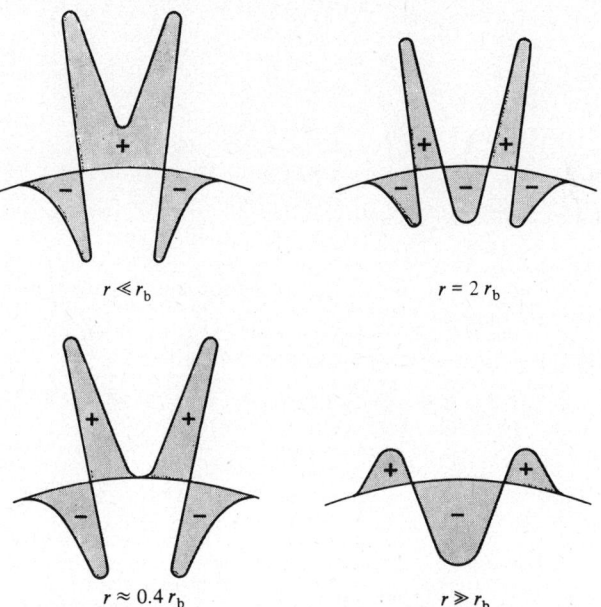

Fig. 11.32. Theoretical angular dependences $\delta N_e(\theta)$ behind a small, slightly charged body ($\rho_b \ll D_e$, $H = 0$, $\phi_b = 0$, $V_b/v_i = 8$). (Bud'ko, 1966.)

decreases as r^{-3}, the perturbation of the electron density in a nonisothermal plasma can be expressed analytically in a simple form in terms of the universal function $F_0((V_b/v_i)\sin\theta)$ and of its derivative, which describe $\delta N_e(\theta)$ for a large body (see Figs. 11.25, 11.28, and 11.31). Thus we have

$$\delta N_e(\theta,r) = -\left(\frac{V_b}{v_i}\right)^2 \frac{T_e}{T_e+T_i} \frac{1}{\pi N r^3} \left[(1-3\cos^2\theta)F_0 - \frac{V_b}{v_i}\sin\theta\cos^2\theta F_0'\right]. \quad (11.48)$$

As noted previously, the main special features of the theoretical angular dependences obtained for a point body were detected in the above-described experiments (Hester & Sonin, 1970a) at relatively small distances from a weakly charged, large body (see Fig. 11.16).

In a magnetized plasma the electron-density perturbation, averaged about a direction normal to the plane of $\mathbf{V}_b$ and $\mathbf{H}_0$, is

$$\delta N_e(\theta,r) = \pi\left(\frac{V_b}{v_i}\right)^2 b^2 \ln\left(\frac{1}{b}\right)\frac{D_e^2}{\rho_{Hi}r}\left[\frac{r_b}{r}B_{1H}\left(\theta,\theta_0,\frac{V_b}{v_i}\right) + B_{2H}\left(\theta,\theta_0,\frac{V_b}{v_i}\right)\right]. \quad (11.49)$$

In formula (11.49)

$$B_{1H} = \frac{\sin\theta_0}{\sin^2(\theta_0-\theta)}F_{1H}(a_H), \quad B_{2H} = \frac{\sin^2\theta_0}{\sin(\theta_0-\theta)}F_{2H}(a_H), \quad (11.50)$$

where

$$F_{1H} = \frac{1}{2\pi^{1/2}}\operatorname{Re}\left\{\frac{W''(a_H)}{[2+i\pi^{1/2}a_H W(a_H)]^2}\right\}, \quad (11.51)$$

$$F_{2H} = \frac{1}{2\pi^{1/2}}\operatorname{Re}\left\{\frac{W''(a_H)}{2+i\pi^{1/2}a_H W(a_H)]}\right\}, \quad (11.52)$$

and where $W''(a_H)$ is the second derivative of the Kramp function (11.21) and $a_H = (V_b/v_i)(\sin\theta/\sin(\theta_0-\theta))$.

Formula (11.49) has a structure similar to that of formula (11.45), but in this case at distances $r \ll r_b$ the disturbance decreases as r^{-2}, and for $r \gg r_b$ it decreases as r^{-1}. Fig. 11.33 shows the corresponding angular dependences at various distances from the body. These exhibit the same features as the angular functions for $\mathbf{H}_0 = 0$, except that in this case they are more pronounced.

The perturbation of the electron density $\delta N_e(\theta)$ described by formula (11.49) is averaged about the Z direction, which is parallel to the vector product $\mathbf{V}_b \times \mathbf{H}_0$. In general, $\delta N_e(\theta,r)$ can be calculated for a small body ($\rho_b \ll D_e$) in the intermediate zone, i.e., at distances

11.3. Effect of an electric field

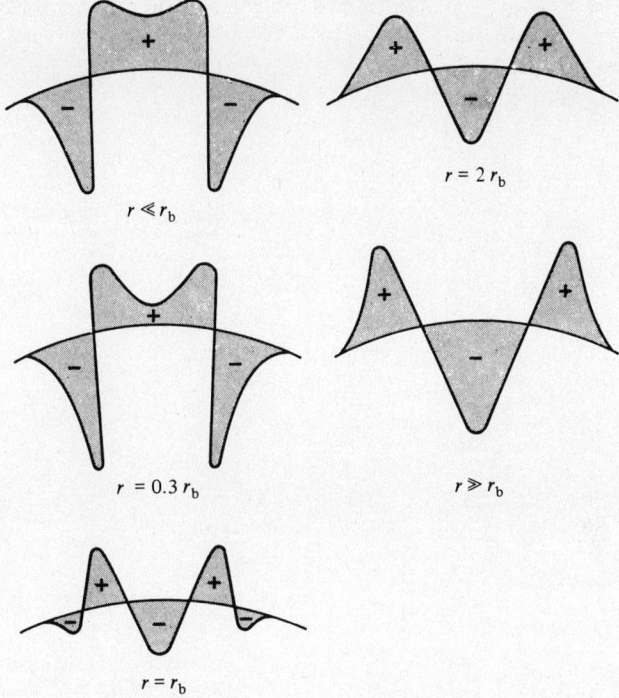

Fig. 11.33. Theoretical angular dependences $\delta N_e(\theta)$ behind a small, weakly charged body ($\rho_b \ll D_e$, $\mathbf{H}_0 \neq 0$, $\theta_0 = \pi/2$, $\phi_b < 0$, $V_b/v_i = 8$). (Vas'kov 1966b.)

$V_b/v_i D_e$, $V_b/v_i \rho_{Hi} \ll r \ll r_b$, using numerical integration. The formulas are:

$$\delta N_e(\theta, r) = \frac{e|\phi_b|}{kT} \frac{V_b}{v_i} \frac{\rho_b}{\rho_{Hi}} \left(\frac{D_e}{r}\right)^2 F(\theta), \quad \text{if} \quad \sin(\theta_0 - \theta) > 0$$

$$\delta N_e(\theta, r) = 0 \quad \text{if} \quad \sin(\theta_0 - \theta) < 0 \qquad (11.53)$$

where

$$F(\theta) = -\operatorname{Re}\left[\frac{1}{\pi^{1/2}}\left(W''(a_H)\int_{-\infty}^{\infty} \frac{[\exp i(s|Z|/\rho_{Hi})]L(s^2/2)\mathrm{d}s}{[2+(s^2/\rho_{Hi}^2)s^2+i\pi^{1/2}a_H W(a_H)L(s^2/2)]^2}\right)\right]. \qquad (11.54)$$

In (11.54) Z is the length of a normal dropped from a point (r,θ) to the $(\mathbf{V}_b, \mathbf{H}_0)$ plane passing through the coordinate origin, W'' is the second derivative of the Kramp function, Re denotes the real part of the expression in braces, ρ_{Hi} is the Larmor radius of the ions, and

$$L(x) = \exp(-x)I_0(x), \qquad (11.55)$$

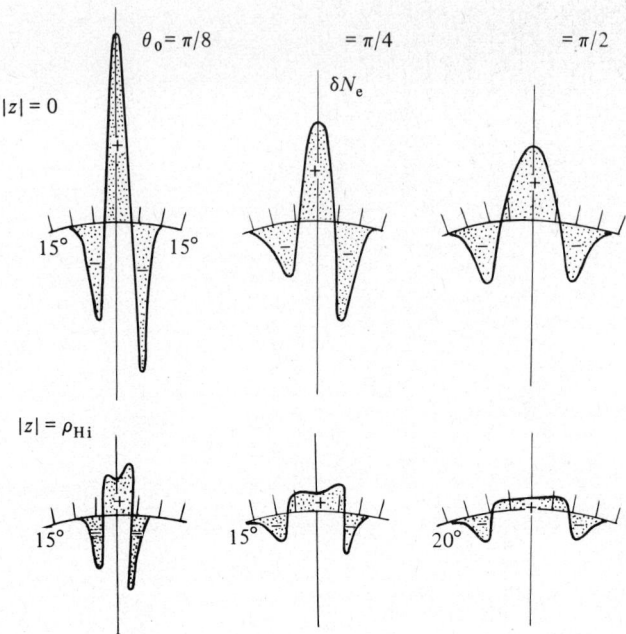

Fig. 11.34. Theoretical angular dependences of the electron-density perturbation $\delta N_e(\theta)$ behind a small, weakly charged body, at various distances from it and for various magnetic-field directions $\mathbf{H}_0$ (angle θ_0) ($\rho_b \ll D_e$, $\mathbf{H}_0 \neq 0$, $\phi_b < 0$, $V_b/v_i = 8$, $r_b \ll r$, $r_b \gg D_e V_b/v_i$). (Doubovoi & Yaroslavtsev, 1976.)

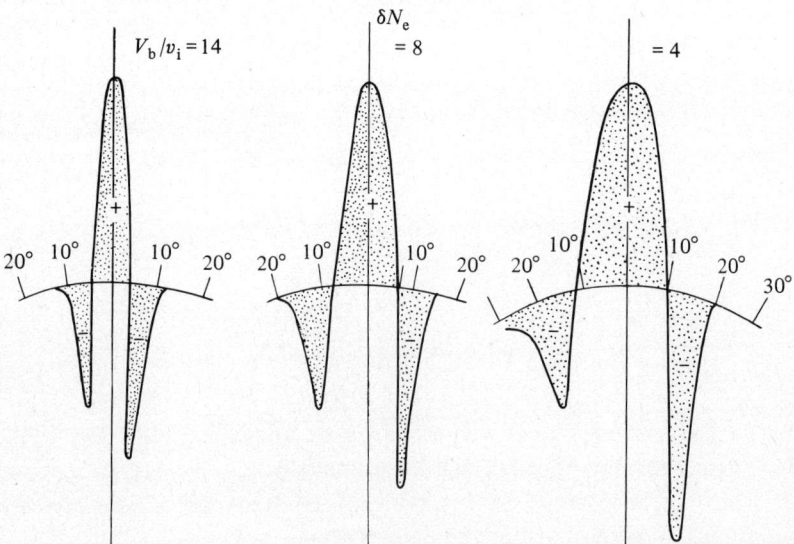

Fig. 11.35. Theoretical angular dependences $\delta N_e(\theta)$ behind a small weakly charged body, for various V_b/v_i ($\rho_b \ll D_e$, $H_0 \neq 0$, $\theta_0 - \theta = \pi/4$, $\phi_0 < 0$, $|z| = 0$, $r_b \ll r$, $r_b \gg D_e V_b/v_i$). (Dubovoi & Yaroslavtsev, 1976.)

11.3. Effect of an electric field

I_0 being a Bessel function with an imaginary argument (Dubovoi & Yaroslavtsev, 1976).

Figs. 11.34 and 11.35 show the angular dependences $\delta N_e(\theta)$ for various directions of the magnetic-field vector $\mathbf{H}_0$ (values of the angle θ_0) and various values of V_b/v_i, calculated using formula (11.53). In general, these curves of $\delta N_e(\theta,\theta_0)$ are similar to the relations given previously. For $\theta_0 = \pi/2$ and $|Z| = \rho_{Hi}$, the results obtained with formula (11.54) are quite close to the curves obtained using formula (11.49) in the $r \ll r_b$ region (see Fig. 11.32). Note that, further from the body (as z increases), the focusing effect is enhanced, but the magnitude of δN_e drops rapidly. With a decrease in V_b/v_i, the angular width of the disturbance region naturally grows (Fig. 11.35). The $\delta N_e(\theta, \theta_0)$ relations presented in Figs. 11.34 and 11.35 were constructed to the accuracy of the second approximation of formula (11.54). Taking the second approximation into account makes the $\delta N_e(\theta, \theta_0)$ variations less smooth, the maximum fluctuations $(\delta N_e)_{max}$ ranging from $0.1\,\delta N_e$ to $0.3\,\delta N_e$. It should also be mentioned here that in the presence of a magnetic field $\mathbf{H}_0$ the perturbation δN_e is in all cases cut off sharply, i.e., $\delta N_e = 0$ for $\theta_0 \leqslant \theta$ (see (11.37), (11.40) and (11.53)).

12

Disturbances of the plasma near quasi-stationary bodies ($V_b \lesssim v_i$)

In the near-Earth plasma (see Tables 2.1 and 2.2 of Volume 1) it sometimes happens that artificial satellites or space probes move through parts of it at velocities V_b not only commensurate with the thermal velocities of ions v_i, but even less than v_i. Under these conditions the physical processes in the vicinity of the body differ radically in many respects from those that were described above, when $V_b \gg v_i$. If $V_b < v_i$, then instead of the body's velocity, the factors determining the plasma processes will be the body potential ϕ_b, the reflecting properties of the body surface, and, naturally, the magnetic field and linear dimensions of the body. Under transition conditions, namely when $V_b \simeq v_i$ or $V_b > v_i$, some phenomena are qualitatively similar to those taking place for $V_b \gg v_i$, although quantitatively, of course, they differ considerably. Problems in plasma diagnostics with the aid of various kinds of sounding measurements assume a unique character when the body moves slowly, making a special analysis necessary. Such problems form an independent major branch of plasma physics. However, this branch lies beyond the framework of this book, and only a few results of the corresponding theoretical studies will be given here briefly, for the sake of completeness.

Just as in the case when $V_b \gg v_i$, for $V_b \lesssim v_i$ the calculated results are most complete when the linear dimensions of the body $\rho_b \gg D_e$ and $\rho_b \ll D_e$. Since we are dealing with low velocities of the body, the processes in the neighbourhood of a stationary body are the starting-point for the study of such cases.

12.1. A small stationary body ($\rho_b \ll D_e$, $V_b = 0$)

For velocities $V_b \ll v_i$, or more precisely when $V_b = 0$, the nature of the charged-particle trajectories and of the particle-density distribution $N(\mathbf{r})$ in the body's vicinity will depend on the total energy of a particle

$$\varepsilon(\mathbf{r}) = \tfrac{1}{2}Mv^2 + e\phi(\mathbf{r}) \tag{12.1}$$

12.1. A small stationary body

and on the sign of the field potential. Here v is the particle velocity, $\mathbf{r}$ is the distance from the arbitrarily selected body centre, and $\phi(\mathbf{r})$ is the potential distribution near the body. In this case two kinds of particles can appear: *trapped* $N_t(\mathbf{r})$ and *free* $N_f(\mathbf{r})$. The density in the vicinity of the body will thus be

$$N(\mathbf{r}) = N_t(\mathbf{r}) + N_f(\mathbf{r}). \tag{12.2}$$

Trapped particles are defined as particles whose orbits are finite, in the sense that they are closed around the body. Such orbits appear if

$$\varepsilon(\mathbf{r}) < 0 \tag{12.3}$$

in the case of an *attractive potential*, positive for electrons and negative for ions. Since in the near-Earth plasma bodies almost always have a negative potential, the *trapped particles* will be predominantly *ions*. However, finite orbits can appear only under the influence of collisions between particles: in order to trap particles spiralling around a body, it is necessary that they lose some of their energy. In the absence of collisions, attracted particles impinging upon the body are absorbed by its surface and are not reflected. Thus, it is obvious that the density of trapped particles may increase considerably around the body, in view of the fact that they gradually accumulate there. In the case of an equilibrium distribution

$$N_t(\mathbf{r}) = N_0 \exp(|e\phi(\mathbf{r})|/kT) \tag{12.4a}$$

when $|e\phi| \gg kT$, the density of trapped particles $N_t \gg N_0$. For electrons attracted to a Coulomb centre, when they collide with neutral particles, at distances $r > (M/m)^{1/2} \rho_b$ we find that

$$N_t(\mathbf{r}) = \frac{4}{3\pi^{1/2}} N_0 \left(\frac{|e\phi(\mathbf{r})|}{kT}\right)^{3/2} \left(\frac{2|e\phi(\mathbf{r})|}{5kT} + 1\right) \tag{12.4b}$$

where $\phi(\mathbf{r}) = Q_b/r$, with Q_b the charge of the Coulomb centre (Gurevich, 1963, 1964; Al'pert et al., 1965). It follows from formula (12.4b) that already for $|e\phi| \simeq 2kT$ the trapped-particle density increases considerably: $N_t/N_0 \simeq 4$. For the majority of cases corresponding to the conditions of interest to us here, the calculations of N_t are very complicated, mainly because collisions have to be taken into account. Thus let us confine ourselves to just a few brief remarks concerning these particles. Note that in most cases where $V_b \ll v_i$, especially at large distances from the Earth, collisions between particles in general play a minor role, so that the time of accumulation of particles in finite orbits is quite long.

If at a given point near the body the energy of the particles, in a Coulomb

field for instance, satisfies the condition
$$\varepsilon(\mathbf{r}) > 0, \tag{12.5}$$
then the particles are free, in the sense that they describe open *infinite trajectories*. In order for there to be free particles, with a density $N_f(r)$, the field can be either *attractive* or *repulsive*. For instance, for ions and a negative body potential $\phi_b < 0$, the field is attractive. The density of the corresponding free particles is denoted N_f^+. For electrons, when $\phi_b < 0$ the field will be repulsive, the density of the corresponding particles being denoted N_f^-.

For small bodies at distances that are not great ($r \lesssim D_e$) simple analytical formulas are obtained for N_f^+, namely

$$N_f^+ = N_0 \left\{ \frac{s}{\pi} \left[1 + \left(1 - \frac{\rho_b}{r}\right)^{1/2}\right] + \tfrac{1}{2}[1 - \Phi(s^{1/2})]\exp\left(S + \tfrac{1}{2}\left[1 - \left(\frac{\rho_b}{r}\right)^2\right]^{1/2}\right)\right.$$
$$\left. \times \left[1 - \Phi\left[\left(\frac{sr}{\rho_b + r}\right)^{1/2}\right]\right]\exp\left(\frac{sr}{\rho_b + r}\right)\right\}, \tag{12.6}$$

where
$$s = (|e\phi_b|/kT)(\rho_b/r), \quad \phi \ll (kT/e)(D_e/\rho_b), \quad (D_e/\rho_b) \gg 1 \tag{12.7}$$
and
$$\Phi(a) = (2/\pi^{1/2}) \int_0^a \exp(-u^2) du \tag{12.8}$$

is the error function (Gurevich, 1963). For a Coulomb centre ($\rho_b \to 0$), with a charge Q_b, when $\phi(r) = Q_b/4\pi\varepsilon_0 r = \phi_b \rho_b/r$ we have

$$N_f^+ = N_0 \left\{ 2\left(\frac{s}{\pi}\right)^{1/2} + [1 - \Phi(s^{1/2})]\exp(s)\right\}. \tag{12.9}$$

In plasma regions where the magnitude of the field is small $e\phi(r)/kT \ll 1$, but the plasma is still quite disturbed, the density of attracted particles is described by an even simpler formula:

$$N_f^+ = N_0 \left\{ 1 + \frac{|e\phi(r)|}{kT} - \left(\frac{\rho_b}{r}\right)^2 \left[\frac{|e\phi(r)|}{kT} + \frac{1}{2}\right]\right\}. \tag{12.10}$$

Formula (12.10) overlaps adequately with formula (12.6), which applies to the region where $|e\phi(r)|/kT \lesssim 1$ and the field is a Coulomb field. Fig. 12.1 shows N_f^+/N_0 curves for the entire disturbed region, for $\rho_b = 0.07 D_e$ and various values of the body potential ϕ_b. It is seen that close to the body the density of attracted particles N_f^+ increases significantly. For a Coulomb centre the values of N_f^+ increase even more. For example, for

12.1. A small stationary body

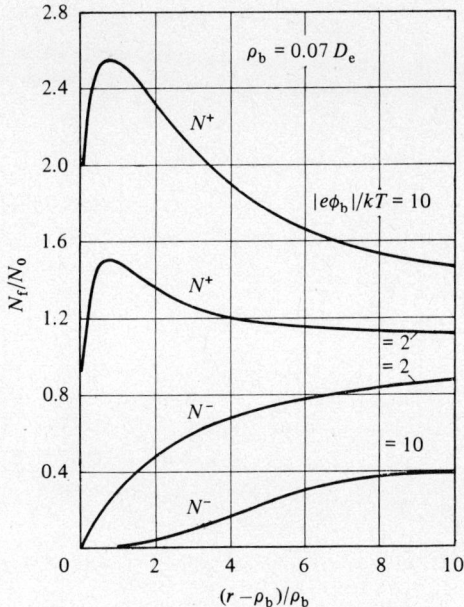

Fig. 12.1. Relative densities of attracted N_f^+ and repelled N_f^- free particles, as functions of $(r - \rho_b)/\rho_b$ for a small charged sphere at rest (Gurevich, 1963).

$|e\phi_b|/kT \gg 1$ it follows from (12.9) that $N_f^+/N_0 = (2/\pi^{1/2})(|e\phi_b|/kT)^{1/2}$ and for $|e\phi_b|/kT = 10$ the ratio $N_f^+/N_0 = 4$.

The function $\phi(r)$ outside the Coulomb zone is described by quite complicated formulas, which will not be presented here (see Al'pert, 1948). In the region of Debye screening $(r < D_e)$ it is evident that $\phi \sim 1/r$. For $D_e < r < D_e \ln[(D_e/\rho_b)(kT/|e\phi_b|)]$, the potential $\phi(r)$ decreases exponentially, and at great distances for $r > D_e \ln[(D_e/\rho_b)(kT/|e\phi_b|)]$ the potential $\phi(r) \sim r^{-2}$. For $\rho_b = 0.07 D_e$, the calculated $\phi(r)$ curves for $|e\phi_b|/kT = 2$ and 3 are given in Fig. 12.2.

The densities of repelled free particles are described not by (12.6) and (12.9) but rather by, respectively, the formulas

$$N_f^- = N_0 \frac{\exp(-s)}{2} \left\{ 1 + \phi \left[\left(\frac{r - \rho_b}{\rho_b} s \right)^{1/2} \right] + \left(1 - \frac{\rho_b^2}{r^2} \right)^{1/2} \right.$$
$$\left. \times \left[1 - \Phi \left[\left(\frac{r^2 s}{\rho_b(\rho_b + r)} \right)^{1/2} \right] \exp \left(\frac{\rho_b s}{\rho_b + r} \right) \right\}, \quad (12.11)$$

$$N_f^- = N_0 \exp(-s). \quad (12.12)$$

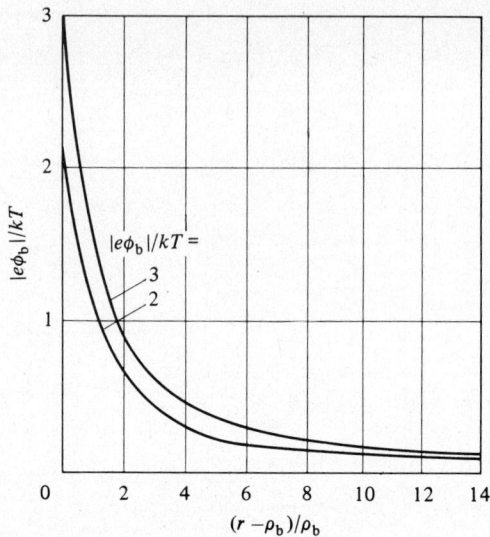

Fig. 12.2. Plasma potential ϕ as a function of $(r - \rho_b)/\rho_b$ for a small charged sphere at rest (Gurevich, 1963).

Fig. 12.1 gives N_f^-/N_0 curves for two values of the body potential. Close to the body surface $N_f^- \ll N_0$, and right out to distances of ρ_b to $2\rho_b$ the density of repelled free particles N_f^- is only half as great as, or less than half as great as, the undisturbed particle density N_0.

12.2. A large stationary body ($\rho_b \gg D_e$, $V_b = 0$)

The *charged boundary layer*, which is formed in a certain zone around the body surface due to the marked effect of Debye screening, is an important feature of the plasma structure in the vicinity of a large body. In this layer the quasi-neutral nature of the plasma is disturbed considerably. For body potentials that are not very large, i.e., if

$$|\phi_b| < (kT/e)(\rho_b/D_e)^{4/3} \tag{12.13}$$

the thickness of the sheath is $(r - \rho_b) \simeq D_e$. If the opposite is true, i.e., if

$$|\phi_b| > (kT/e)(\rho_b/D_e)^{4/3} \tag{12.14}$$

then the thickness is $(r - \rho_b) \simeq \rho_b$. The criteria (12.13) and (12.14) are thus a measure of whether the body potential is *small* or *large*, and they define the boundary between the near and far zones of a stationary or slowly moving large body. On the two sides of this boundary the radial variation of the particle density is quite different in nature. Nevertheless, the position of the boundary is described by the very same formula, even for

12.2. A large stationary body

considerably different laws of radial variation of the potential $\phi(r)$. For example, the densities of attracted and repelled free particles, when $\rho_b \gg D_e$, are respectively

$$N_f^+ = \frac{N_0}{2}\left\{\exp(\phi^*)\left[1 - \Phi(\phi^{*1/2}) + \frac{2}{\pi^{1/2}}\phi^{*1/2}\right] + \left(1 - \frac{\rho_b^2}{r^2}\right)^2\right.$$
$$\left. \times \exp\left[\left(\phi^*\frac{\rho_b^2}{r^2} - \phi_1^*\right)\frac{r^2}{\rho_b^2 - r^2}\right]\right\}, \tag{12.15}$$

$$N_f^- = N_0 \frac{\exp(\phi^*)}{2}\left\{1 + \Phi[(\phi_b^* - \phi^*)^{1/2}] + \left(1 - \frac{\rho_b^2}{r^2}\right)^{1/2}\right.$$
$$\left. \times \left[1 - \Phi\left[\left(\frac{\rho_b^2\phi_b^* - r^2\phi^*}{r^2 - \rho_b^2}\right)^{1/2}\right]\right]\exp\left(\frac{\rho_b^2\phi_b^* - r^2\phi^*}{r^2 - \rho_b^2}\right)\right\}, \tag{12.16}$$

where

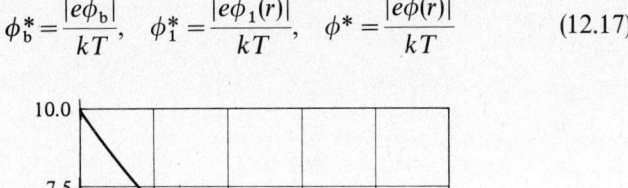

$$\phi_b^* = \frac{|e\phi_b|}{kT}, \quad \phi_1^* = \frac{|e\phi_1(r)|}{kT}, \quad \phi^* = \frac{|e\phi(r)|}{kT} \tag{12.17}$$

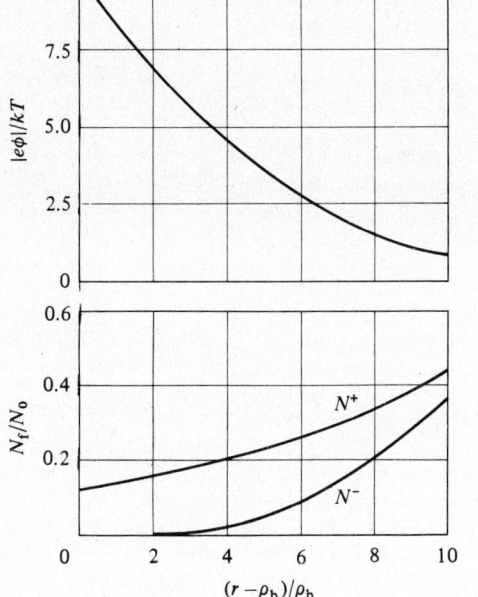

Fig. 12.3. The normalized potential $|e\phi|/kT$ and normalized density of free particles N_f/N_0 as functions of $(r - \rho_b)/\rho_b$ for a large charged sphere (Al'pert et al., 1963; Gurevich, 1964).

and $\phi_1(r)$ is the potential at the boundary of the sheath (Gurevich, 1963).

The potential $\phi(r)$ can be calculated only with the aid of numerical integration. The corresponding curves of the potential $\phi(r)$ for $e\phi_b/kT = 10$ and $\rho_b \gg D_e$ in the far and near zones of the body, together with the curves (obtained from these data) of the densities of attracted and repelled free particles, are given in Figs. 12.3 and 12.4. Naturally, the curves in Fig. 12.3 'match up' quite well with the beginning of the curves in Fig. 12.4, i.e., at the boundary of the sheath, which in Fig. 12.4 arbitrarily corresponds to the origin of the coordinates, since $D_e \gg \rho_b$. A decrease in the densities of particles other than repelled free particles is an important physical feature of the plasma structure in the vicinity of a large body. This is because the velocity of the attracted particles in the sheath is greatly increased, while their flux remains unchanged. The density of free particles may, however, also increase around a body if its surface is a good reflector. For perfect reflection of particles, as is evident from Fig. 12.5, which gives the corresponding calculated results for $\phi_b = 5kT/e$ and $= 10kT/e$, the

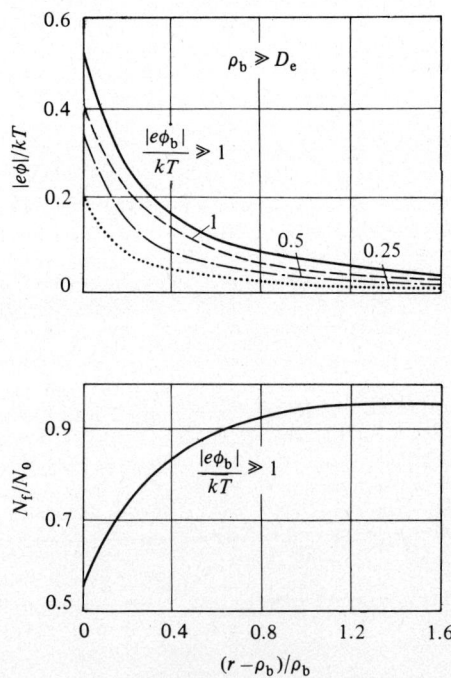

Fig. 12.4. Curves of the normalized potential $|e\phi|/kT$ and of the normalized density of free particles N_f/N_0 in the near zone of a charged sphere (Al'pert et al., 1963; Gurevich, 1964).

12.3. Slowly moving bodies

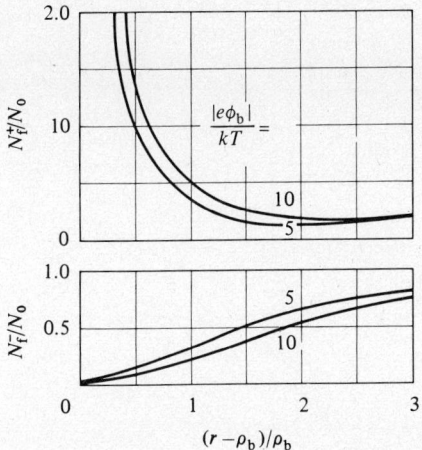

Fig. 12.5. Curves of the normalized densities N_f^+/N_0 and N_f^-/N_0 in the vicinity of a large perfectly reflecting sphere (Al'pert et al., 1963; Gurevich, 1964).

density $N_f^+ > 10N_0$ to $20N_0$. However, these conditions are seldom realized in experiments with bodies in space. In individual cases, on the other hand, local effects of almost total reflection of particles are observed, complicating processes in the vicinity of bodies of complex structure, such as artificial satellites or space probes. This has to be kept in mind when formulating and analysing the results of the various experiments.

12.3. Slowly moving bodies ($V_b \simeq$, $<$, or $\ll v_i$; $\rho_b \ll D_e$ or $\rho_b \gg D_e$)

When the body moves, even slowly, the effects on the plasma in its vicinity already differ radically from the case $V_b = 0$ for low body potentials. First let us consider the corresponding phenomena in the vicinity of a small spherical body of radius $\rho_b \ll D_e$, which have been investigated theoretically in considerable detail (Moskalenko, 1964a, b; Knyazyuk & Moskalenko, 1966).

For low body potentials

$$|\phi_b| \ll (kT/e)(D_e/\rho_b) \qquad (12.18)$$

the electric field in the plasma is weak, and also we can assume that at all distances it decreases according to a Coulomb law, since for $r > D_e$ the potential energy of charged particles is lower than the kinetic energy of their thermal motion. This simplifies the calculations considerably. In this case the density of attracted particles (i.e., ions in the case of $\phi_b < D_e$)

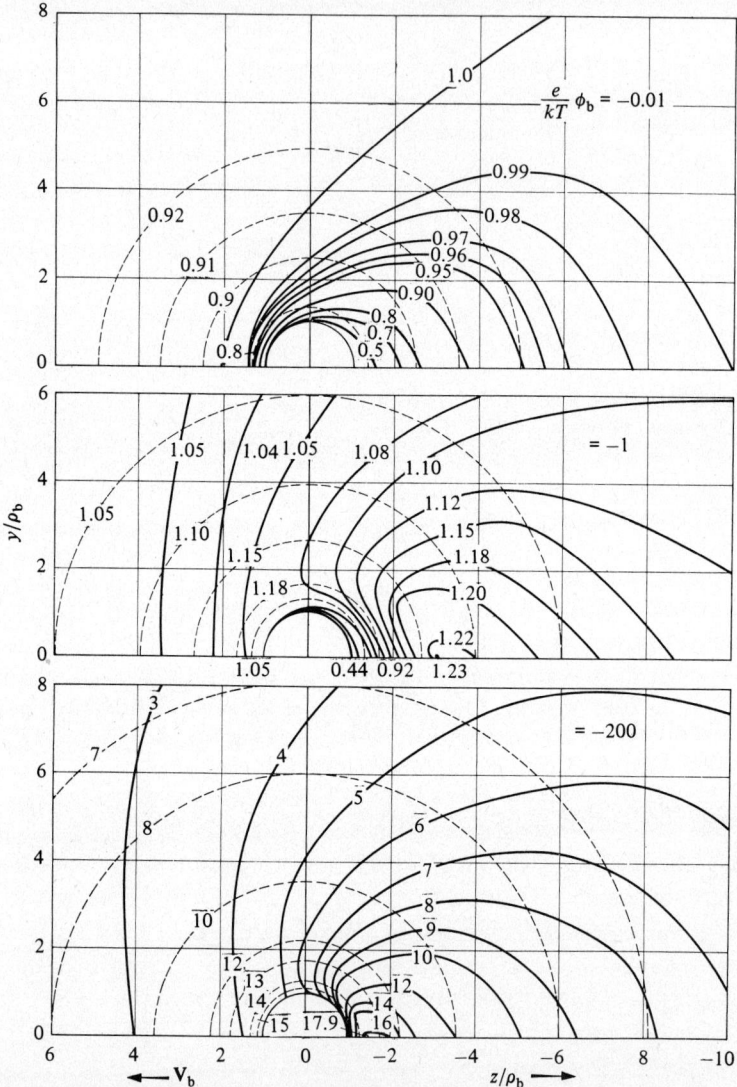

Fig. 12.6. Curves of equal values of the normalized density $N(\theta,r)/N_0$ of attracted free particles in the vicinity of a small sphere ($\rho_b \ll D_e$) moving at a velocity $V_b \simeq v_i$, for different values of the body potential ϕ_b (solid curves). The dashed curves show N_f^+/N_0 for $V_b = 0$ (Knyazyuk & Moskalenko, 1966).

12.3. Slowly moving bodies

can be expressed as

$$N(\theta,r) = N_f^+(r) + \frac{N_0}{\pi^{1/2}}\frac{V_b}{v_i} f\left(\theta, r, \phi_b, \frac{V_b}{v_i}\right), \qquad (12.19)$$

where $N_f^+(r)$ is the density for $V_b = 0$ (the corresponding formulas were given in the preceding section), and the function f, expressed in terms of quadratures, can be calculated numerically. Fig. 12.6 shows the results of calculations for $V_b \simeq v_i$ and different values of ϕ_b. The solid curves correspond to equal values of $N(\theta,r)/N_0$, and the dashed curves to $N_f^+(r)/N_0$ for a stationary body ($V_b = 0$). The calculations pertain to a body surface that completely neutralizes ions impinging upon it. Thus, when the body potential is low ($\phi_b \simeq 10^{-2} kT/e$), the ion density in both cases $V_b = 0$ and $V_b \simeq v_i$ will be lower than the unperturbed particle density

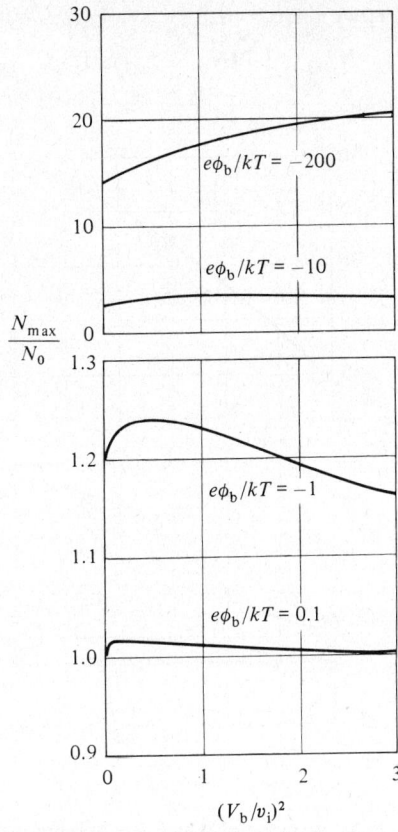

Fig. 12.7. Ratio of the maximum particle density N_{max} to N_0 along the axis of a small sphere ($\rho_b \ll D_e$), as a function of $(V_b/v_i)^2$ (Knyazyuk & Moskalenko, 1966).

(ions) and repelled (electrons) particles. Figs. 12.9 and 12.10 show the curves of the functions f_N and f_ϕ calculated by Moskalenko, 1969, using numerical methods, for various values of $|e\phi_b|/kT$.

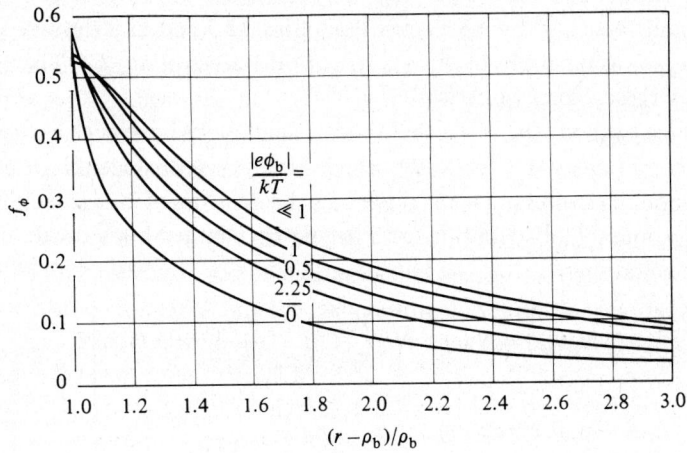

Fig. 12.10. Curves of the function f_ϕ (see (12.20)). (Moskalenko, 1969.)

12.3. Slowly moving bodies

can be expressed as

$$N(\theta, r) = N_f^+(r) + \frac{N_0}{\pi^{1/2}} \frac{V_b}{v_i} f\left(\theta, r, \phi_b, \frac{V_b}{v_i}\right), \quad (12.19)$$

where $N_f^+(r)$ is the density for $V_b = 0$ (the corresponding formulas were given in the preceding section), and the function f, expressed in terms of quadratures, can be calculated numerically. Fig. 12.6 shows the results of calculations for $V_b \simeq v_i$ and different values of ϕ_b. The solid curves correspond to equal values of $N(\theta, r)/N_0$, and the dashed curves to $N_f^+(r)/N_0$ for a stationary body ($V_b = 0$). The calculations pertain to a body surface that completely neutralizes ions impinging upon it. Thus, when the body potential is low ($\phi_b \simeq 10^{-2} kT/e$), the ion density in both cases $V_b = 0$ and $V_b \simeq v_i$ will be lower than the unperturbed particle density

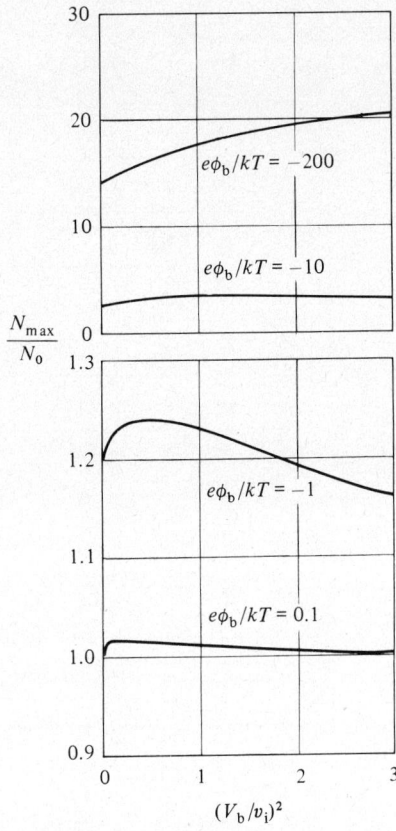

Fig. 12.7. Ratio of the maximum particle density N_{max} to N_0 along the axis of a small sphere ($\rho_b \ll D_e$), as a function of $(V_b/v_i)^2$ (Knyazyuk & Moskalenko, 1966).

N_0. Behind the body, near its axis, a rarefaction region appears; near its surface ($r/\rho_b = 1, \theta = 0$) the value of $N(1,0) \simeq 8 \times 10^{-2} N_0$.

However, with a rise in body potential the structure of the plasma becomes altered significantly. For instance, already for $\phi_b = kT/e$ (see Fig. 12.6), due to the effect of the electric field, the concentration, or bunching, of particles will everywhere predominate: $N(\theta, r) > N_0$. Here the maximum focusing (the maximum value $N_{max}(\theta, r)$) occurs on the axis behind the body, at a distance z_{max} from its surface which depends on the potential ϕ_b. For $\phi_b = kT/e$, we have $N_{max}/N_0 = 1.23$ and $z_{max}/\rho_b = 3$. However, in the immediate vicinity of the body surface, for low body potentials ϕ_b a small rarefaction region also forms. For example, for $\varphi_b \simeq kT/e$ at a distance $z/\rho_b = 1$ the value of $N(1,0) \simeq 0.21 N_0$. Then, with an increase in body potential the rarefaction region gradually disappears, the degree of particle concentration increases, and the maximum of N approaches the body. When $\phi_b = 200 kT/e$, we have

$$N_{max} \simeq 1.79 N_0, \quad z_{max} \simeq 1.3 \rho_b.$$

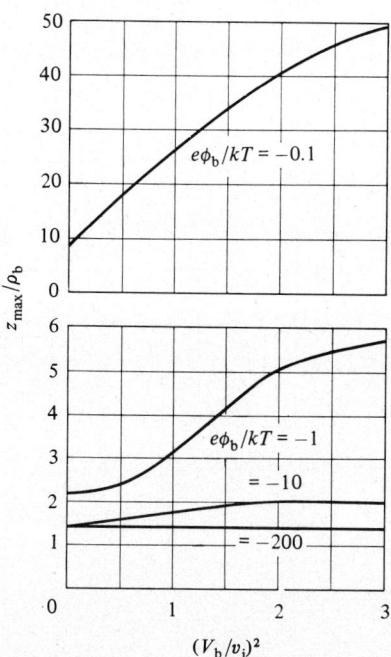

Fig. 12.8. Ratio of z_{max}, the distance at which the maximum density N_{max} occurs, to ρ_b, as a function of $(V_b/v_i)^2$ behind a small sphere (Knyazyuk & Moskalenko, 1966).

12.3. Slowly moving bodies

With an increase in body velocity the position of N_{max}, on the other hand, recedes from the body surface, a result associated with an enlargement of the rarefaction region (the 'shadow'). Consequently, as in the case $V_b \gg v_i$, two 'concurrent' effects determine the structure of the plasma disturbance in the vicinity of a slowly moving body: 'shading' of particles behind the body and 'focusing', by the electric field. Figs. 12.7 and 12.8 show N_{max}/N_0 and z_{max}/ρ_b as functions of $(V_b/v_i)^2$ for various values of $e\phi_b/kT$. The figures portray these effects quite well.

For a large body ($\rho_b \gg D_e$), theoretical calculations have been carried out quite comprehensively for body velocities $V_b \ll v_i$ (Moskalenko, 1969).

Beyond the region of Debye screening, the densities of ions and electrons and the potential distribution for a negatively charged body are described by the equations

$$\left. \begin{array}{l} N_i(\theta, r) = N_i(r) + N_{i0} \dfrac{V_b}{v_i} f_N\left(\dfrac{r}{\rho_b}, \dfrac{|e\phi_b|}{kT}\right) \cos\theta, \\[8pt] N_e(\theta, r) = N_e(r) + N_{e0} \dfrac{V_0}{v_i} f_N\left(\dfrac{r}{\rho_b}, \dfrac{|e\phi_b|}{kT}\right) \cos\theta, \\[8pt] \phi(r, \theta) = \phi(r) + \dfrac{kT}{e} \dfrac{V_b}{v_i} f_\phi\left(\dfrac{r}{\rho_b}, \dfrac{|e\phi_b|}{kT}\right) \cos\theta, \end{array} \right\} \quad (12.20)$$

where $N_i(r)$, $N_e(r)$ and $\phi(r)$ are the values of these quantities for $V_b = 0$; the formulas for these were given in the preceding section for free attracted

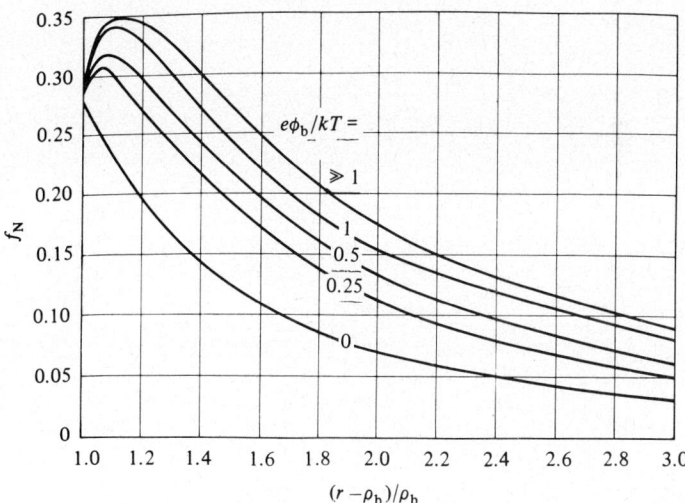

Fig. 12.9. Curves of the function f_N (see (12.20)). (Moskalenko, 1969.)

(ions) and repelled (electrons) particles. Figs. 12.9 and 12.10 show the curves of the functions f_N and f_ϕ calculated by Moskalenko, 1969, using numerical methods, for various values of $|e\phi_b|/kT$.

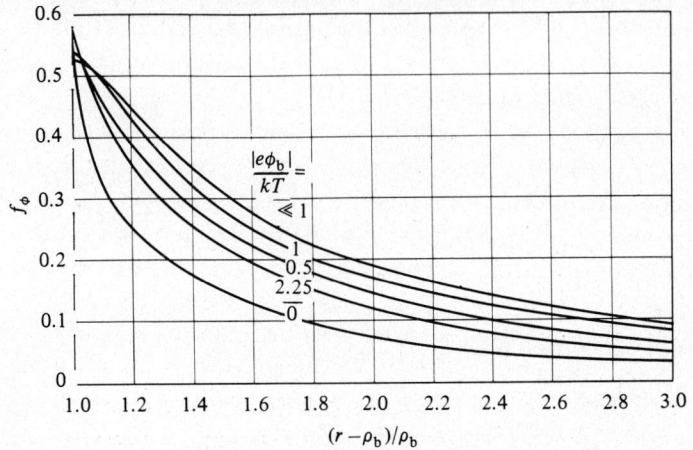

Fig. 12.10. Curves of the function f_ϕ (see (12.20)). (Moskalenko, 1969.)

13

Scattering of radio waves in the wake of a fast-moving body

The wake of a fast-moving body, as we saw in Chapter 11, is an inhomogeneous cloud extended behind the body, and which is 'carried along' by the latter. This cloud is characterized by the perturbation of the electron density, and thus by the dielectric permittivity of the plasma around the body. The extent of the body's wake, understandably enough, is of the order of the mean free path of the particles: at such distances $\delta N(r)$ gradually becomes dissipated. The mean free path, which in the ionosphere increases with the altitude, may amount to hundreds of metres, several kilometres, or more, because of the decrease in the frequency of particle collisions. Therefore, despite the fact that $\delta N(r)$ is small in the far zone of the wake, which constitutes its main part, it is to be expected that under certain conditions there must be considerable scattering of electromagnetic waves from the long inhomogeneous cloud produced behind the body. The first simple evaluations of this effect (Al'pert, 1960) indicated that, if the influence of the external magnetic field of the Earth is taken into account, which gives a cylindrical wake structure, the differential effective cross-section $d\sigma$ for radio-wave scattering at the wake may be as great as $100\,m^2$ or more, being tens or hundreds of times greater than the scattering cross-sections of the bodies themselves.

Further, more rigorous, detailed theoretical studies of this effect (Al'pert & Pitaevskii, 1961; Al'pert *et al.*, 1965) indeed verified that under certain conditions the scattering cross-section of a wake may be quite large, and, if the body is moving through the ionosphere, below the F-region maximum close to the caustic, then $d\sigma$ may even be of the order of $10^4\,m^2$ or more (Gurevich & Pitaevskii, 1966; Vas'kov, 1969a). On the other hand, as yet there are no sufficiently convincing results that experimentally confirm this effect. In particular, as will be seen below, this is because of the lobe structure of the wake. The width of the part of the principle lobe which corresponds to large $d\sigma$ is only of the order of some fractions of a degree. Therefore, due to the rapid motion of the body, the scattering field can be detected at the Earth's surface only in the form of brief bursts lasting a second or less. Consequently, to study this effect experimentally, special quite thorough experiments have to be set up, and there is no certainty that the various reports in the literature concerning observations of high magnitudes of the radio-wave scattering fields of artificial satellites (see, for instance, Kraus et al., 1960; Kraus, 1965) are actually a result of experimental detection of this phenomenon. However, the effect of radio-wave scattering from a body's wake is undoubtedly of interest not only with regard to the physics of interactions of moving bodies with a plasma, but also for a number of practical reasons. The results of theoretical studies of this effect will be outlined briefly below.

The scattering at the wake of a body can be calculated with the aid of perturbation theory, since the values of $\delta N(\mathbf{r})$ for the part of the wake which plays a role in the radio-wave scattering are small in comparison with the electron density N_0 of the unperturbed plasma. Ultimately, using the calculation method given by Landau & Lifshitz, 1957, the following expression is found for the differential effective scattering cross-section $d\sigma$ of the wake:

$$d\sigma = \frac{|E_s|}{|E_0|^2} r^2 do = \frac{1}{16\pi^2} \left(\frac{\omega_0}{c}\right)^4 \frac{|\delta N_q|^2}{N_0^2} \sin^2 \psi \, do. \qquad (13.1)$$

This quantity gives the fraction of the wave energy that is backscattered in a solid-angle element do. In formula (13.1) ψ is the angle between the electric-field vector $\mathbf{E}$ of the incident wave and the wave vector $\mathbf{K}_s$ of the scattered wave, ω_0 is the Langmuir frequency of the electrons, and δN_q is the Fourier component of the spatial distribution of the electron-density perturbation $\delta N(\mathbf{r})$. Thus the differential scattering cross-section is defined in terms of the perturbation of the electron density in $\mathbf{q}$-space, namely in terms of the quantity

$$\delta N_q = \int \delta N(\mathbf{r}) \exp(-i\mathbf{q} \cdot \mathbf{r}) d^3 r. \qquad (13.2)$$

Here

$$\mathbf{q} = \mathbf{K}_s - \mathbf{K}_0, \qquad (13.3)$$

where $\mathbf{K}_s$ and $\mathbf{K}_0$ are the wave vectors of the scattered and incident waves. Solution of the steady-state $(\partial f/\partial t = 0)$ kinetic equations (see Volume 1 (3.7)) in the quasi-neutral approximation, taking (3.12) into account, enables us to obtain the perturbation δN_q in the form of finite equations (Pitaevskii, 1961). This facilitates the precise calculations of $d\sigma$, which, however, can be completed only with the aid of numerical integration.

Scattering of radio waves by the wake

Finally, for a sphere of radius ρ_b we get

$$d\sigma = \frac{1}{16}\left(\frac{\omega_0}{c}\right)^4 \left(\frac{\rho_b V_b}{\Omega_{Hi}}\right)^2 F(\tau,\zeta,\delta,\eta)|G(q\rho_b,\chi)|^2 \sin^2\psi \, do, \qquad (13.4a)$$

where χ is the angle between the vectors $\mathbf{q}$ and $\mathbf{V}_b$;

$$\left.\begin{aligned}\tau &= \frac{qV_b}{\Omega_{Hi}}\cos\chi = \frac{qV_b}{\Omega_{Hi}}(\cos\theta_0\sin\theta_q + \sin\theta_0\cos\theta_q\cos\phi_q) \\ \zeta &= \frac{1}{4}\left(\frac{v_i q}{\Omega_{Hi}}\right)^2 \sin^2\theta_q, \; \delta = \frac{1}{2}\left(\frac{v_i q}{\Omega_{Hi}}\right)^2 \cos^2\theta_q, \; \eta = \frac{v_{ii}}{\Omega_{Hi}},\end{aligned}\right\} \qquad (13.4b)$$

θ_q is the angle between $\mathbf{q}$ and the normal to $\mathbf{H}_0$, while ϕ_q is the angle between the ($\mathbf{q}, \mathbf{V}_b$) and ($\mathbf{V}_b, \mathbf{H}_0$) planes, θ_0 is the angle between $\mathbf{V}_b$ and $\mathbf{H}_0$, and v_{ii} is the effective frequency of collisions between ions.

The functions F and G, defining the differential effective scattering cross-section $d\sigma$ (13.4a), have to be calculated using numerical integration. The function G is a smoothly, gradually varying function of the angle χ. Moreover, it will be seen below that $d\sigma$ is a maximum for low values of τ, so that in the calculations we can use the value of G for $\chi = \pi/2$. In this case

$$G(q\rho_b, \chi = \pi/2) = [J_1(q\rho_b)/q\rho_b]^2, \qquad (13.5)$$

where J_1 is a Bessel function.

The function F governs the main properties of $d\sigma$. It can thus be called the *scattering function*. It depends on four parameters, which in turn vary with the velocity of the body V_b and the angle that the velocity vector makes with the Earth's magnetic field $\mathbf{H}_0$, as well as the thermal velocity v_i, the gyrofrequency Ω_{Hi}, and the collision frequency v_{ii} of the ions in the ionosphere. The characteristics of the function F will become evident from the figures and tables to be presented below. The latter were calculated for three altitudes in the ionosphere: $z = 300$, 400, and 700 km, where the effect of scattering from a body wake in the ionosphere is apparently quite pronounced.

For a narrow region in the vicinity of the maxima of the differential cross section $d\sigma$ (see below), instead of (13.4) we get the following approximate analytical formula (Vas'kov, 1969b):

$$d\sigma = \tfrac{1}{16}(\omega_0/c)^4(\rho_b V_b/v_{ii})^2|P_n|^2 do, \qquad (13.6)$$

where

$$P_n = \frac{\pi^{1/2}}{a}\frac{W(d)e^{-\mu}I_n(\mu)}{2 + i\pi^{1/2}[n(\Omega_{Hi}/av_{ii}) + (d + i/a)]W(d)e^{-\mu}I_n(\mu)}, \qquad (13.7)$$

and

$$W(d) = \exp(-d^2)\left[1 + \frac{2i}{\pi^{1/2}} \int_0^d \exp(t^2)dt\right] \quad (13.8)$$

is the Kramp function. For $d \gg 0$, we have $W(d) \simeq d^{-1}$, and

$$\left. \begin{array}{c} d = \left(\dfrac{b}{a} + \dfrac{i}{a}\right), \quad a = |q_\parallel| v_i/v_{ii}, \\[2mm] b = \dfrac{\mathbf{q} \cdot \mathbf{V}_b - n\Omega_{Hi}}{v_{ii}}, \quad \mu = \tfrac{1}{2} q_\perp^2 (v_{ii}^2/\Omega_{Hi}^2), \end{array} \right\} \quad (13.9)$$

where $n = 1, 2, \ldots$ is the number of the corresponding maximum of $d\sigma$, $q_\parallel$ and $q_\perp$ are the components of $\mathbf{q}$ parallel and normal to vector $\mathbf{H}_0$, and I_n is a Bessel function with an imaginary argument. Although the expression (13.7) is rather complicated, nevertheless it enables us to determine $d\sigma$ without having to resort to numerical integration. However, the comprehensive analysis of the properties of the scattering cross-section was carried out on the basis of a study of the accurate formula (13.4), some numerical results of which will also be given below.

The main special feature of the scattering function F is its oscillating (multi-lobed) character as a function of the angle χ (Fig. 13.1). If $\theta_0 = 0$, i.e., if the body moves along the Earth's magnetic-field vector $\mathbf{H}_0$, or if the angle $\theta_\mathbf{q} = 0$, then the principal maximum of the scattering function (zero-order maximum (0)) will correspond to a value $\tau = 0$ (or $\chi = \pi/2$), its lateral maxima ($\pm 1, \pm 2, \ldots$ order) lying symmetrically about the principal one. This is shown by Fig. 13.1a, the curves of which correspond to different values of the parameters ζ and δ. An analysis of the general properties of the scattering function indicate that its principal maximum has the largest amplitudes for $\theta_0 = \theta_\mathbf{q} = 0$. If $\theta_0 = 0$, the principal maximum lies in the direction of 'specular reflection' of the wave from the magnetic-field direction $\mathbf{H}_0$. This means that the bisector of the angle between $\mathbf{K}$ and $\mathbf{K}_s$, i.e., the vector $\mathbf{q}$, is perpendicular to $\mathbf{H}_0$. Here, since for $\theta_0 = 0$ the function F is independent of the angle $\phi_\mathbf{q}$, the function $F(\theta_\mathbf{q})$ forms a surface of revolution about the vector $\mathbf{K}_s$, along which its principal 'spatial lobe' is directed. For $\theta_0 \neq 0$, the principal maximum corresponds to values $\theta_\mathbf{q} = -\theta_0$ (Fig. 13.1b). At this same angle the vector $\mathbf{q}$ turns about a line normal to $\mathbf{H}_0$. The number of lateral maxima and their amplitudes diminish rapidly with an increase in the lobe number; they depend considerably on the values of the parameters ζ, δ, and n determining the convergence of the corresponding integrals. The function F, as is easily shown (see (13.4b)), is a function of the angle $\phi_\mathbf{q}$ only if $\theta_0 \neq 0$.

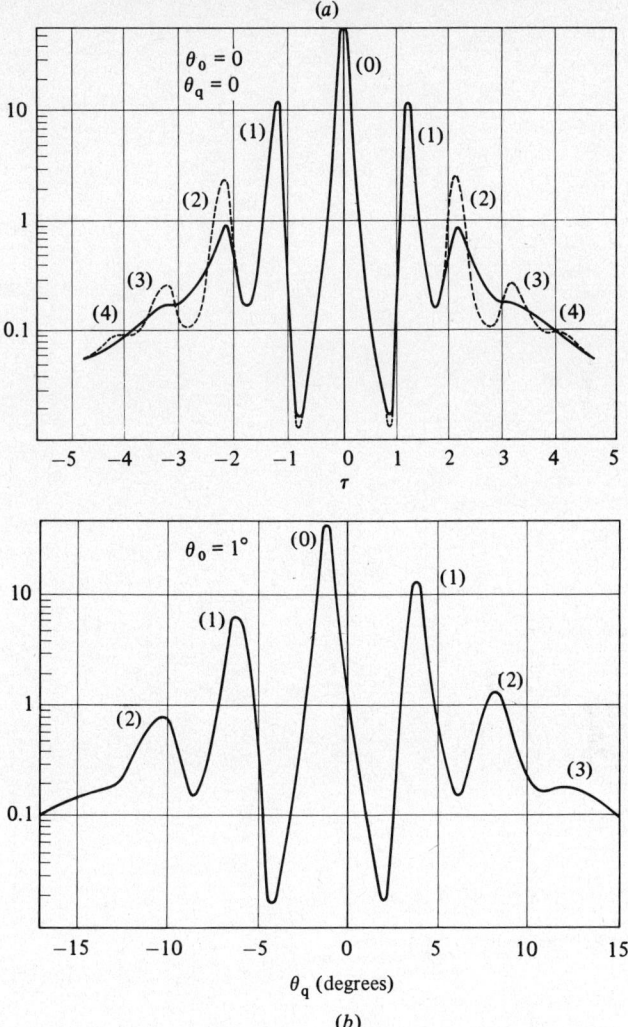

Fig. 13.1. Curves of the scattering function $F(\tau, \zeta, \delta, \eta)$, determining the effective scattering cross-section of the wake of a rapidly moving body. The solid and dashed curves correspond to different values of ζ and δ (Al'pert & Pitaevskii, 1961).

Consequently, the greatest effect of the scattering will be observed when the body moves along the magnetic-field vector, and *the major role is played by the principal maximum of F*. It can take on quite high values in the angle ranges $\Delta\theta_0$ or $\Delta\theta_q$, amounting to only some fractions of a degree. Fig. 13.2 shows the principal maximum $F_{0,\max}$ (zero-order maximum) of the scattering function as a function of the wavelength λ for

Table 13.1. Zero-order and first-order maxima of the function F for $\theta_0 = 0$

	$F_{0,max}$									$F_{1,max}$		
	z (km)											
	$\theta_q = 0$			$\theta_q = \pm 0.3°$			$\theta_q = \pm 1°$			$\theta_q = -5°$		
λ(m)	300	400	700	300	400	700	300	400	700	300	400	700
15	5	14	241	0.5	0.4	—	0.01	0.05	0.001	—	—	—
20	11	31	479	1.8	0.5	0.7	0.1	0.07	0.03	—	—	—
30	53	134	1535	13	11	45	1.3	1.0	0.3	10.3	12.6	3.4

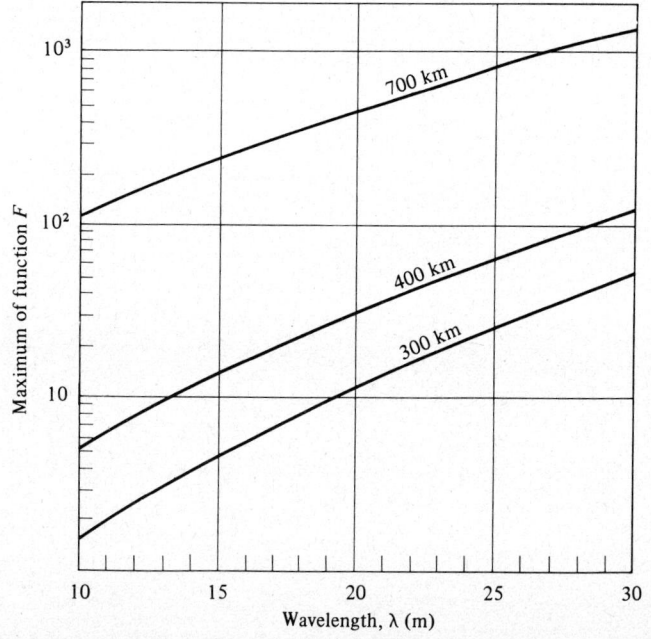

Fig. 13.2. Dependence of the principal maximum of the scattering function on wavelength for different altitudes in the ionosphere ($\theta_0 = \theta_q = 0$). (Al'pert & Pitaevskii, 1961.)

$\theta_0 = 0$, while Table 13.1 gives the maximum values for different altitudes z and wavelengths λ in the vicinity of the principal maximum. In addition, the table gives the first-order maximum values $F_{1,max}$ for $\lambda = 30$ m and $\theta_0 = 0$, which in the given case correspond to an angle $\theta_q = -5°$.

Inspection of Table 13.1 and Fig. 13.1 shows that the field of the scattered wave should be quite large only in a small range of angles, in which the zero-order maximum of the scattering function has high values. Calculations show that we can take a corresponding angular width of the lobe

$$\alpha = 2\Delta\theta_0 \quad \text{or} \quad 2\Delta\theta_q \simeq (0.6\text{--}0.8)° \simeq 10^{-2}\,\text{rad}. \tag{13.10}$$

Thus, for a velocity of the body $V_b \simeq 8\,\text{km}\,\text{s}^{-1}$, the scattering field at the Earth's surface for motion of the body along $\mathbf{H}_0$ at the above-indicated altitudes z 'irradiates in a burst' an area on the Earth's surface having a linear size $\simeq (\alpha z)$, the irradiation lasting, at the most, only for a time

$$\Delta t \simeq \alpha z/V_b \simeq 0.4 \text{ to } 1\,\text{s}. \tag{13.11}$$

The total angle interval, covering several lobes, amounts to about 15 or 20°, but the bursts of ± first, ± second, and higher orders are considerably weaker. Here, within the given time intervals Δt the effect of the scattering field recorded by a receiver naturally is determined not by the maximum values of the scattering function $F_{0,\text{max}}$, given in Fig. 13.2 and Table 13.1, but rather by some average value $\bar{F}$ in the indicated intervals of angle α. Such 'average' values of the function depend on a number of factors, in particular on the characteristics of the receiver and on the angular variation of F. Approximate estimates show that we can assume

$$\bar{F} \simeq \tfrac{1}{2} F_{0,\text{max}}. \tag{13.12}$$

With the aid of formula (13.4a) the dependences of the differential effective scattering cross-section $(d\sigma/do)_{\text{max}}$ on the wavelength λ and on the altitude z were calculated (Fig. 13.3), on the basis of the $F_{0,\text{max}}$ values given in Fig. 13.2. Since $d\sigma \sim \omega_0^4 \sim N_0^2$ (see (13.4)), i.e., it depends markedly on the electron density, the effective cross-section $d\sigma$ of the body's wake will be small during the nighttime hours. Therefore, the corresponding curves in Fig. 13.3 pertain to a mean model of the ionosphere in the daytime. It is also interesting to compare the effective scattering cross-section of the body's wake with the scattering $d\sigma_0/do$ from the body itself. For such a comparison we consider scattering from a smooth metal sphere of radius ρ_b. Other bodies, with similar linear dimensions but with more complicated (in particular, rough) surface structures, will have lower values of $d\sigma_0/do$. Fig. 13.3 shows the plots of $d\sigma/d\sigma_0$, i.e., the ratio of the differential effective scattering cross-section $d\sigma/do$ of the body's wake to the differential effective cross-section $d\sigma/do$ of the body itself, for $d\sigma/d\sigma_0 > 2$, namely for those wavelengths at which the effect of scattering from the wake is twice or more as great as the effect of the scattering from the sphere itself. It should be noted, however, that since the action time of the effect of the scattering from the sphere

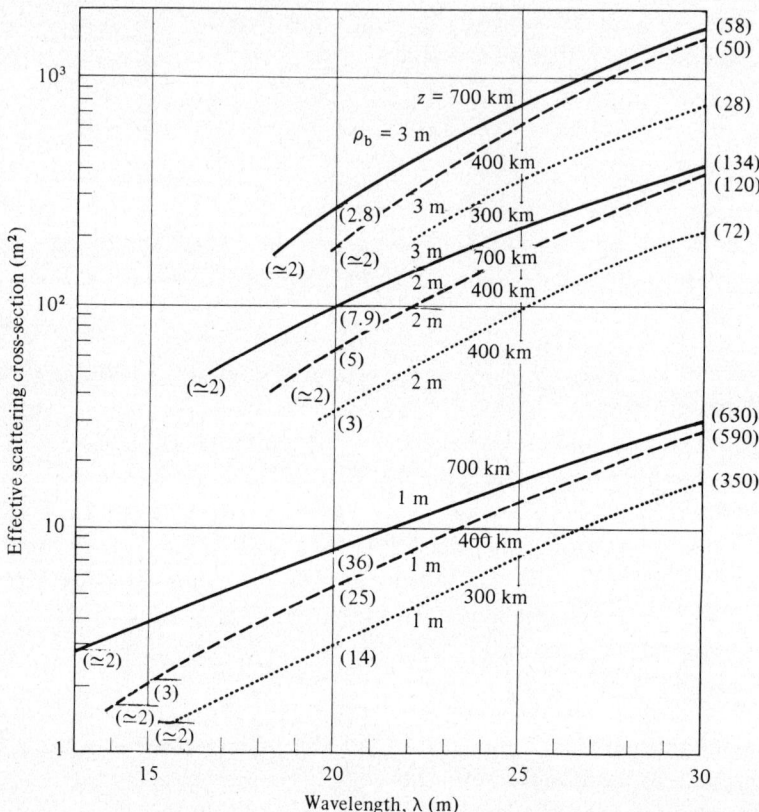

Fig. 13.3 Plots of the effective scattering cross-section $(d\sigma/do)_{max}$ for the wake of a rapidly moving sphere of radius ρ_b at various altitudes z in the ionosphere. The numbers by the curves give the values of $d\sigma/d\sigma_o$ for $\lambda = 20$ and 30 m (Al'pert & Pitaevskii, 1961).

is practically equal to the 'transit' time of the field of its scattering through the angular range of the receiving antenna, it follows that this time may exceed considerably the duration of a burst Δt (see (13.11)).

Note also that, if the region of transit of the body is irradiated from several points $(s_1, s_2, s_3, \ldots)$ situated on the Earth's surface at different angles, several scattered waves will be picked up at reception point E (Fig. 13.4), and the total time of action $\sum \Delta t$ of the effect of the scattering from the body's wake may be substantially longer.

Inspection of Fig. 13.3 shows that the differential effective scattering cross-section $d\sigma/do$ increases rapidly with an increase in wavelength, approximately according to an exponential law; the $d\sigma/do$ variations plotted on a logarithmic scale are close to straight lines. Here the ratios

Scattering of radio waves by the wake

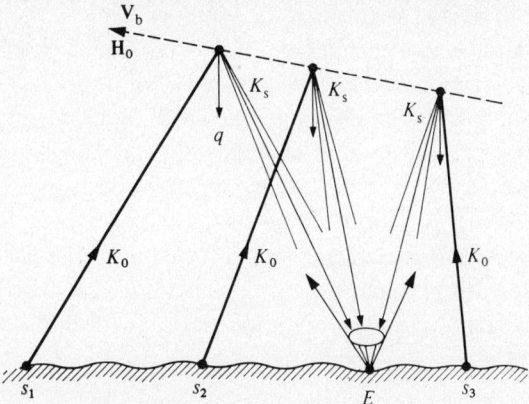

Fig. 13.4. Diagram illustrating reception at the point E of satellite-scattered radio waves transmitted from the points S_1, S_2 and S_3.

$d\sigma/d\sigma_o$ (the numbers by the lines) are also seen to increase considerably with an increase in the wavelength λ. In the λ range being considered, the scattering field from the body's wake is hundreds of times more intense than the scattering field from the sphere itself. The differential cross-section $d\sigma/do$ also increases with the altitude, but a great deal more slowly than with an increase in λ. The greater $d\sigma/do$ values at higher z can be explained by the fact that the drop in the collision frequency v_{ii} with altitude is compensated by a drop in the electron density N_0; these two factors have opposite effects on $d\sigma$.

All the foregoing theoretical results were obtained assuming that the incident waves scattered by the wake of the body are plane waves. Taking the sphericity of an incident wave into account reveals some additional peculiarities (Vas'kov, 1969c), which, however, do not alter too radically the above-considered scattering characteristics or the results of quantitative calculations of the differential effective cross-section, provided the inhomogeneity of the ionosphere can be neglected and the body moves through regions where the refractive index is not close to zero. In this case the effect of the sphericity of the wave will reduce to the following: the principal maximum of the scattering cross-section will also be smaller, because of the sphericity of its front (for a plane wave the corresponding divergence of the formulas for $d\sigma$ is removed only by the collision frequency v_{ii}). This is connected with the fact that the effective length of the wake, which determines the main part of the scattering-field intensity, is limited by the first Fresnel zone (radius $\rho_F \simeq (\lambda s)^{1/2}$, where s is the distance between the source of the incident waves and the body). However, the role of the sphericity of the wave becomes very important when the body moves through a homogeneous medium and traverses regions of reflection of the incident wave where the refractive index $n^2 \to 0$. In this case, as has been shown (Gurevich & Pitaevskii, 1966), the value of $d\sigma/do$ may increase by two orders of magnitude or more, especially for small angles of the incident wave. This effect starts to be very strong when the body moves close to the caustic formed upon reflection of the spherical wave from the inhomogeneous medium. In this case the focusing of the waves will be a maximum. Here, in order to obtain this effect, it is enough to take into account just the inhomogeneity of the ionosphere with altitude, as was done by Gurevich & Pitaevskii, 1966, and Vas'kov, 1969a.

The caustic, the envelope of the family of rays emitted by the source, is produced because of refraction of waves in an inhomogeneous medium. The caustic separates the region irradiated by the waves from the shadow region. Naturally, therefore, the nature of the variation in the scattering cross-section changes when the body crosses from the irradiated zone into the shadow zone, and it depends on the position of the body relative to the caustic. Detailed calculations of this effect (Vas'kov, 1969a,d) gave the following results. The intensity of a scattered wave is a maximum, if the source of the incident waves and the observer are at the same point, when the body moves tangent to the surface of the caustic and the Earth's magnetic field is normal to this surface. Fig. 13.5 shows the differential effective cross-section of a sphere of radius $\rho_b = 1$ m, as a function of

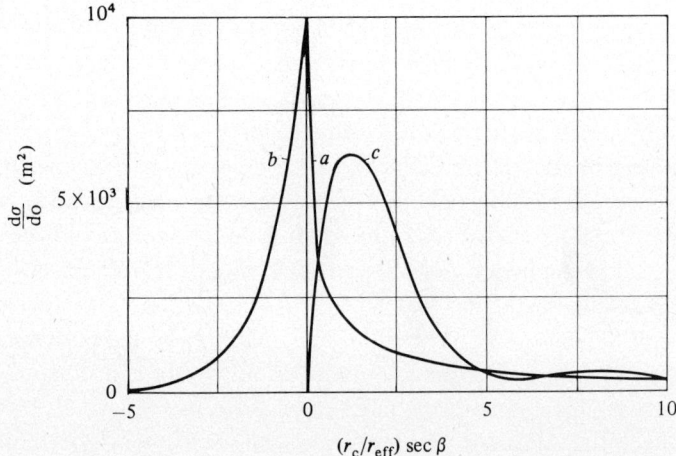

Fig. 13.5. Differential effective scattering cross-section $d\sigma/do$ of a rapidly moving body passing through the region of a caustic (Vas'kov, 1969d).

the distance r_c from the caustic, for an ionosphere region at an altitude $z = 250 \text{ km} (N_0 \simeq 10^6 \text{ cm}^{-3})$.

Curve a in Fig. 13.5 corresponds to the approach of the moving body to the caustic from the irradiated side ($r_c > 0$); curve b corresponds to the actual crossing of the caustic into the shadow region ($r_c < 0$); curve c describes the $d\sigma/do$ relation when the body moves only in the irradiated region. If the body crosses the caustic into the shadow region, scattering takes place only from the part of the body's wake lying in the irradiated region. The quantity $r_{\text{eff}} = V_b/v_{ii}$ is called the 'effective length' of the wake, and β is the angle between V_b and the normal to the caustic. Fig. 13.5 shows that in the given case the cross-section reaches 10^4 m^2 when $r_c = 0$. Note that the curves of $d\sigma/do$ versus $(r_c/r_{\text{eff}}) \sec \beta$ in Fig. 13.5 were obtained by Vas'kov, 1969d, for the case when $\beta \neq \pi/2$, i.e., under the condition

$$r_{\text{eff}} \cos \beta > [(\lambda/2\pi)^2 (dn^2/dz)^{-1}]^{1/3}. \tag{13.13}$$

Since for $z \simeq 250 \text{ km}$, $r_{\text{eff}} \simeq 2 \times 10^2 \text{ m}$, and for $N \simeq 10^6 \text{ cm}^{-3}$ the reflection condition is satisfied for $\lambda \simeq 33 \text{ m}$, by assuming $dn^2/dz \simeq 10^{-4} \text{ m}^{-1}$, we find that the curves in Fig. 13.5 correspond to the case where $|\cos \beta| > 0.25$ and thus $\beta < 76°$. With an increase in the angle β, as was shown above, $d\sigma/do$ becomes greater; however, accurate calculations for $\beta \to \pi/2$ are rather complicated and thus were not carried out.

For the indicated value $r_{\text{eff}} \simeq 200 \text{ m}$ and a body moving near the caustic, intense scattering is observed at distances of the order of 1 or 2 km. Thus the duration of a burst of the intense scattering field at the observation

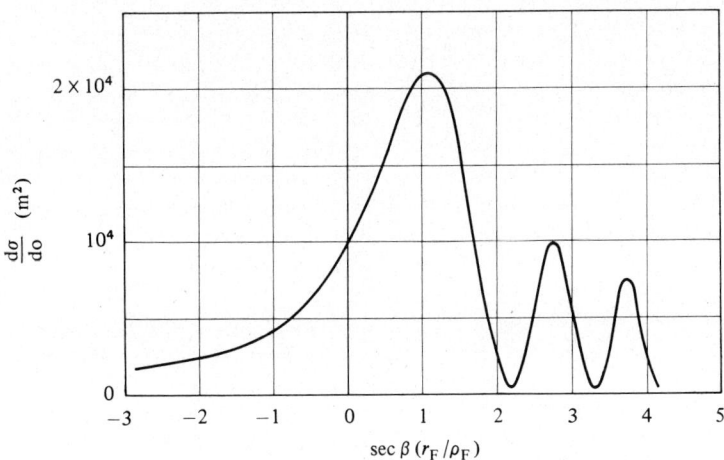

Fig. 13.6. Oscillatory nature of the $d\sigma/do$ curve for a rapidly moving body (Vas'kov, 1969d).

point is short: $\Delta t \simeq 0.1$ to 0.2 s. It is interesting that in some cases of body movement close to the caustic $d\sigma/do = 0$, so the scattering field recorded near the Earth's surface has an oscillatory nature. This effect occurs for a certain geometry of the wave-front position relative to the caustic, when the sphericity of the wave may limit the scattering cross-section and waves arriving from the caustic region interfere with waves reflected from the Fresnel zone. Fig. 13.6 gives the corresponding curve, calculated for the above-indicated conditions, of $d\sigma/do$ as a function of the distance r_F of the body from the centre of the Fresnel zone (ρ_F is once again the radius of the first Fresnel zone).

14

Some remarks about the excitation of waves and plasma instability near a fast-moving body

A very important subject related to studies of the flow around a body moving rapidly through a plasma ($V_b \gg v_i$) is the instability of the plasma region around the body. This subject is associated with a consideration of the conditions for excitation in the body's vicinity of oscillations and waves, which either travel with the body or are emitted by the disturbed plasma region and go beyond it.

In its general formulation this problem involves the solution of the kinetic equations (3.7) (see Volume 1), taking into account the dependence on the time t, i.e., including the terms $\partial f_e(\mathbf{r},\mathbf{v},t)/\partial t$ and $\partial f_i(\mathbf{r},\mathbf{v},t)/\partial t$. It appears to be more convenient to solve such problems in a system of coordinates that is at rest with respect to the body (i.e., the plasma moving and body at rest). However, as yet no final results of studies of this type have been obtained, even for simple particular cases. All the above-considered data were arrived at by solving the kinetic equations in a coordinate system moving with the body, i.e., by solving a number of steady-state problems in which the time dependences of the distribution functions f_e and f_i were ignored. Moreover, it is clear that, first, since the motion of the body is supersonic, very low frequency (VLF) waves dependent on the motion of ions must in some way play a role and appear during the interaction of the body with the plasma, even in the steady-state case. Second, it is to be expected that an inhomogeneous cloud carried with the body, interacting with the particle fluxes or waves impinging upon the body, may cause the appearance in the plasma of various kinds of oscillations and waves. Moreover, due to the influence of the Earth's external magnetic field, all these phenomena take on a specific, more complex character. Consequently, under certain conditions individual cases can be distinguished, i.e., conditions favourable to the excitation of specific wave spectra.

As a whole, this problem constitutes a complicated, little-studied area of plasma theory. Investigating it systematically is one of the most interesting, and at the same time most important, tasks of the theory of plasma flow around bodies. The role of wave processes in the vicinity of a rapidly moving body, in particular, is illustrated by the following results of wake studies.

14.1. The relationship between the perturbation of the electron density $\delta N_e(\mathbf{r})$ in the wake and ion-acoustic waves

The 'wave' nature of the perturbation of the electron density in the far zone of a body's wake is revealed immediately by an analysis of its Fourier component $\delta N_\mathbf{q}$ (see formula (13.2)). In the preceding chapter it was shown that, precisely for $\delta N_\mathbf{q}$, a self-consistent solution of the kinetic equation and the Poisson equation in the form of finite equations could be obtained, taking into account the effect of the electric field appearing in the wake. This enabled a detailed study to be made of the wake structure, and the corresponding results were given in Chapter 11. For this, an inverse Fourier transformation of $\delta N_\mathbf{q}$ was used, i.e., numerical integration of the formula

$$\delta N(\mathbf{r}) = \int \delta N_\mathbf{q} \exp(-i\mathbf{q}\cdot\mathbf{r}) d^3q. \qquad (14.1)$$

For simplicity, let us illustrate here for an isotropic plasma, i.e., when $\mathbf{H}_0 = 0$, the corresponding properties of $\delta N_\mathbf{q}$ and its 'wave' features. In this case for a nonisothermal plasma, we have

$$\delta N_\mathbf{q} = -\frac{\pi^{3/2}}{q} N_0 \left(\frac{V_b}{v_i}\right) \rho_b^2 \frac{W((V_b/v_i)\cos\chi)}{1 + (T_e/T_i)[1 + i\pi^{1/2}(V_b/v_i)\cos\chi\, W((V_b/v_i)\cos\chi)]}, \qquad (14.2)$$

where, as previously (see formula (13.8)), W is the Kramp function and χ is the angle between $\mathbf{V}_b$ and $\mathbf{q}$. Note that if the denominator of (14.2) is equated to zero, we get the dispersion equation for longitudinal ion-acoustic waves $\omega = \mathbf{q}\cdot\mathbf{V}_b$, where $qD_e = 2\pi D_e/\lambda \ll 1$ (and D_e is the Debye length). The formula for $\delta N_\mathbf{q}$ and also for the perturbations of the electric and magnetic fields in the wake, can, however, be obtained in a different way. The appropriate calculations (Bud'ko, 1969b) show that the nature of these perturbations is related to the excitation of ion-acoustic waves by the body. Thus the presence of a 'dispersion denominator' in (14.2) has a quite obvious physical meaning, and the theoretically obtained maximum perturbation of the electron density in a cone with an opening angle $\theta \simeq 2\sin^{-1}(v_i/V_b)$ (see Figs. 11.22 and 11.23) is a consequence of *Čerenkov excitation of ion-acoustic waves* by the body. In a certain sense, there is a similarity between this cone and the Mach cone of aerodynamics. But the remarks made previously with regard to this should be kept in mind.

In an isothermal plasma, due to the considerable attenuation of ion-acoustic waves, the cone of the disturbed region becomes blurred and lacks sharp boundaries (see Fig. 11.22). In a nonisothermal plasma with $T_e \gg T_i$, where the attenuation of ion-acoustic waves is slight, this effect becomes more pronounced. The Čerenkov excitation of ion-acoustic waves causes the regions of particle rarefaction to acquire sharp boundaries and to become much narrower (see Figs. 11.25 and 11.26). On the other hand, the appearance of concentration regions behind the body, which take on a narrow 'lobelike character' with an increase in T_e/T_i (Fig. 11.26), can, as noted previously in Chapter 11, be attributed to an enhancement of the effect of the electric field because of particle focusing, as well as to the slight attenuation of the ion-acoustic waves.

In a magnetized plasma, where $\mathbf{H}_0 \neq 0$, the formula for δN_q has a structure like that of formula (14.2) for an isotropic plasma (see, for instance, Al'pert et al., 1965). As in the case where $\mathbf{H}_0 = 0$, the quantity δN_q has a 'dispersion denominator', the properties of which are determined by the dispersion equations for the various kinds of magnetoacoustic waves. A study of the structure of the perturbation when $\mathbf{H}_0 \neq 0$ will naturally, from this point of view, be more complicated than in the isotropic case.

14.2. The interaction between incident electromagnetic waves and the wake

It is to be expected that under certain conditions 'resonances' of various kinds may be observed in the inhomogeneous cloud accompanying a body, because of the interaction between the currents generated in the cloud and the field of incident electromagnetic waves. Plasma oscillations arising in the wake itself can apparently also lead to modulation of the incident waves. Moreover, in the wake of a body there may be a simultaneous coherent excitation of oscillations at frequencies satisfying various resonance conditions of the plasma, for example, in the vicinity of the upper and lower hybrid resonances, making the total field of the oscillations complex in structure. The different experimental facts which have been observed using satellites (for instance, those to be discussed in the next chapter), but which have not been reliably verified so far by theory, may well be caused by wave processes arising due to the interaction of the wake with the natural electromagnetic radiation of the plasma surrounding it. Such data, requiring further theoretical explanation, include, for instance, those obtained with OGO 1. Distinctive features of such experimental findings may be their correlation with the period of rotation of the body, or with the orientation of the wake relative to the external magnetic field H_0, correlation of the velocity of the waves incident on the body with the velocity V_b of the latter, etc. But, as emphasized above, none of these phenomena has as yet been properly studied.

As in the previous section, by way of an illustration let us cite the results of a theoretical analysis of the 'resonant' interaction of a wave packet incident upon the body's wake with the perturbed region of the plasma; these results appear to explain some of the experimental data.

It was shown (Bud'ko, 1969), that if the 'resonance' condition

$$d\omega/d\mathbf{K} = \mathbf{V}_b \qquad (14.3)$$

is satisfied, i.e., if the group velocity of the incident wave packet equals the velocity of the body and if $V_b \perp H_0$, then in an inhomogeneous medium oscillations corresponding to the resonance branch in the vicinity of the lower-hybrid frequency ω_L will be excited. Physically, condition (14.3) means that the wave packet moves along with the wake of the body, promoting a strong interaction between them. Finally, it polarizes the inhomogeneous cloud, slow longitudinal waves are excited in it, and this effect lasts for a long time. These theoretical calculations were stimulated by some experimental results obtained with Alouette (Brice & Smith, 1965; McEwen & Barrington, 1967), indicating that such effects occurred. Whistling atmospherics recorded aboard the satellite (Fig. 14.1, branch a) excited plasma oscillations (branch b), which were cut off at the lower-hybrid frequency. The theoretical results for this effect are plotted schematically in Fig. 14.1b, as a time dependence of the frequency of the packet of oscillations excited in the vicinity of the lower-hybrid frequency; the fit with the experimental data is seen to be quite good.

The theoretical results given in Fig. 14.1, correspond to times

$$t \gg t_0 \simeq \lambda_L/V_b \qquad (14.4)$$

for which simple, usable formulas are obtained; see the figure for the definition of t_0. In (14.4) λ_L is the wavelength of the excited oscillations. The width Δf_L of this oscillation packet and its amplitude E_L vary with time as follows:

$$\Delta f_L \sim t^{-1}, \quad E_L \sim t^{-3/2}. \qquad (14.5)$$

For the experimental results given in Fig. 14.1, data of various measurements aboard Alouette give $t_0 \simeq 0.1$ s. Note, however, that a quantitative comparison of theory with the indicated experimental data could not be carried out, since in Brice & Smith, 1965, and McEwen & Barrington, 1967, the necessary data for the oscillation amplitude E_L, its dependence

14.3. Emission from the wake and plasma instability

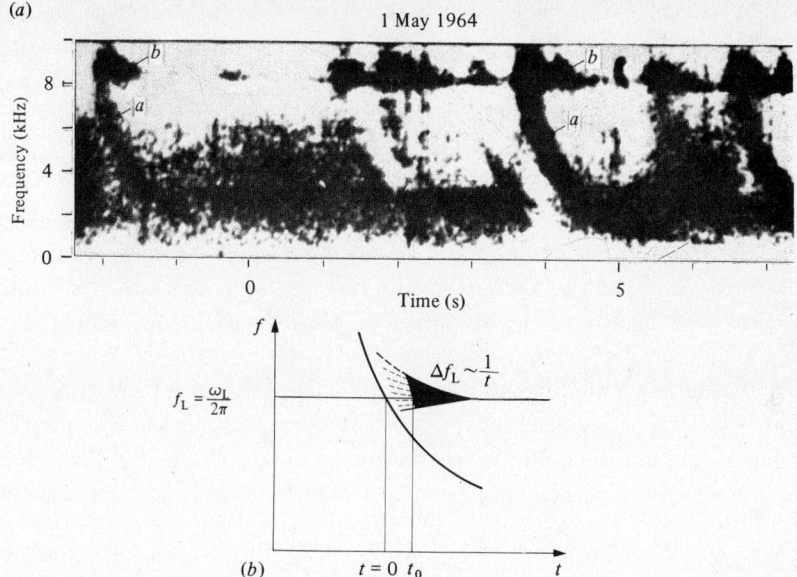

Fig. 14.1 Spectrogram of a plasma emission (branch b) excited near the lower-hybrid resonance by whistling atmospherics (branch a) (Alouette, 1 May 1964). A schematic plot of theoretical calculations of this effect is shown below (Bud'ko, 1969b).

on time, the width Δf_L of the packet of excited waves, etc., were lacking. Nevertheless, the agreement obtained between theory and experiment appears to be quite interesting with regard to the development of the theory describing the effects being discussed in this chapter.

14.3. Emission from the wake and plasma instability

It is advisable to conclude this chapter with a few brief remarks. We saw above that in the vicinity of a body waves may be excited that are determined by the motion of ions, since the velocity of the body $V_b \gg v_i$. For the attenuation of these oscillations to be low, their wavelengths have to be smaller than the mean free path of the ions and the plasma has to be nonisothermal ($T_e \gg T_i$).

The excitation of electron waves calls for more favourable special conditions than does the excitation of ion waves, since the electron velocity $v_e \gg V_b$, so the phase velocity of the anticipated electron waves will also be higher than V_b. However, 'electron resonances' may be excited, for instance, by the scattering of electrons at the inhomogeneous (quasi-periodic) structure of the body's wake, by a favourable orientation of the vector $\mathbf{H}_0$ relative to $\mathbf{V}_b$, and by other factors. Thus there is as yet no good basis for ruling out the possibility of their excitation. With regard to this, it should be noted that in the above-cited experiments aboard Gemini/Agena and Explorer 31 the effective electron temperature $T_{e,\text{eff}}$ close to the body (behind it and on its axis) was found to be 1.5 to 2 times the electron temperature T_e of the surrounding undisturbed plasma. This is apparently associated with a disturbance of the nearby Maxwellian electron distribution. However, whatever the cause of this effect, this disturbance of the equilibrium distribution of the particles will promote plasma instability and the excitation of plasma oscillations around the body.

Conditions promoting plasma instability and the growth of disturbances in the vicinity of various kinds of 'resonances' may appear in the wake of a body due to a nonuniform distribution of the density and temperature of the plasma. In the presence of an external magnetic field such a disturbance may make conditions favourable for the enhancement of certain kinds of oscillations. Since in the immediate vicinity of the body the effective ion temperature $T_i \ll T_e$, two particle fluxes may well form around the body, i.e., beam instability may appear, promoting the growth of small deviations from the steady state of the plasma and exciting wave processes in it. Under actual conditions, of course, the growth rates of such oscillations will be small.

With regard to problems of this kind, of course, a major role, and in a number of cases a decisive role, may be played by the interaction of the body's wake with the particle fluxes incident upon it. In this case, as in many other problems, the properties of the body surface have considerable influence (the nature of its interaction with the particle fluxes, see Chapter 10 above), as does the body potential ϕ_b (Gurevich, 1964). Studies have also been made of the possibility of the appearance of plasma instability in the vicinity of a moving body as a consequence of the photoemission from its surface (Smirnova, 1967, 1969). An instability mechanism created by the presence in the ionosphere of different species of ions is also of interest. This leads to the appearance of an electric field in the disturbed zone behind the body, accelerating the motion of light ions, as well as to the possibility of the excitation of ion-acoustic waves in the plasma (Gurevich et al., 1972).

In conclusion, let us mention once again that, in our opinion, detailed theoretical studies of this group of problems, together with a search for similar wave processes in the vicinity of bodies in various experiments, are of great interest.

Part 2

Waves and oscillations in the near-Earth and interplanetary plasma

15

Introductory remarks

During the last 10 or 15 years some very intensive studies of wave processes in the near-Earth plasma and in the interplanetary medium, out to distances of a million or more kilometres from the Earth, have been performed successfully. Every issue of the specialized journals includes new experimental results and/or attempts to give them a theoretical explanation. It is, however, becoming increasingly difficult to interpret the plethora of different data in the literature and to present them systematically. Here experiments occupy the predominant position. Although some theoretical investigations have enabled us to broaden our views concerning the excitation mechanisms of various kinds of observed waves, as well as concerning their mode of travel through the medium, still many facts remain unexplained, and some of these even appear to be rather curious.

It is not an easy task to analyse such a comprehensive, many-faceted field of study or to generalize the vast amounts of information obtained from the numerous experiments. This is the job of individual monographs. Within the framework of the present book, however, the author will attempt to give the reader, briefly but as comprehensively as possible, some conception of the majority of the wave processes observed in the vicinity of the Earth and in interplanetary space. Naturally, to do this, we have had to limit the contents of this part, as follows:

First of all, mainly experimental results will be presented. The principal types of wave processes observed over a wide frequency range in the near-Earth plasma and the interplanetary medium will be described. Secondly, we consider data collected mainly out in the plasma itself (*in situ*), i.e., during measurements with satellites and rockets. Naturally, however, no complete picture of the waves generated in the near-Earth plasma can be obtained without making use of some results of ground-based observations. Thus data of this type are considered too.

15.1. A general outline of the results of various experiments

The multitude of data gathered in experiments conducted under diverse conditions lead to the following general conclusions.

(1) In the considered regions of the natural plasma *all types of waves* and oscillations predicted by the linear theory are observed, namely: branches describing waves associated with the ion gyroresonances and with the lower-hybrid frequency, gyroresonance electron oscillations, Langmuir waves and waves at the upper-hybrid frequency, and finally the longitudinal electrostatic waves inherent to a warm plasma, in particular ion-acoustic and magnetoacoustic waves, and multiple ion and electron gyroresonance oscillations.

(2) Wave processes of *identical* physical nature have been recorded both in the plasma region close to the Earth, at altitudes $z \simeq \ldots 10^2 \ldots 10^3$ km, and further out, at distances of hundreds of thousands or millions of kilometres: in the geomagnetic tail and in the solar wind; these can even be assumed to take place close to the Sun, at distances from the solar centre $R_\odot \simeq 0.3$ AU (AU = astronomical unit).

(3) An important, fundamental feature of the results of numerous experiments is that the observed waves are often *transverse* ($\mathbf{K}_0 \perp \mathbf{E}, \mathbf{H}$). The data obtained show a good fit with the relation $n = \mu_0 cH/E$ between the electric and magnetic components of the transverse electromagnetic waves, n being the refractive index of a wave, μ_0 the permeability of free space, and c the velocity of light. On the other hand, in many cases the observed radiation can be excited only in the form of electrostatic longitudinal waves ($\mathbf{K}_0 \parallel \mathbf{E}$). This indicates that an active *transformation of longitudinal waves into transverse waves* is going on; frequently, however, the specific transformation mechanisms are unknown. This has been indicated for a long time by a great many results of ground-based observations of waves generated in the near-Earth plasma.

(4) Resonances appear actively at the *lower-hybrid frequency*, and this frequency plays a role in various phenomena taking place in the near-Earth plasma. At the lower-hybrid frequency, cutoff of *generated* oscillations is observed, as well as reflection of waves emitted by sources far away from the observation point. In the frequency interval between the lower-hybrid frequency and the ion-cyclotron frequency, *trapping* of waves in the near-Earth plasma occurs, making the wave trajectories quite complex.

(5) Experiments using artificial satellites revealed some new resonance oscillations of the plasma which had not been predicted by theory. Some of these have already been explained theoretically, albeit not definitively.

15.1. An outline of experimental results

Certain of the data are of considerable interest, and they probably play a major role with regard to odd half-integral gyroresonances of electrons: narrow bands of longitudinal waves excited at the frequencies $(s + \frac{1}{2})\omega_H$, where s is a whole number, and especially resonance at the frequency $\frac{3}{2}\omega_H$. The excitation of second and third harmonic resonance oscillations $2\omega_U$ and $3\omega_U$ (of the upper-hybrid frequency) and second harmonic resonance $2\omega_0$ (of the Langmuir electron frequency) were observed, in particular, under the influence of radio waves transmitted from a satellite. The plasma oscillations at combination frequencies $(\omega_U - \omega_H)$, $(\omega_0 - \omega_H)$ recorded during these experiments, as well as those at other frequencies not simply identifiable with the typical resonance frequencies, cannot be explained by the linear theory of a plasma.

(6) During recent years *active experiments* have come to play an important part in studies of various processes in the near-Earth plasma. Regular experiments of this type were initiated aboard Alouette, Explorer, and ISIS, the ionosondes of which sent out radio waves that had a strong effect on the medium. Some of the data from these experiments will be presented in §19.1 and referred to later in the present chapter. There have also been some quite successful attempts at *active modification of the ionosphere* in the altitude region $z \gtrsim 200$ to 300 km, with the aid of intense HF radio waves transmitted from the Earth's surface. A number of *nonlinear* effects discovered during these experiments were described in §7.1. They included some things that were very interesting and unexpected, such as the generation of an artificial sporadic layer with a cloudlike structure, the creation of small-scale inhomogeneities, certain properties of the spectra of scattered radio waves, etc. Attempts to excite artificially discrete wave packets in the whistler mode in the magnetosphere, using LF waves, led to some quite important, interesting results. For instance, an approximately thousandfold increase in radio-wave intensity was detected in the vicinity of the geomagnetic equator. Some results of these experiments will be described in §18.2.

Experiments using high-altitude rockets, designed to *inject* very intense electron fluxes, occupy a special position among active experiments in the near-Earth plasma. In addition to providing a study of the various kinds of phenomena occurring in the plasma close to the rocket, these experiments also occasioned a whole complex of simultaneous ground-based observations, both in the launch region and at its magnetically conjugate point. The results of this cycle of studies will not be considered here, but accounts of them can be found, for instance, in: Cartwright & Kellogg, 1971, 1974; Winckler, 1974; Gendrin, 1974; Miyatake *et al.*, 1974;

Matsumoto et al., 1971, 1974; Monson et al., 1976; Kellogg et al., 1976; Monson & Kellogg, 1978.

(7) Many oscillatory processes observed in the plasma regions of interest to us are *nonlinear in nature*. Some of these have been explained satisfactorily by theory. For instance, the role of the parametric decay instability has been clarified, and quite detailed studies have been made of nonlinear effects of the heating type in the ionosphere, or, for example, radiation at twice the Langmuir frequency $2\omega_0$ in the magnetosphere. These and other thoretical results constitute an important achievement in the given field of plasma physics.

15.2. Classification of the observed wave processes

So far, during the description of the various wave processes, the terminology used to classify them according to frequency range has not usually been too definite. This sometimes prevents a correct physical representation of the nature of these processes and leads to some confusion. The situation was, of course, complicated 'historically' during the development of this new branch of physics, in part because the location of the source of the waves being observed was often not known. Oscillations with frequencies higher than the gyrofrequency ω_H or the Langmuir frequency ω_0 of electrons, recorded at great distances from the earth, are called very-low-frequency (VLF) or low-frequency (LF) waves by some authors. However, these are actually by nature high-frequency (HF) waves, since they are caused by *electron oscillations*. In other publications, VLF or LF waves are defined as waves with frequencies less than the lower-hybrid frequency ω_L or waves with frequencies between the ion gyrofrequency Ω_H and ω_L, i.e., waves whose behavior is determined largely by *ion oscillations*. Similarly, the definition of the term extremely-low-frequency (ELF) waves is by no means clear either. These are frequently defined as waves with frequencies $\omega > \Omega_H$ or even $\omega \gtrsim \omega_L$.

To the author, it seems to be physically justifiable to classify the various wave processes on the basis of the resonance branches of the oscillations of a cold plasma, with the aid of their characteristic frequencies (see formulas (4.49)–(4.55) Volume 1). Thus we have the following definitions:

ELF waves pertain to the frequency range $0 \lesssim \omega \lesssim \Omega_H$;

VLF   "   "   "   "   "   " $\Omega_H < \omega \simeq \omega_L$

and $> \omega_L \ll \omega_H$;

LF   "   "   "   "   "   " $\omega_L \ll \omega \lesssim \omega_H$ or $\lesssim \omega_0$;

HF   "   "   "   "   "   " $\omega > \omega_H, \omega > \omega_0$.

Note that the ELF range as defined here is equivalent to what many authors call the ELF and ULF ranges. In the existing literature, however, the terms ULF, ELF, and VLF are by no means used consistently, in plasma physics or elsewhere. Any terminology and classification employed will naturally always involve some arbitrariness, so it is difficult, and often even essentially impossible, to define precise boundaries between the frequency ranges, say, when considering waves whose mechanisms are due to the nonisothermal nature or nonlinearity of the plasma. For instance, in a nonisothermal plasma, where $T_e \gg T_i$, the characteristic boundary between the VLF and LF ranges will be the Langmuir frequency of the ions Ω_0, rather than the lower-hybrid frequency ω_L. On the other hand, the above classification facilitates a physical approach to the various experimental data and makes possible a more orderly exposition of the corresponding experimental results.

15.3. Generation mechanisms for waves of different types

The phenomena under review will not be considered theoretically here in detail. This is in fact often completely impossible, since the proposed theories are still sometimes quite unrealistic. It would be advisable, however, in some cases to mention which processes can explain, or describe quite satisfactorily, the various types of waves and oscillations excited in the near-Earth and interplanetary plasma. The following brief remarks of general character will, in particular, serve this purpose. More detailed discussions of the corresponding theoretical subjects are given in, for instance: Kimura, 1974; Matsumoto & Kimura, 1971; Fredricks, 1975; Gendrin, 1975; Shawhan, 1977; and also in the references cited in these works.

The *energy sources* of the wave processes of natural origin to be described below are:

1. the solar wind;
2. the large-scale dynamics of the near-Earth plasma, regulated by the Earth's rotational energy;
3. the magnetic field of the Earth.

The conversion of the energy of these sources manifests itself in the plasma in the form of the kinetic energy of the electric fields. A further conversion into oscillatory energy of the plasma takes place as a result of a disturbance of the Maxwellian equilibrium distribution of the plasma particles, which becomes unstable relative to various types of wave processes. An *unstable state* in turn can appear for any of a number of reasons. The main ones with regard to the processes being considered here are:

1. diffusion of charged particles along the magnetic field and other transport processes in a plasma, pitch angle scattering of particles;

2. plasma drift as a result of the effect of gradients of plasma density and magnetic field;

3. heating of the medium and particle acceleration in the electric field;

4. disturbances arising in a mixture of a cold (ionosphere) and hot (solar wind) plasma.

Finally, in the media being considered two quite general forms of instabilities will be predominant; *gyroresonance* instabilities and *flow* (beam) instabilities.

An *anisotropic particle distribution*, i.e., a distribution for which the temperature $T_\perp$ in the direction perpendicular to the static magnetic field is higher than the longitudinal temperature $T_\parallel$ of the particles, or else a particle distribution that is anisotropic in pitch angle, represents a situation where the particles have free energy, which leads to instability (see §§6.2.3 and 6.3.5, Volume 1). Here the free energy of the particles may be liberated if the conditions of *gyroresonance* (see (3.18), Volume 1), relative to the cyclotron frequencies of electrons ω_H or ions Ω_H, are satisfied. For instance, the ion-cyclotron waves which appear within the bounds of the plasmapause (see §16.3) or the hydromagnetic whistlers generated beyond the plasmapause in the transition region of the magnetosphere (see §16.1) are apparently explained by instabilities of this type. In these cases the energy sources are hot ions, injected, in particular, from the region of the geomagnetic tail. Random noise-like waves such as hiss (see §17.2) are apparently caused by electron gyroresonance; they are presumably generated at the inner boundary of the plasmapause at the expense of electrons trapped in the radiation belt. Similarly, in the transition region beyond the plasmapause, narrow-band chorus radiation is generated, due to gyroresonance during periods of heightened magnetic activity on the dayside of the Earth, following the injection into this region of electrons with energies from 5 to 150 keV (Burton & Holzer, 1974). The 'explosive' nature of these oscillations and their high intensity attest to the nonlinearity of the process. The so-called 'polar HF radiation', a source of the Earth's radiation which goes beyond its limits (see §19.3), is presumably excited by a Doppler-shifted gyroresonance mechanism (Melrose, 1976; see also Gurnett, 1974; Alexander & Kaiser, 1976).

The fundamental condition for *beam* (flow) instability is the presence of a 'bump' in the tail of the particle distribution function, i.e., an increased phase-space density (an additional maximum) in an interval around some velocity v_{max} (see §§6.2.1 and 6.2.2, Volume 1). If under the influence of

15.3. Generation mechanisms for waves of different types

this increased particle flux the instability condition is satisfied, then the energy of the oscillations generated is derived from the part of the distribution function just below v_{max} where the function has a positive slope, i.e., where the phase-space density increases with increasing particle velocity. The lower end of this range, close to the velocity v_{min} where the phase-space density is minimum, is the most favourable for the excitation of oscillations, provided that the resonance condition $v_{min} = v_\phi = c/n_b$ is satisfied (where v_ϕ is the phase velocity of the excited oscillations), since in the regions of gyroresonances ω_H and Ω_H, of the hybrid resonances ω_U and ω_L, and of the plasma resonances ω_0 and Ω_0, the kinetic refractive indexes n will reach in this case their highest values (see Chapters 4 and 5, Volume 1). Consequently, interaction even with low-energy particles of the beam may result in the generation of plasma oscillations.

In the given plasma regions beam instabilities of various kinds are observed. For instance, along polar magnetic field lines, in the vicinity of a shock front, and in the plasma sheet of the geomagnetic tail (see Fig. 2.9, Volume 1), directed ion fluxes will appear, i.e., current systems, and these give rise to *current-driven instability*. Presumably a number of the wave processes detected in the indicated plasma regions, and due to ion-cyclotron or ion-acoustic instability, can be attributed to the influence of ion fluxes. However, the beam instability that occurs due to *flow of electrons* along polar field lines apparently leads to excitation of the hybrid resonances ω_L and ω_U observed here. In the vicinity of a shock front the mechanism of so-called *two-stream instability* is operating. This is caused by the presence of two *opposing* beams of ions and electrons, arising because of reflection of beams of solar-wind particles in the vicinity of the shock front. The gyroresonance oscillations of ions and electrons observed above the shock-front region, as well as the Langmuir electron oscillations there, may well be related to reflected electrons and ions. The *drift instability* can also be classified as a kind of beam instability. It is a result of the appearance of bulk particle motion caused by large-scale gradients of the plasma density or temperature. In this case an additional source of free energy appears in the plasma due to plasma drift. This energy may be converted into vibrational energy of the particles. The ion-cyclotron waves in the polar cap and the micropulsations of the magnetic field are presumably associated with this instability mechanism.

Nonlinear mechanisms of wave generation are operating along with the indicated 'linear' and 'quasi-linear' mechanisms in the plasma regions being considered. Some nonlinear instabilities playing a role here were described in Chapter 7, Volume 1, and will be mentioned below, in Chapters

16–19. An important trait of nonlinear processes is their 'individual' nature, meaning that they manifest themselves differently under different conditions. The nonlinear effects observed during the experiments described can often not even be assigned a role, or identified with the aid of an adequate theory and the appropriate calculated results. However, in addition to the clear-cut evidence for nonlinearity of a number of processes (see Chapter 7), their major role in the regions of natural plasma of interest to us is connected with the following idiosyncrasies. First of all, in many cases strong quasi-monochromatic waves are present in the plasma. These fields may either be of artificial origin (created by radio-wave transmitters, see §§ 7.1.3 and 18.2) or of natural origin (radiation from lighting flashes or, for instance, high-amplitude plasma waves generated in the magnetosphere). Another factor considerably influencing the nonlinearity of oscillatory processes is the trapping of particles by monochromatic waves (see Chapter 7), which affects the dynamics of a 'wave-particle' interaction. For large trapping 'lengths' and the presence of a quite large inhomogeneity, even for small field amplitudes the nonlinear effects may turn out to be considerable. Naturally, the nonlinear nature of various processes of 'wave–wave' interaction will also manifest itself.

16

Results of studies of ELF waves

The ELF range will here be assumed to include only wave processes whose frequencies are lower than the proton gyrofrequency ($0 < \omega \lesssim \Omega_H(H^+)$). In the region of the ionosphere where different species of ions play a role, we will also consider some phenomena caused by the behavior of waves in the frequency ranges lying between the gyroresonances of the different ions. Insofar as the author was able to determine on the basis of experimental results obtained by a number of investigators, who referred to the recorded plasma oscillations as ELF noise or waves, the corresponding processes were excited in plasma regions where their frequencies were much higher than $\Omega_H(H^+)$. For instance, so-called ELF noise was recorded in the range $f = 100$ to 800 Hz predominantly at distances from the Earth $R \gtrsim 3.5R_0$ (where $L=6$, magnetic latitude $45°$), for $\Omega_H(H^+) <$ or $\ll 100$ Hz (Russell et al., 1969). These data should, however, pertain to the region of VLF waves, since the corresponding frequencies lie in the interval $\Omega_H < \omega < \omega_L$.

16.1. Hydromagnetic whistlers

Trains of successive discrete wave packets of magnetospheric origin, with frequencies from a fraction of a hertz to a few hertz, have been observed at the Earth's surface. These were initially called 'pearl-type micropulsations' (Benioff, 1960; Troitskaya, 1961; Saito, 1962). It was soon realized, however, that these wave packets are propagated like whistling atmospherics (electron-whistler waves), being guided along the Earth's magnetic force lines that pass through the source location (Tepley, 1961). Thus they are called *hydromagnetic whistlers* (they are also knows as PC-1 micropulsations; see the following section). Under different conditions the frequencies of these wave packets lie in the following ranges:

$$0.1 \text{ to } 0.2 \text{ Hz} < f \simeq 2 \text{ to } 5 \text{ Hz}.$$

Hydromagnetic whistlers correspond to the ion branch of the waves

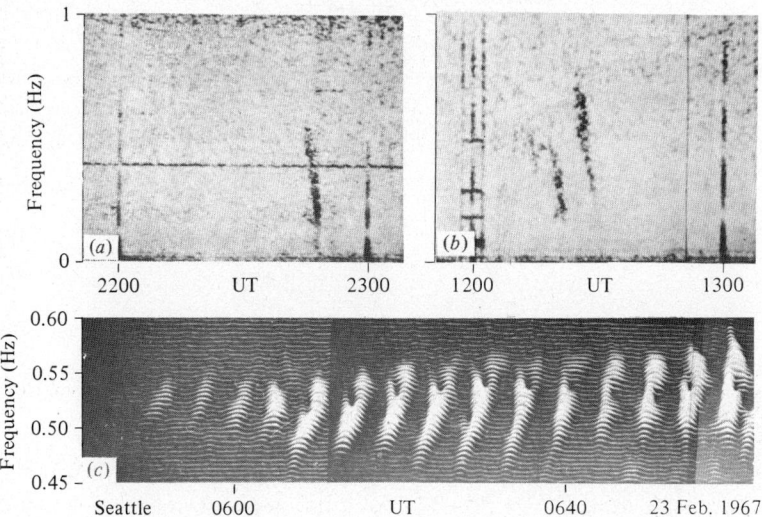

Fig. 16.1. Spectrograms: (a) and (b) rarely observed packets of ELF electron waves $d\omega/dt < 0$ (Qanaq: (a) = 21 Dec. 1967; (b) = 30 Jan. 1968); (c) trains of ELF ion-wave packets (hydromagnetic whistlers) $d\omega/dt > 0$ (Seattle, 21 Feb. 1967). (Kenney & Knaflich, 1968; Liemohn, 1969.)

$n_1(\omega)$, which in the limit where $\omega \to 0$ becomes an Alfvén wave, and as $\omega \to \Omega_H$ becomes an ion-cyclotron wave (See Fig. 4.3, Volume 1). Since they are multiply reflected at the magnetically conjugate points, these wave packets form a train of discrete signals at the observation point. The frequency–time characteristic of these signals will depend on the sign of the derivative $dn_1/d\omega$, and their frequency increases with time: $d\omega/dt > 0$ (Tepley & Wentworth, 1962; Gendrin & Stefant, 1962; Mainstone & McNicol, 1962; Jacobs & Watanabe, 1963; Campbell & Stilner, 1965). Figs. 16.1 and 16.2 show some typical spectrograms of hydromagnetic whistlers, i.e., frequency ω of a component of a wave packet as a function of the group propagation time t (along a magnetic force line) (Kenney & Knaflich, 1968; Liemohn, 1969). The upper diagrams in Fig. 16.1 depict spectrograms of rarely observed wave packets with an opposite sign of the time dependence $d\omega/dt < 0$. These correspond to the electron branch of the waves $n_2(\omega)$, which in the limit (as $\omega \to 0$) becomes a fast modified Alfvén wave, while for $\omega \gg \Omega_H$ it becomes an electron whistler.

On the basis of numerous experiments, it was initially decided that hydromagnetic whistlers are generated in the equatorial region of the magnetosphere, predominantly at distances $R \simeq 3R_0$ to $10R_0$ from the Earth, where the plasma density varies under different conditions from

16.1. Hydromagnetic whistlers

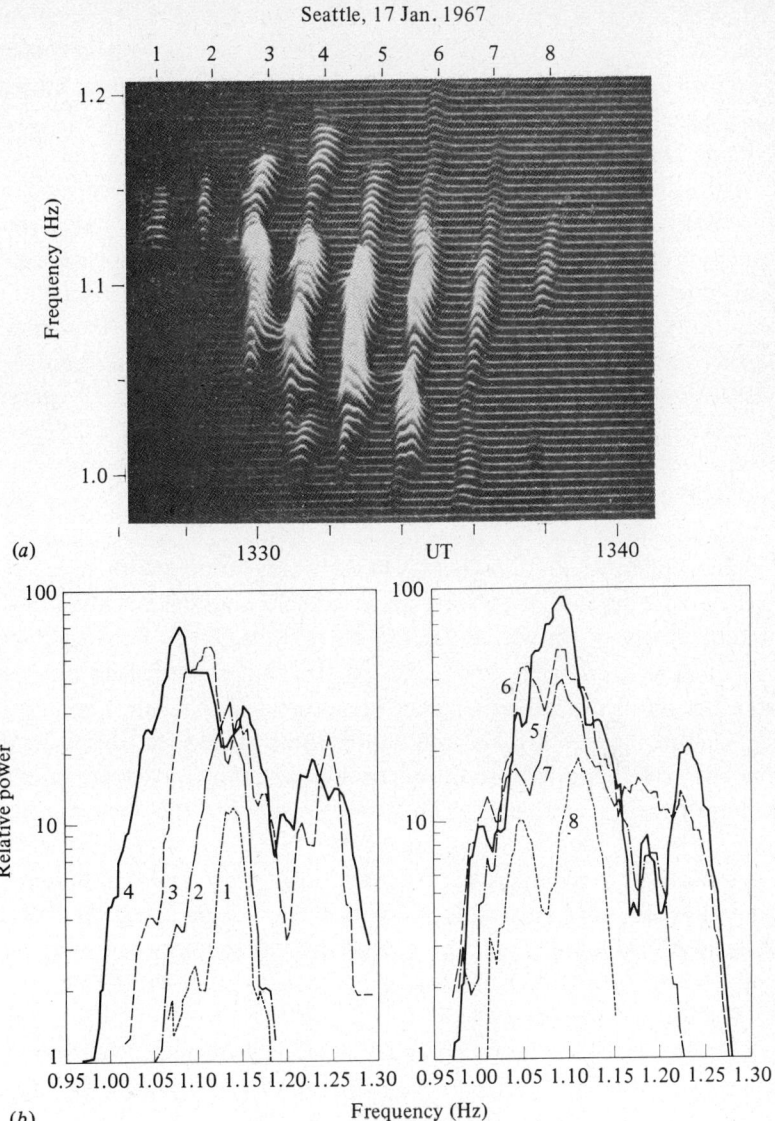

Fig. 16.2. Spectrogram of a train of eight hydromagnetic whistlers and their energy (power) spectra (Seattle, 17 Jan. 1967). (Kenney & Knaflich, 1968; Liemohn, 1969.)

$N \simeq 0.1$ to $10^3 \, \text{cm}^{-3}$ (Kenney et al., 1968; Higuchi & Jacobs, 1970). However, later it was shown that the sources of the wave packets forming hydromagnetic whistlers, the frequencies of which lie in the interval $f \simeq 0.5$ to 2.5 Hz, apparently lie predominantly in narrower regions, namely at distances $R \simeq 3.5R_0$ to $4R_0$, close to the boundary of the plasmapause, where $N \simeq 10^2$ to 10^3cm^{-3}. On the other hand, signals with frequencies $f \simeq 0.2$ to 0.5 Hz are apparently generated in the intermediate zone between the plasmapause and the magnetopause, where $N \simeq 1 \text{cm}^{-3}$ (see Fig. 2.1, Volume 1) (Al'pert & Fligel, 1977).

Hydromagnetic whistlers are generated because of the gyroresonance instability of the plasma, which is associated with a gyroresonant enhancement of the waves as they repeatedly traverse the generation region. In this process a major role is apparently played by the temperature anisotropy $(T_\perp > T_\parallel)$ and possibly by the anisotropy of the particle distribution function connected with the loss cone. The excitation of these waves presumably takes place via the action of protons with energies $E_p \simeq 30$ to 120 keV. As regards the wave packets with $d\omega/dt < 0$ depicted in the upper diagrams of Fig. 16.1, these are observed extermely rarely and their excitation mechanisms are not clear.

A study of trains of hydromagnetic whistlers (Figs. 16.1 and 16.2) explains an important idiosyncrasy of these wave trains. With an increase in the number of individual wave packets, there is at first an enhancement of their intensity. Inspection of the energy spectra of eight successive hydromagnetic whistlers in Fig. 16.2 shows that the intensity increased from the 1st to the 5th signal, and then decreased. The 8th (last) and 1st signals were the weakest. This can be attributed to the effect, over a certain time interval, of a gyroresonant intensification of the wave packets in the region of their generation, which is located close to the apogee of their trajectories of motion (Liemohn, 1969). Similar descriptions of the properties of hydromagnetic whistlers, and pearl micropulsations in general, which have been observed with the aid of an extensive network of stations (Liemohn, 1971), can be found in many sources (see, for instance, Pope, 1964; Jacobs, 1970; Tartaglia, 1970; Gulel'mi & Troitskaya, 1973; Troitskaya & Gulel'mi, 1969).

Calculations of the group delay time for hydromagnetic whistlers

$$\tau(\omega) = \int \frac{ds}{u_1} = \frac{1}{c} \int \frac{\Omega_0(\Omega_H - \tfrac{1}{2}\omega)}{[\Omega_H(\Omega_H - \omega)^3]^{1/2}} ds \qquad (16.1)$$

where u_1 is the group velocity of these waves (see (8.14), Volume 1) and ds is an element of the path along a magnetic force line, enable one to locate the source of these signals, as well as to determine the parameters of the corresponding plasma regions. A general qualitative fit of these calculated results with the frequency–time characteristics $\omega(\tau)$ of hydromagnetic whistlers was already obtained in the first studies of this type (Obayashi,

16.1. Hydromagnetic whistlers

1965; Hultqvist, 1966). For the diagnostics of the regions of generation of these waves, it is naturally important to use a model of the outer ionosphere which is closer to actual conditions. Fig. 16.3 shows the results of such calculations of the group delay time as a function of the geomagnetic latitude, for three ionospheric models corresponding to the most frequently encountered altitude dependences of the electron density and temperature in the outer ionosphere (middle band), as well as the maximum and minimum values of these quantities (top and bottom bands). The figure

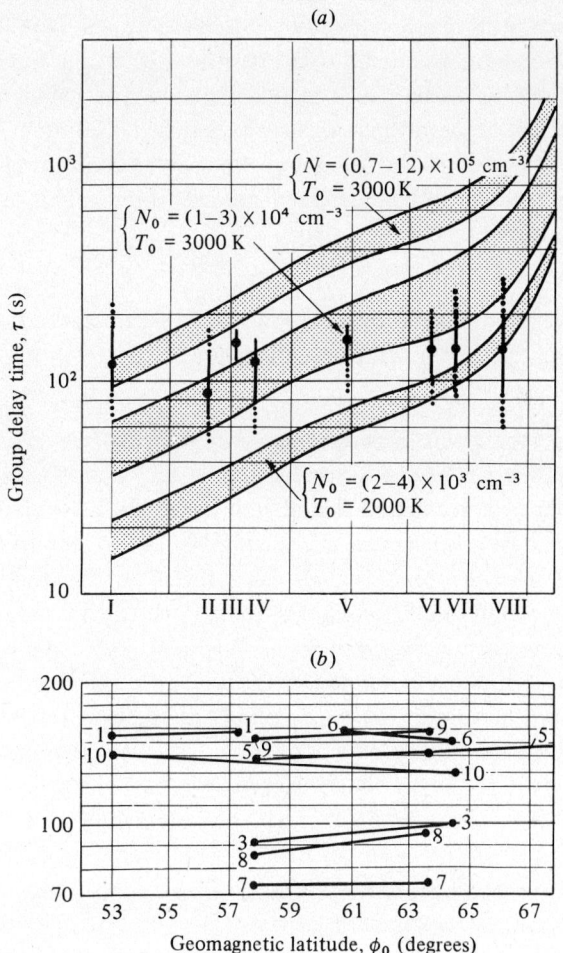

Fig. 16.3. Theoretical dependences of the group delay time of hydromagnetic whistlers $\tau(\phi_0)$ on geomagnetic latitude (shaded bands), for various models of the outer ionosphere. The results of simultaneous observations at the points I, II,..., VIII are indicated by dots and straight lines (Al'pert & Fligel, 1977).

also gives the mean values (dots) of the group delay time τ of hydromagnetic whistlers in the range $f \simeq 0.5$ to 2.5 Hz, determined on the basis of extensive series of observations in a network of stations I, II, ... VIII, situated at geomagnetic latitudes ϕ_0 from 53° to 66° (Al'pert & Fligel, 1977). These observation points correspond to the apogees of the Earth's magnetic field lines, at altitudes $z_a \simeq 11 \times 10^3$ to 33×10^3 km. In Fig. 16.3b lines 1, 2, ... 10 correspond to the values of τ obtained on the basis of simultaneous observations of hydromagnetic whistlers at various, quite separated, points. Line 5 goes out of the figure; values close to the values of $\tau = 135$ and 141 s shown on line 5, namely $\tau = 141$ and 134 s, were determined at a geomagnetic latitude $\phi_0 = 89°$ in the Northern Hemisphere (Qanaq) and at $\phi_0 = -89°16'$ (Antarctica). It is seen that the time values τ are quite close for different observation points, well separated in latitude. An analysis of these data also led to the foregoing conclusions concerning the region of generation of hydromagnetic whistlers and to the values of N and E_p given above.

16.2. Hydromagnetic waves (pulsations of the magnetic field) in the magnetosphere

ELF transverse waves at frequencies $f \simeq 10^{-3}$ to $1 \text{ Hz} < F_H(H^+)$ (where $F_H(H^+)$ is the local value of the proton gyrofrequency) have been observed in many experiments at distances $R \gtrsim 4R_0$ to $6R_0$ from the Earth, in the magnetosphere, the transition region, and the geomagnetic tail. They are apparently almost always present in these parts of the near-Earth plasma. However, these waves are quite diverse in nature. They are mainly recorded as pulsations (often quasi-periodic) of the magnetic field, being observed predominantly during the periods of heightened geomagnetic activity associated with magnetic storms (McPherron et al., 1972).

The earliest, quite regular, observations of these waves included, for instance, the data obtained with Pioneer 5 (Coleman, 1964) and Mariner 4 (Siscoe et al., 1967). In the *magnetosphere and transition zone*, at distances $R \simeq 5R_0$ to $15R_0$, Pioneer 5 recorded, predominantly at 1500 to 1700 hours Local Time (LT), bands of ELF waves; these were picked up with a magnetic antenna at frequencies $f \simeq 0.3$ Hz, less than the proton gyrofrequency $F_H(H^+)$, the intensity being $H^2 \simeq 1\,\gamma^2\,\text{Hz}^{-1}$ to $10^2\,\gamma^2\,\text{Hz}^{-1}$. *In the interplanetary medium*, the space probe Mariner 4 picked up the same kind of ELF noise, in the range $f \simeq 3 \times 10^{-4}$ to $0.5 \text{ Hz} < F_H(H^+)$. The intensity varied in this frequency range from $H^2 \simeq 10^{-1}\,\gamma^2\,\text{Hz}^{-1}$ to $3 \times 10^3\,\gamma^2\,\text{Hz}^{-1}$, i.e, approximately according to an $H^2 \sim f^{-3/2}$ law.

16.2. Hydromagnetic waves in the magnetosphere

Further studies of ELF pulsations of the magnetic field under various conditions indicated that their properties differ considerably within quite narrow frequency intervals and depend on the nature of the geomagnetic activity. Systematic observations of ELF waves in the range $f \simeq 0.4 \times 10^{-2}$ to 1.5×10^{-2} Hz were carried out during a period of *quiet geomagnetic activity* at the ATS 1 satellite (Cummings et al., 1969, 1972). Such waves were found to be generated mainly during the daylight hours: maximum activity at 1400 LT, with a rapid disappearance of this radiation at twilight. The amplitude of the magnetic-field pulsations, the so-called *PC-4 and PC-5 pulsations*, with $f \simeq 10^{-2}$ Hz, varied from $H \simeq 1$ to 10γ, the mean being $H \simeq 3\gamma$. Earlier observations with Explorer 12 at distances $R \simeq 7R_0$ to $11R_0$ and at 0700 to 1200 LT revealed ELF waves with $f \simeq 0.5 \times 10^{-2}$ to 10^{-2} Hz, the amplitude being $H \simeq 6$ to 11γ (Patel, 1965).

In the equatorial plane, at frequencies $f \simeq 3 \times 10^{-2} \pm 1 \times 10^{-2}$ Hz *PC-3 pulsations*, and at $f \simeq 0.3$ Hz *PC-1 pulsations*, were recorded by the geostationary satellite Dodge, at a distance $R \simeq 6.3R_0$ from the Earth. These were quasi-periodic ELF transverse waves with amplitudes of, respectively, 1 and 3γ (Dwarkin et al., 1971).

The packets of ELF waves known as *PC-1 pulsations* are, as mentioned previously, the source of the hydromagnetic whistlers described in the previous section. These were regularly observed in the magnetosphere at the geostationary satellite ATS 1, close to the region of their generation. During a *strong magnetic storm* (substorm) on 10 Oct. 1969 such wave packets were recorded simultaneously at ATS 1 and at an observation point situated at the foot of the magnetic field line on the Earth's surface (McPherron & Coleman, 1970), at a frequency $f \simeq 0.2$ to 0.3 Hz. During the onset stage of the storm the maximum amplitude of the packets corresponded to $f \simeq 0.27$ Hz, and later to $f \simeq 0.18$ Hz. An important consequence of these observations was recognition that at the Earth's surface the structure of the signal was more complicated. Moreover, its intensity near ATS 1 was at first three times as high, and then about nine times as high, as at the Earth's surface. This difference can be attributed to effects arising as the wave packets propagate, as well as to changes in the position of ATS 1 relative to the magnetic force line along which the wave packets recorded at the observation point were ducted.

During *geomagnetic storms* ELF waves at frequencies $f \simeq 0.2$ to 0.3 Hz with amplitudes $H \simeq 1$ to 2γ were also recorded on OGO 5 (see, for instance, Barfield & Coleman, 1970). It is interesting that in the polar-cap zone, at geomagnetic latitude $\phi_0 \simeq 45°$ and at a distance $R \simeq 5R_0$, during a storm on 1 Nov. 1968 quasi-periodic ELF waves with $f \simeq 7$ Hz $\simeq$

$0.8F_H(H^+)$ and an amplitude $H = 5\gamma$ were recorded (where $F_H(H^+)$ is again the local value of the proton gyrofrequency) (McPherron et al., 1972).

An analysis of the observations of PC-1 magnetic-field pulsations at the geostationary satellite ATS 1 indicated that their appearance is connected with geomagnetic activity (Bossen et al., 1976). The PC-1 activity is sharply enhanced during the course of about an hour as the main phase of the magnetic storm develops. They are observed most frequently in the pre-evening sector of the Earth, with a maximum at 1600 to 1700 LT. In these experiments the PC-1 wave packets observed had frequencies f and amplitudes H in the ranges:

$$f \simeq 0.2 \text{ to } 2 \text{ Hz}, \quad H = 1 \text{ to } 7\gamma.$$

The lifetimes of the PC-1 emissions usually did not exceed half an hour, but sometimes they lasted as long as an hour or more. An important result of these experiments was establishment of the fact that the frequency of the PC-1 emission $f \simeq 0.1F_H(H^+)$ to $0.4F_H(H^+)$, and that predominantly $f \simeq 0.1F_H(H^+)$ to $0.2F_H(H^+)$. It should be noted that the frequency of the electron gyroresonance LF wave packets (chorus) generated in the magnetosphere, expressed as a fraction of the electron gyrofrequency, varies over approximately these same limits (see §18.3 below). The emission of the PC-1 signals detected in these experiments is apparently due to cyclotron resonance involving protons with energies of the order of 30 keV, in the plasma region where the density $N \simeq 30 \text{ cm}^{-3}$ (Bossen et al., 1976).

During a period of *heightened geomagnetic activity*, ELF waves with frequencies lower than those of PC-1 waves are also observed. For instance, quasi-periodic transverse waves with $f \simeq 5 \times 10^{-2}$ Hz and an amplitude $H = 0.3\gamma$ were recorded during a strong magnetic storm on 17 Aug. 1967 at OGO 5, close to the magnetic equator at distances $R \simeq 5R_0$ to $7R_0$ (McPherron & Coleman 1971). During this same period ELF waves with $f \simeq 4 \times 10^{-2}$ Hz and $H \simeq 1\gamma$ were recorded at ATS 1 (McPherron et al., 1972).

16.3. Ion-cyclotron whistlers and waves

In the ionosphere the ion branch of the ELF waves $n_1(\omega)$ (see Fig. 4.3, Volume 1) was first observed with Alouette 1 and Injun 3, in the form of a so-called ion whistler close to the proton gyroresonance (Smith et al., 1964). ELF waves emitted by a lighting flash are the source of such a proton whistler. When it enters the ionosphere, a packet of these waves is divided into ordinary and extraordinary components. One of these, the extraordinary wave, forms the electron-whistler mode $n_2(\omega)$. It is guided

16.3. Ion-cyclotron whistlers and waves

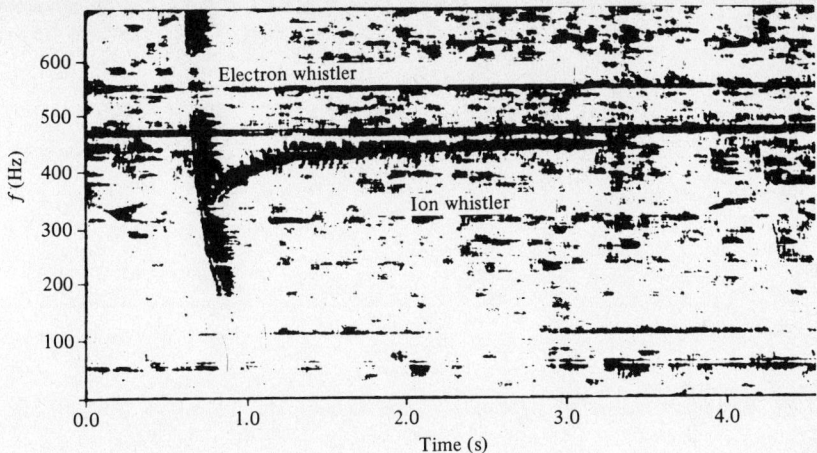

Fig. 16.4. Spectrogram of a proton whistler (Injun 3, altitude $z = 1308$ km, 5 March 1963). (Gurnett & Brice, 1966.)

along the Earth's magnetic field and reaches the magnetically conjugate point. After being reflected near this point, the extraordinary wave returns and is recorded as a whistling atmospheric. The $\omega(t)$ relation for this signal is extended in time; $d\omega/dt < 0$. Electron whistlers will be described briefly below, in Chapter 18; more detailed studies can be found in, for example, Helliwell, 1965; Gendrin, 1971; Russell et al., 1972).

The other wave, the ion ordinary wave, is observed directly in the ionosphere only above the radiation source, since it is cut off at the gyrofrequencies of the different ions. Thus it does not reach the magnetically conjugate point and is not observed at the Earth's surface. Fig. 16.4 gives a spectrogram of the ion-proton whistlers recorded aboard Injun 3 (Gurnett & Brice, 1966). Inspection of the figure reveals the tail of the electron whistler (short fractional-hop whistler), which because of the short propagation path from the Earth's surface to the satellite is only slightly dispersed. On the other hand, the proton whistler is highly dispersed as ω approaches $\Omega_H(H^+)$, and it usually becomes extended over several seconds. Subsequently, helium whistlers were detected as well (Fig. 16.5; Barrington et al., 1966). As yet no whistlers of heavier ions (O^+, N^+, etc.) have been recorded. They are cut off in the region of extremely low frequencies (a few hertz or tens of hertz), and thus are apparently poorly resolved by present-day apparatus. Interestingly enough, three ion whistlers were recorded simultaneously by Injun 5: a proton whistler, a helium whistler, and a whistler whose frequency put its mass per unit charge at 8 proton masses (Gurnett & Rodriguez, 1970). Presumably this was an

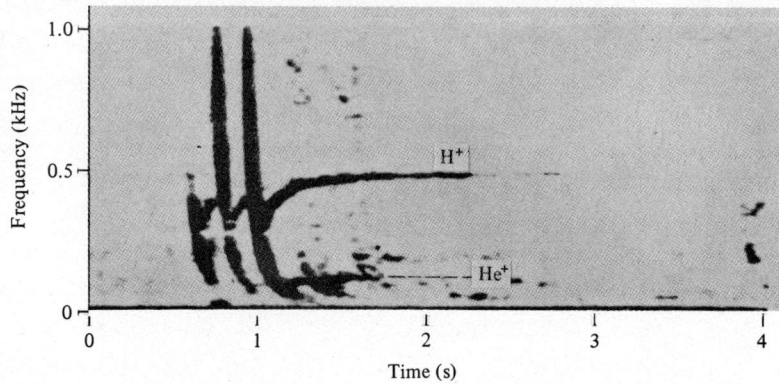

Fig. 16.5. Spectrogram of proton H$^+$ and helium He$^+$ whistlers (Alouette 2, altitude $z = 1900$ km, 4 Oct. 1968). (Barrington et al., 1966.)

ion whistler formed by a doubly ionized oxygen atom O^{++} or a singly ionized helium molecule He$_2^+$.

Ion-cyclotron whistlers (mainly proton whistlers) have been studied in detail (Gurnett et al., 1965; Shawhan, 1966). A theoretical analysis of the corresponding experimental data made it possible to carry out diagnostics of the outer ionosphere. To do this, first the crossover frequency of the tail of the electron and ion whistlers was used; this is found from the equation

$$n_1^2 = n_2^2. \tag{16.2}$$

Secondly, Gurnett et al., 1965, determined the signal amplitude as a function of frequency as $\omega \to \Omega_H$. Finally, the ion and electron densities, the proton gyrofrequency, and also the plasma temperature in the vicinity of the satellite, were calculated (Shawhan & Gurnett, 1966; Gurnett & Shawhan, 1966; Gurnett & Brice, 1966). In addition to the relation between the ionic composition of the plasma and the crossover frequency, the corresponding formulas for the indicated plasma parameters were obtained by calculating the group delay time $\tau(\omega)$ and the spatial damping $\beta(\omega)$ of the signal amplitude from the integrals

$$\tau(\omega) = \int \frac{ds}{u_1}, \quad \beta(\omega) \simeq \int \exp\left(\frac{\omega}{c}\kappa\right) ds, \tag{16.3}$$

where u_1 is the group velocity. The damping β as a function of the frequency difference $\Delta\omega = \Omega_H(H^+) - \omega$ or of the time t shows a close fit with experiment if we assume that $\kappa = \kappa_{Hi}$, i.e., if we use the formulas for ion-cyclotron damping (Gurnett & Brice, 1966). This shows the kinetic

16.3. Ion-cyclotron whistlers and waves

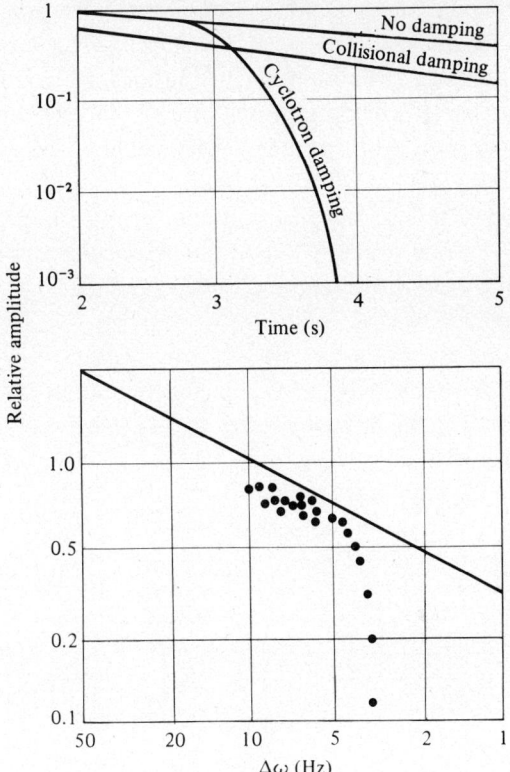

Fig. 16.6. Amplitude–frequency characteristics of a proton whistler. Dots indicate the experimental data, while the solid lines are theoretical curves (Gurnett & Brice, 1966).

nature of the attenuation of proton whistlers. A study of Fig. 16.6, in particular, verifies this conclusion. This figure (upper diagram) shows the theoretical relationships for the relative amplitude as a function of time, calculated (1) without taking damping into account, (2) taking collisions between charged particles into account, and (3) using the cyclotron attenuation coefficient κ_{Hi}. The lower diagram in Fig. 16.6 shows the experimental dependence of the proton-whistler amplitude on $\Delta\omega$ (dots), as compared with the expected variation of the relative amplitude ignoring damping.

The variation in the damping of the proton-whistler amplitude as a function of time (or of $\Delta\omega$) obtained from these studies is, however, more gradual than the theoretical variation taking into account just cyclotron damping (Lucas & Brice, 1971). These authors assumed that this is because the effect of irregularities in the distribution of the proton density $N(H^+)$ in the vicinity of the satellite has to be taken into account. It is shown that, by taking into consideration the fluctuations in electron density δN, the temperature obtained from an analysis of the amplitude–frequency characteristics of proton whistlers may even be about twice as high as the value obtained by ignoring δN.

It should be noted here, too, that in a number of cases proton whistlers apparently have a fine structure due to the gyroresonances of the oxygen ions O^+, i.e., the proton-whistler amplitude is modulated at the frequency $\Omega_H(O^+)$ (Stefant, 1970).

Gyroresonant excitation of *ion-cyclotron waves* was observed on Explorer 45 close to the *plasmapause* in the vicinity of the geomagnetic equator (Taylor et al., 1975; Taylor & Lyons, 1976). Most of these instances are associated with the presence of a ring current of ions, formed in the magnetosphere during a magnetic storm; the lifetime of such a ring current varies from a few hours to several days (Joselyn & Lyons, 1976). On

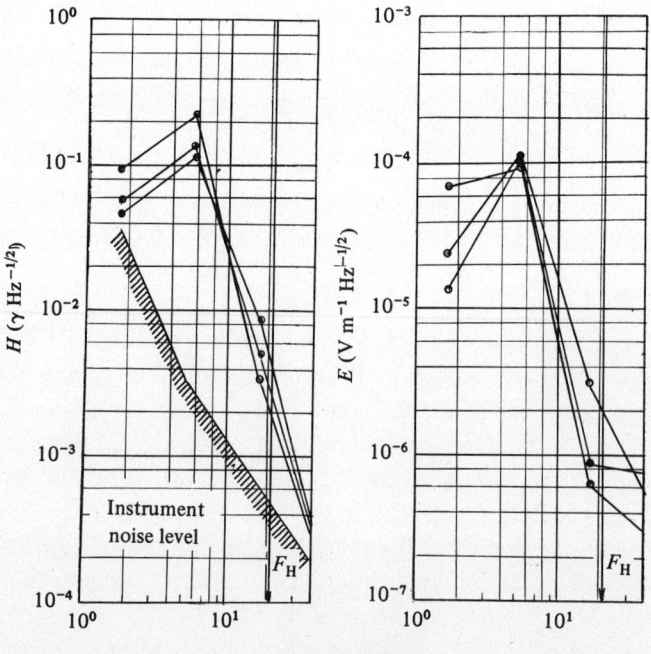

Fig. 16.7. Spectra of $E(V\,m^{-1}\,Hz^{-1/2})$ and $H(\gamma\,Hz^{-1/2})$ for ELF waves. Results of measurements close to the plasmapause (Hawkeye 1, $L \simeq 2.8$ to 3.9, $\phi \simeq 2$ to $9°$, 8 July 1975). (Kintner & Gurnett, 1977.)

16.3. Ion-cyclotron whistlers and waves

Hawkeye 1 instruments more sensitive than those on Explorer 45 were used to observe the magnetic, H, and electric, E, fields of these waves at geomagnetic latitudes $\phi \simeq 2$ to $9°$ and force lines $L \simeq 2.8$ to 3.3 (Kintner & Gurnett, 1977). Cases of simultaneous recording of the E and H fields of ion-cyclotron waves were very rare. Fig. 16.7 shows the corresponding spectra obtained during three successive series of measurements. These spectra were used to calculate the refractive index of the plasma $n = \mu_0 c H / E$, assuming that the observed waves were transverse and propagated along $\mathbf{H}_0$. The results of simultaneous measurements of the Earth's magnetic field $\mathbf{H}_0$, together with the corresponding values of the electron density N, were used to calculate the theoretically predicted $n_1(\omega)$ and $n_2(\omega)$ curves for the ordinary and extraordinary waves (see (4.61), Volume 1). The values of N were selected by fitting the $n_1(\omega)$ and $n_2(\omega)$ curves to the experimental data. The individual experimental values of n_1 and n_2 (dots in Fig. 16.8) show a good fit with the $n_1(\omega)$ and $n_2(\omega)$ curves (solid curves in Fig. 16.8). This indicates that the observed radiation (Fig. 16.7) really consists of ion $n_1(\omega)$ (for $\omega < \Omega_H$) and electron $n_2(\omega)$ electro-

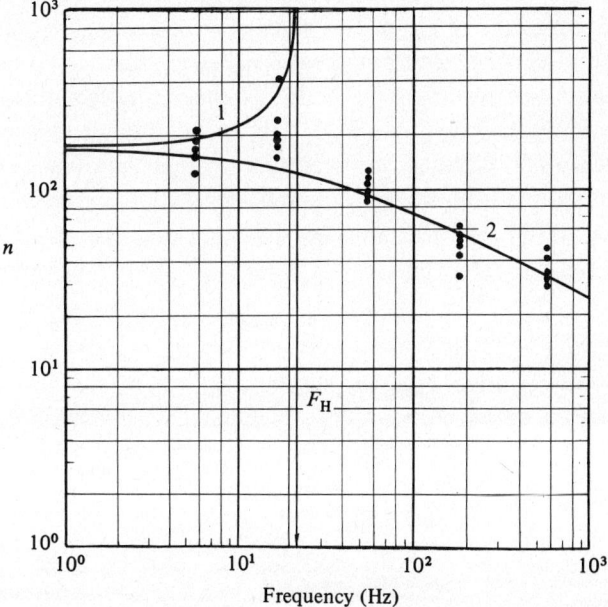

Fig. 16.8. Curves of the refractive indexes $n_{1,2}(\omega)$ for the ordinary and extraordinary waves, according to data from measurements (close to the plasmapause) of the field components $\mathbf{E}$ and $\mathbf{H}$ of ion-cyclotron waves (Hawkeye 1, 8 July 1975). (Kintner & Gurnett, 1977.)

magnetic waves. This is also evidence that, in a number of cases in the corresponding plasma regions, longitudinal ion-cyclotron waves may apparently be transformed into transverse waves.

In the *polar zone* low-amplitude ion-cyclotron waves in the extraordinary mode were recorded on OGO 5 during a magnetic storm on 1 Nov. 1968 at frequencies $f \simeq 0.67 F_H(H^+)$ and $0.87 F_H(H^+)$ (Fredricks & Russell, 1973). The magnetic field H of these waves was normal to the Earth's magnetic field $\mathbf{H}_0$ (the three components of H were measured), but the direction of wave propagation made an arbitrary angle with $\mathbf{H}_0$. An analysis of the nature of these waves did not lead to any definite conclusion as to their origin. It was found, however, that their activity is associated with drift currents and plasma gradients.

16.4. Hiss and chorus ELF radiation. Cutoff and intensification of emission as $n^2 \to 0$

In the outer ionosphere, as at the Earth's surface, in discrete frequency ranges radiation bands of the hiss type and also narrow radiation bands of the chorus type are observed. The latter are a succession of discrete signals appearing at random, and lasting for about a few tenths of a second; their frequency usually increases with time. Spectrograms of these kinds of emissions, pertaining to the ELF frequency range, were obtained during observations with Injun 3 (see Fig. 16.9; Taylor & Gurnett, 1968). The frequencies of the emissions referred to by these authors as 'ELF hiss and chorus' varied from several hundred Hz to approximately 2 kHz. At first sight it would appear from these data that the lower portion of this frequency range corresponds to $\omega < \Omega_H(H^+)$, i.e., to the region below the

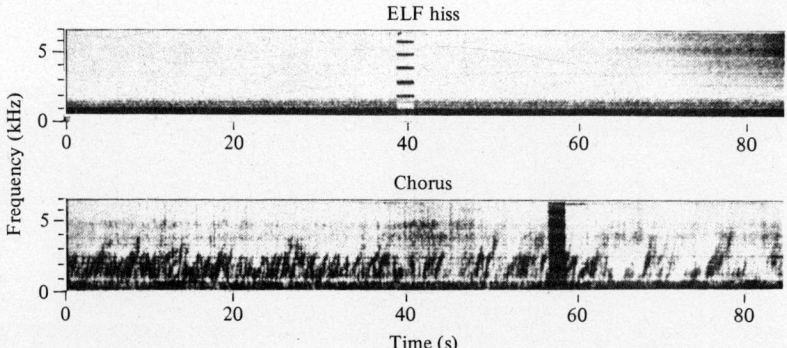

Fig. 16.9. Spectrograms of ELF hiss and chorus waves (Injun 3, 1963). (Taylor & Gurnett, 1968.)

16.4. Hiss and chorus ELF radiation

proton gyrofrequency; in the ionosphere this is 600 to 650 kHz at medium latitudes. The upper limit of this radiation lies, however, at a value $\omega > \Omega_H(H^+)$. On the other hand, further studies indicated that this emission apparently consists entirely of VLF waves and is only cut off in the ELF range. It is generated at great heights in the outer ionosphere or magnetosphere, at frequencies ω higher than the local values of the gyrofrequency $\Omega_H(H^+)$ for the corresponding plasma region. This can be deduced from the following data.

The Injun 3 experiments showed that the emission in question is often cut off sharply at a specified altitude, at the frequency for which $n_1^2 \to 0$, which lies between the proton and helium gyrofrequencies. Such results are presented, for example, in Fig. 16.10 for hiss and in Fig. 16.11 for chorus (Gurnett & Burns, 1968). At lower altitudes the cutoff frequency decreases, since the gyrofrequencies $\Omega_H(H^+)$ and $\Omega_H(He^+)$ are higher there. Therefore, the waves in question are excited above the satellite and they propagate initially like a packet of extraordinary waves (branch n_2^2 in Fig. 4.8, Volume 1). At the altitude where ω becomes equal to the crossover frequency ω_{12}, the sign (sense) of the polarization of the extraordinary wave changes. Thus, subsequently the propagation of waves with a frequency $\omega < \omega_{12}$ will be described by the n_1^2 branch, and these waves are cut off at the frequency for which $n_1^2 = 0$. Below the height at which $n_1^2 = 0$, however, only a part of the wave spectrum coming down from above can be propagated, namely waves having a frequency lower than the cutoff frequency. Inspection of Fig. 16.10 shows that at an altitude $z \simeq 1000$ km the corresponding frequency is 250 Hz. Consequently, there is reason to assume that in the given case the lower boundary of the hiss spectrum is actually below the minimum values of ω at which these waves are observed at an altitude $z \simeq 1000$ km.

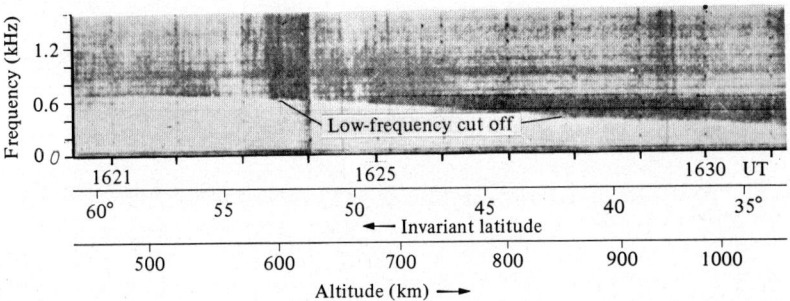

Fig. 16.10. Spectrogram illustrating the cutoff of ELF hiss emission (Injun 3, 1963). (Gurnett & Burns, 1968.)

118   *Results of studies of ELF waves*

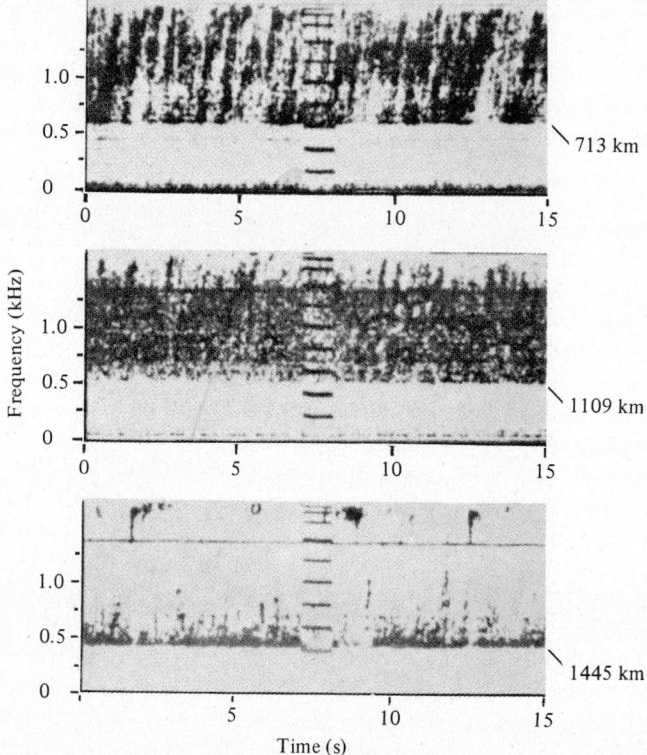

Fig. 16.11. Spectrograms illustrating the cutoff of ELF chorus emission (Injun 3, 1963). (Gurnett & Burns, 1968.)

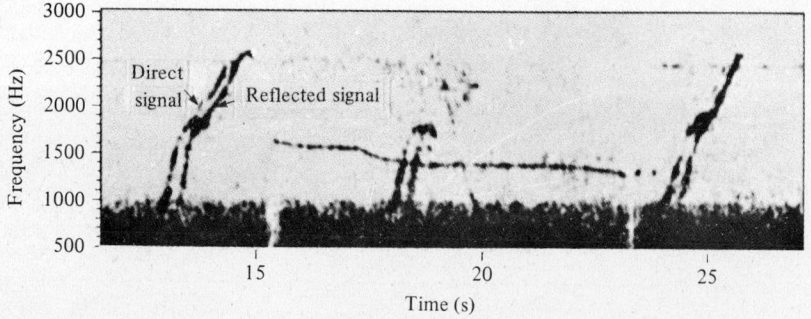

Fig. 16.12. Spectrogram illustrating the reflection of a discrete packet of ELF waves from regions of the ionosphere lying below the orbit of Injun 5 ($z = 1336$ km, 15 Jan. 1969). (Mosier & Gurnett, 1969a.)

16.4. Hiss and chorus ELF radiation

Later experiments on Injun 5 included direct measurements of the component of the Poynting vector parallel to the Earth's magnetic field. Fig. 16.12 show a typical recording, with successive discrete wave packets. One of these is seen to come down to the satellite, while another is reflected from the lower-lying regions of the ionosphere. In this series of experiments it was found that in general the waves depicted in Figs. 16.10 and 16.11 propagate predominantly in a downward direction (Mosier & Gurnett, 1969a; Mosier, 1970, 1971). However, in a narrow frequency interval at the signal-cutoff boundary, less intense, upward-moving waves were also observed, i.e., waves reflected from lower-lying regions of the ionosphere. This is apparently connected with the fact that in some parts of the ionosphere below the satellite the cutoff frequency ω_c is lower (rather than higher) than the value of ω_c near the satellite; thus these waves can propagate downward and be reflected upward. In the experiments being described, a method making it possible to obtain colour spectrograms was subsequently used.

Fig. 16.13 shows a black and white reproduction of such a colour spectrogram on which a mixture of ELF hiss and chorus in the range 0.25 to 1 kHz was recorded (Gurnett et al., 1971). On the original spectrograms, waves moving downward are red and waves moving upward are green. In Fig. 16.13 the wave direction varied at a certain time: after first being propagated downward, the wave started to move upward. This occurred when the height of the satellite decreased from $z \simeq 2480$ km to $z \simeq 2464$ km. Here the upper and lower boundaries of the emission frequencies remained nearly unchanged, but the emission took on a more discrete structure (chorus). The change in the direction of arrival of these waves may possibly have occurred because the satellite crossed the region where they were being generated.

It should be noted here that reflection of downward-moving electron waves (whistling atmospherics), at a frequency between the proton and helium gyrofrequencies, which corresponds to $n^2 = 0$, was observed on

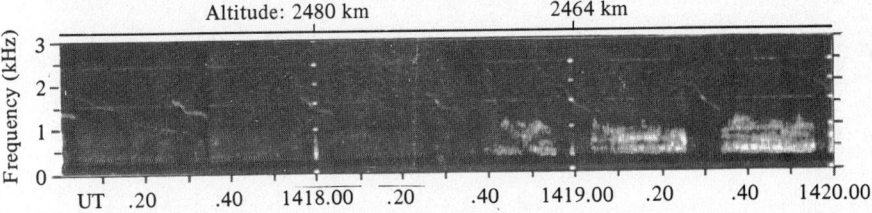

Fig. 16.13. Spectrogram of ELF hiss and chorus (Injun 5, $z = 2480$ km, 17 Jan. 1969). (See Gurnett et al., 1971, for the original version in colour.)

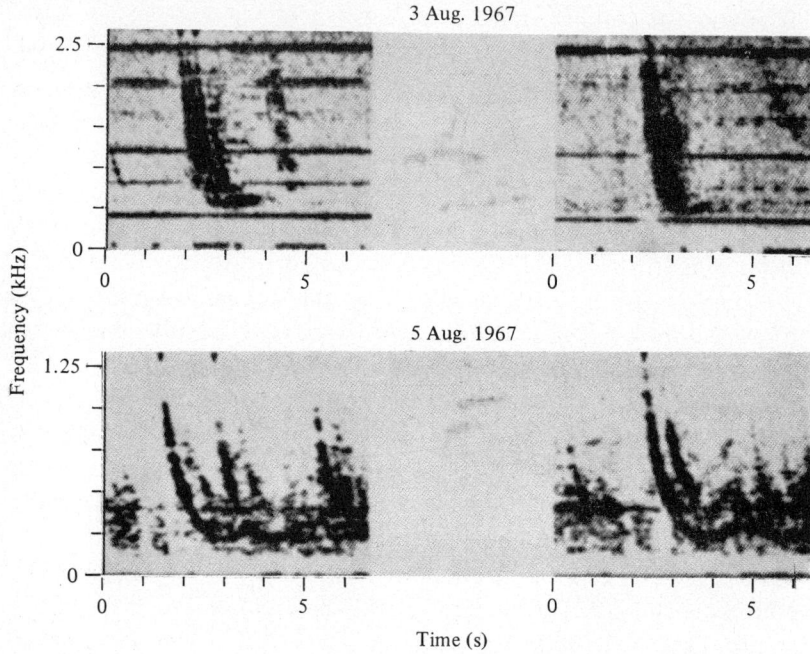

Fig. 16.14. Spectrogram illustrating the reflection of a short-fractional-hop whistler in the ionosphere at the cutoff frequency ω_c (OGO 4, z = 900 km, 3 Aug. 1967). (Muzzio, 1968.)

OGO 2 and OGO 4 (Fig. 16.14; Muzzio, 1968). Thus it would appear that in cases where the two modes (electron and ion waves) do not interact (see, for instance, Rodriguez & Gurnett, 1971), whistling atmospherics do not descend through the ionosphere to the ground, but are reflected back upwards. This may partly explain why the number of whistling atmospherics recorded on the satellite in the hemisphere opposite to the source was considerably higher than the number of whistlers observed simultaneously at the Earth's surface under the satellite (Gurnett & O'Brien, 1964).

As mentioned already, ELF hiss consists of waves in the electron-whistler mode. Simultaneous measurements of their magnetic and electric components revealed that these waves are actually transverse electromagnetic waves. Their components are related by the following expression derived from Maxwell's equation

$$n = Z_0 H/E \qquad (16.4)$$

where $Z_0 = \mu_0 c \simeq 377$ ohms is the wave impedance of free space (Gurnett et al., 1969). This situation can be illustrated by an example: in one of

16.4. Hiss and chorus ELF radiation

the Injun 5 experiments absolute measurements of E and H gave, with the aid of formula (16.4), a refractive index $n \simeq 80$ for frequencies $f \simeq 200$ to 600 Hz. Independent measurements of the electron density at this same satellite gave $n \simeq 95$ for $f \simeq 400$ Hz. These two values agree quite well (with an accuracy of $\pm 20\%$).

We should mention here one effect that was observed in the Injun 3 experiments (Gurnett & O'Brien, 1964). In the vicinity of the frequency $\omega \lesssim \Omega_H(H^+)$ at which the ELF hiss is cut off, an intense narrow-band enhancement of the electric field was recorded. This phenomenon can be attributed to the increase in the electric field $E \sim n^{-1}$ (see formula (16.4)) as $n \to 0$. Since the group velocity decreases rapidly, as $n \to 0$ in the region of wave reflection, the energy density of a wave increases, intensifying the indicated effect. Fig. 16.15 shows a spectrogram illustrating this phenomenon; the labels 'Alt.' and 'Inv.' refer to the altitude and invariant latitude. In these experiments the field was recorded using both electric and magnetic antennas.

In higher regions of the near-Earth plasma *in the polar zone*, intense bands of hiss waves were observed with a magnetic antenna in a frequency range from a few hertz to several hundred hertz, i.e., even up to $f \simeq f_H$. These waves are unique, in that they were detected only in the 'polar cusp' and thus serve as a kind of indicator of this region. This radiation may be ELF, VLF, or even LF. It is observed, as shown by an analysis of the experimental data, during periods of migration of plasma from the

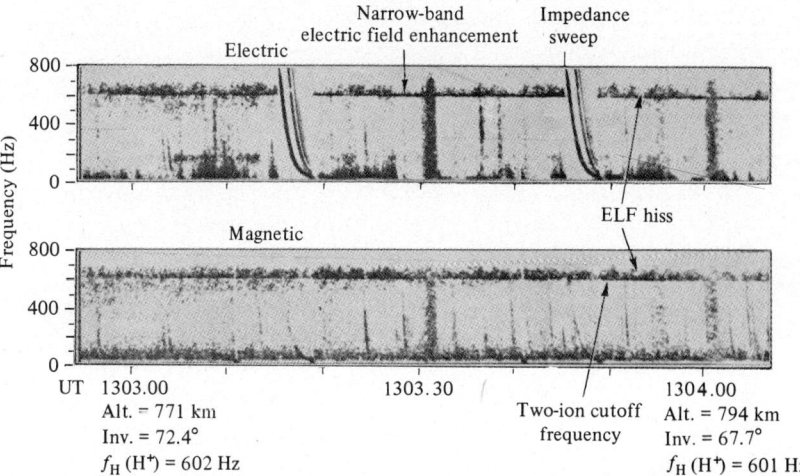

Fig. 16.15 Spectrogram of a narrow band of ELF hiss waves close to the cutoff frequency ω_c (see Fig. 16.10) (Injun 5, 9 Dec. 1968). (Gurnett & O'Brien, 1964.)

transition zone (magnetosheath, see Fig. 2.9) to the polar cusp. Such processes were detected in the near-Earth plasma from an analysis of the data of the IMP 5, ISIS 1, and Injun 5 experiments (Frank, 1971; Frank & Ackerson, 1971). Intense bands of ELF waves at frequencies lower than the proton gyrofrequency were apparently first recorded in this region at great heights on OGO 5 (Scarf et al., 1972, 1974b). Later, on Geos 2 a magnetic antenna recorded here some bands of emission in the range $f \simeq 20$ to 236 Hz (D'Angelo et al., 1974). However, the properties of these waves were investigated the most thoroughly and systematically on Hawkeye 1, since the observations covered a wider frequency range (Gurnett & Frank, 1978b).

These experiments revealed that this emission is almost always present in the vicinity of the *polar cusp*, and that its appearance is associated with bursts of energetic protons, the energies of which are $E_p \simeq 0.2$ to 2 keV. At geocentric distances $R \simeq 6R_0$ intensities $H^2/\Delta f \simeq 10^{-2} \gamma^2$ Hz^{-1} at frequencies $f \simeq 2$ to 5 Hz were observed, and intensities $H^2/\Delta f \simeq 10^{-6}$ γ^2 Hz^{-1} close to the electron gyrofrequency $f_H > f \simeq 10^2$ Hz. The emission in question was in these experiments recorded in the Northern Hemisphere at geocentric distances $R \simeq 5R_0$ to $30R_0$. However, for $R > 10R_0$, they naturally pertain to the transition region of the near-Earth plasma. In these experiments, the satellite orbit did not cross the polar cusp in the Northern Hemisphere at distances $R > 5R_0$. In the Southern Hemisphere, on the other hand, few observations could be carried out. The electrostatic noise in this frequency range made it impossible to carry out measurements with an electric antenna and to demonstrate that the emission being considered actually consists of transverse waves in the whistler mode n_2 (see Fig. 4.2). On the other hand, the fact that the waves are cut off at the electron gyrofrequency f_H is evidence that this is indeed the case.

At lower altitudes, in the polar cap, narrow-band emission of the hiss type was also observed, at frequencies $f \simeq 100$ to 300 Hz, i.e., lower than or of the order of the local proton gyrofrequency (Gurnett & Frank, 1972b). These emission bands were recorded with both electric and magnetic antennas, i.e., the waves were transverse electromagnetic waves. Their appearance was associated with precipitation into the polar zone of low-energy electrons (energies < 10 keV). However, it is still not clear whether these are waves of the electron-whistler mode, characterized by the refractive index n_2, or ion-cyclotron waves, characterized by the refractive index n_1.

17

Results of studies of VLF waves

In the range of frequencies $\Omega_H \simeq \omega \lesssim \omega_L$ a cold plasma does not have resonance branches. In a warm plasma, on the other hand, ion-acoustic waves may be excited at the frequencies $\Omega_H \lesssim \omega \lesssim \Omega_0$. In this frequency range the fact that several different species of ions are present manifests itself forcefully. A number of interesting and important effects are observed that cannot arise in a plasma consisting of electrons and just one species of ion. These effects were first discovered during the course of experiments performed directly in the near-Earth plasma on satellites and space probes.

17.1. Ion-acoustic waves; radiation at the proton gyrofrequency and its harmonics

Ion gyroresonances (i.e. longitudinal ion-cyclotron waves having **K** parallel to **E**, at the frequencies $s\Omega_H$, where s is an integer) are known to be excited readily if **K** is perpendicular to H_0, in which case their Landau damping

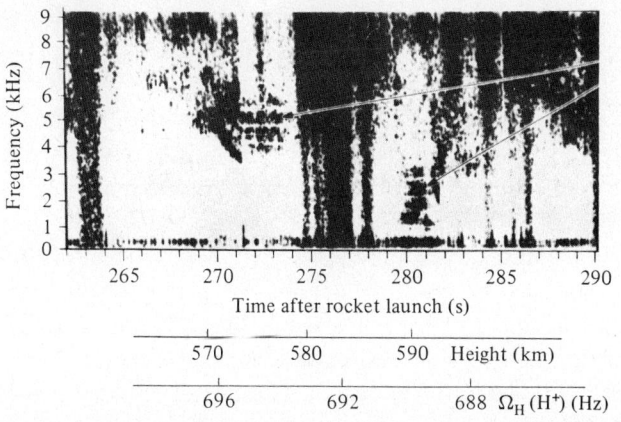

Radiation bands at harmonics $s\Omega_H$ of the proton (H⁺) gyrofrequency

Fig. 17.1. Bands of proton gyroresonance waves recorded on the sounding rocket Javelin 8.46, 25 May 1968 (Mosier & Gurnett, 1969b; Gurnett & Mosier, 1969).

is small. Such proton waves were apparently first recorded in the ionosphere on Injun 5 (Muzzio, 1968) and on a Javelin sounding rocket (Mosier & Gurnett, 1969b; Gurnett & Mosier, 1969). Fig. 17.1 gives a spectrogram obtained on Javelin; on this spectrogram, at altitudes $z \simeq 570$ to 590 km, narrow bands of emission were observed at harmonics of the proton gyrofrequency, up to the eighth.

Cyclotron resonances at the proton gyrofrequency and its harmonics have been studied in a fair amount of detail from the results of observations carried out on the S 3-3 satellite (Kintner et al., 1978). In these experiments, in the polar zone at a geocentric distance $R \simeq 2R_0$, the electric field E of the plasma emissions was recorded in a wide frequency band and was transformed directly into frequency–time spectrograms. The energy spectra $E^2(\omega)$ of the received waves were obtained on the basis of these. It was found that the $E^2(\omega)$ spectra have maxima close to the local values of the proton gyrofrequency $F_H(H^+) \simeq 75$ to 100 Hz, and of its harmonics up to the 3rd or 4th. The frequency spectra of the perturbation $\Delta N/N$ of the plasma density obtained on the basis of these same data exhibit similar maxima, which confirms the electrostatic nature of $E(\omega)$. From the Doppler broadening and shift of the resonance oscillations at $f \simeq 1.2$ to $1.4 F_H(H^+)$ and from the dispersion properties of the $E^2(\omega)$ spectra, the electron temperature $T_e \approx 3.5$ eV and $T_e/T_i \simeq 1$ over the region of the auroral oval were determined. The strong waves observed in these experiments correlate well with presence of strong perpendicular and parallel quasi-static electric fields $E_{dc} \gtrsim 120$ mV m^{-1}, while the weaker field correlates with streams of low-energy electrons $E_c \simeq 0.74$ to 5 keV, indicating the presence of parallel flows along the force lines of the magnetic field H_0, these apparently being the cause of the electrostatic ion-cyclotron instability of the plasma in the polar zone. The maximum field intensities of the observed waves were $E \simeq 50$ mV m^{-1}.

In the experiments on OGO 2, broad bands of emission with maximum amplitude at the proton gyrofrequency $F_H(H^+)$ were discovered in the altitude interval $z = 415$ to 1507 km. The frequency range of the observed VLF waves varied in different cases within the limits $f \simeq 300–700$ Hz to 18 kHz, which correspond to the limits of variation of the proton gyrofrequency $F_H(H^+)$ and of the ion Langmuir frequency Ω_0 at these altitudes. Since only a loop antenna was used in these experiments, just the magnetic field H of these waves was recorded, the wave amplitude diminishing rapidly with increasing frequency (Guthart et al., 1968). These researchers concluded that the emission they had detected represented a spectrum of *ion-acoustic* waves, i.e., a branch describing a fast ion-acoustic

17.1. Ion-acoustic waves and ion-cyclotron waves

wave excited in a warm plasma (see Fig. 5.4 and 5.5 and formulas (5.58) to (5.64), Volume 1). The observed effect on the loop antenna was therefore apparently the result of a magnetic field **H** induced in the antenna as a consequence of its motion relative to the electric field **E** of the electrostatic longitudinal wave. In that case

$$|\mathbf{H}| = \varepsilon_0 |\mathbf{V}_b \times \mathbf{E}| \qquad (17.1)$$

Use of (17.1) showed that, on a number of occasions, with $V_b \simeq 7 \,\text{km s}^{-1}$, the intensity of the field of the longitudinal waves at the proton gyrofrequency was $E \simeq 40 \,\text{V m}^{-1}$! This corresponds to an energy density of a wave $\frac{1}{2}\varepsilon_0 E^2 = 7 \times 10^{-8} \,\text{erg cm}^{-3}$, a value comparable with the kinetic energy density NkT_e of the region of the ionosphere under consideration! The maximum expected intensity of the ion-acoustic waves should theoretically be of the order of $\frac{1}{2}NkT_e$ (Rostoker, 1961; Scarf et al., 1965). Thus, the interpretation given above of the experimental results described seems to be plausible. However, the maximum value of the field intensity E appears to be too high, since the recorded emission covers a broad band of frequencies and the integrated density of their energy will be greater than $7 \times 10^{-8} \,\text{erg cm}^{-3}$. Of course, since the density N and temperature T of the plasma were not determined at the same time in these experiments, it is hard to judge exactly how much the measurement results deviate from the theoretical estimates.

Fast ion-acoustic waves were also apparently observed before by the

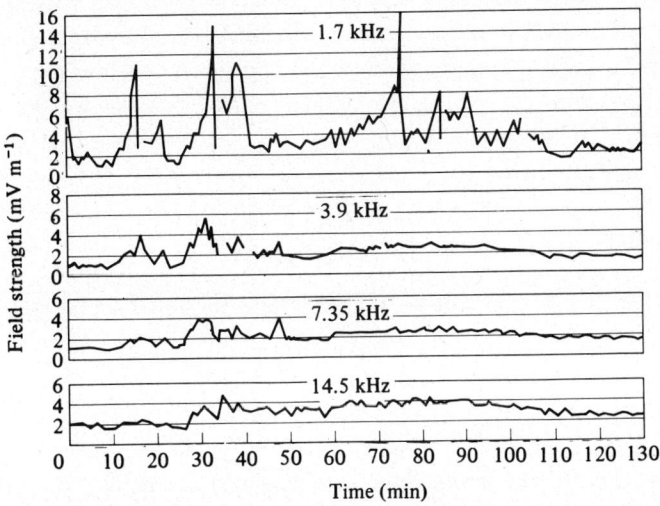

Fig. 17.2. Recording of the field strength of fast ion-acoustic waves observed in the ionosphere; P 11, $z = 268$ to $3720\,\text{km}$, 1965 (Scarf et al., 1965).

P 11 satellite in the altitude interval $z \simeq 270$ to 3700 km at the discrete frequencies $f = 1.7, 3.9, 7.35$, and 14.5 kHz (Scarf et al., 1965). Recordings of the average values of the electric field of these waves, obtained on one of the orbits of P 11, are shown in Fig. 17.2. The mean field level of these waves varied over limits $E \simeq 1$ to $2\,\text{mV}\,\text{m}^{-1}$; it rarely happened that $E_{\min} < 600$ to $800\,\mu\text{V}\,\text{m}^{-1}$; field bursts with $E \simeq 20$ to $100\,\text{mV}\,\text{m}^{-1}$ lasting $\Delta t \simeq 3$ to 10 min. were often observed on the nightside of the Earth; there were even cases where $E \simeq 1\,\text{V}\,\text{m}^{-1}$. Since the lower boundary of the frequency range used ω was greater and sometimes much greater than $\Omega_H(H^+)$, cyclotron excitation of this branch of the ion-acoustic waves could not be observed, just as in the experiments described above (Guthart et al., 1968). Naturally, therefore, the field strength E did not attain very high values, although $E \simeq 1\,\text{V}\,\text{m}^{-1}$ is already quite an intense field. The energy density of the recorded waves in general varied in the limits $E^2 \simeq 10^{-16}$ to $10^{-13}\,\text{erg}\,\text{cm}^{-3}$, which is several orders of magnitude less than NkT_e in the region of the outer ionosphere under investigation. Note that the minimum values of the field $E_{\min}$ observed in these experiments in general agree well with the theoretical estimates of $E_{\min}$. For instance, according to Rostoker, 1961, for $\omega \ll \Omega_0$ and a receiver bandwidth $\Delta\omega/\omega \ll 1$, we have

$$E_{\min}^2 \simeq kT_e \frac{2\pi}{D_i} \frac{4\pi^2}{\Lambda^2} \frac{\Delta\omega}{\omega}, \qquad (17.2)$$

where D_i is the Debye length for the ions and Λ is the wavelength. At $f = 1.7\,\text{kHz}$ and $\Lambda \simeq 90\,\text{cm}$ (such an estimate is given by Rostoker, 1961), when we use the average values of the parameters for the region of the ionosphere through which the satellite orbit passed, we get $E_{\min} \simeq 360\,\mu\text{V}\,\text{m}^{-1}$. This value of $E_{\min}$ is smaller than the values observed, which may constitute further proof of the ion-acoustic nature of the recorded waves. The same conclusion was reached by Scarf et al., 1968a, in experiments that they performed later on OV3 3, which, technically speaking, were more sophisticated. In these experiments measurements were made at four frequencies: $f = 80$ and $400\,\text{Hz}$ and $f = 1.65$ and $7.3\,\text{kHz}$ (Fig. 17.3). In this case both a magnetic loop antenna and a linear electric antenna were used, which made it possible to discern cases where transverse electromagnetic waves rather than longitudinal electrostatic waves were recorded. In the altitude range covered by the OV3 3 orbit, $z = 354$ to $4460\,\text{km}$, two of the frequencies at which measurements were made, 80 and $400\,\text{Hz}$, were often lower than practically all the ion gyrofrequencies in the outer ionosphere. Hence, along with fast ion-acoustic VLF waves,

17.1. Ion-acoustic waves and ion-cyclotron waves

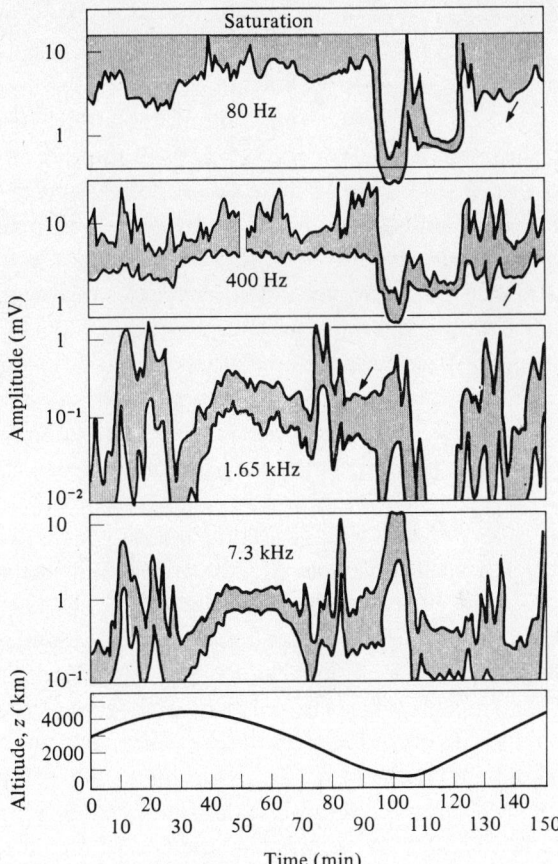

Fig. 17.3. Recordings of the amplitude of slow and fast ion-acoustic waves observed on the satellite OV3 3 in 1966 at altitudes $z = 354$ to 4460 km. The orbit is shown in the lower part of the figure (Scarf et al., 1968a).

slow ion-acoustic ELF waves with $\omega < \Omega_H$ could also be recorded (see Fig. 5.5, Volume 1). The experimental results (Fig. 17.3) show that both branches of ion-acoustic waves are always observed, the slow ELF waves (frequencies 80 and 400 Hz) apparently usually being more intense than the fast ion-acoustic waves. However, this cannot be stated with certainty, since it was not the field strength that was determined here, but rather the potential induced by the field on the antenna: from this potential, the field strength E was estimated, using the theoretical value of the electrical length of the antenna, which was somewhat uncertain (Fig. 17.3 shows to what extent E varied).

Ion-acoustic waves have also been recorded at altitudes lower than in the above-described experiments. For example, on the Javelin rocket 8.45 (Shawhan & Gurnett, 1968) intense longitudinal waves (characterized by high values of E, but no effect on a magnetic antenna) were observed at frequencies $f < 1$ kHz. Here the amplitude of the electric field E was a maximum at the lowest altitudes $z \simeq 250$ to 280 km at which measurements were carried out. It may be that these waves correspond in part to the branch of slow ELF ion-acoustic waves ($\omega < \Omega_H(H^+)$); no precise data on the frequency range are given by Shawhan & Gurnett.

Ion-acoustic waves are also generated in remoter regions of the outer ionosphere (plasmasphere) and beyond its bounds. Results of similar observations, made in space close to the *plasmapause* for $L \simeq 3.4$ in the range $F_H(H^+) < f = 20$ to 500 Hz $< F_0$ on Explorer 45 (S^3-A), are shown in Fig. 17.4 (Anderson & Gurnett, 1973); F_0 is the Langmuir frequency of the ions. This spectrogram, containing a band of ion-acoustic waves, was recorded during a magnetic storm on 16–17 December 1971. Excitation of these particular oscillations was accompanied by a rapid decrease in the number of low-energy protons $E_p \simeq 20$ to 50 keV in the ring current of energetic ions which forms above the equator in the magnetosphere during a magnetic storm. Electrostatic longitudinal waves are seen to have been recorded, since at the left-hand side of the spectrogram no effect on the magnetic antenna is visible. The spectrum of these oscillations consists of vertical bursts with a duration $\Delta t \lesssim 0.25$ s. Their frequency sometimes attains values $f \simeq 600$ Hz, but they are most intense at $f \simeq 200$ to 300 Hz. An analysis of the results of these observations showed that in the frequency range $f = 20$ to 300 Hz near the plasmapause ($\simeq 3.4$) the intensity of the

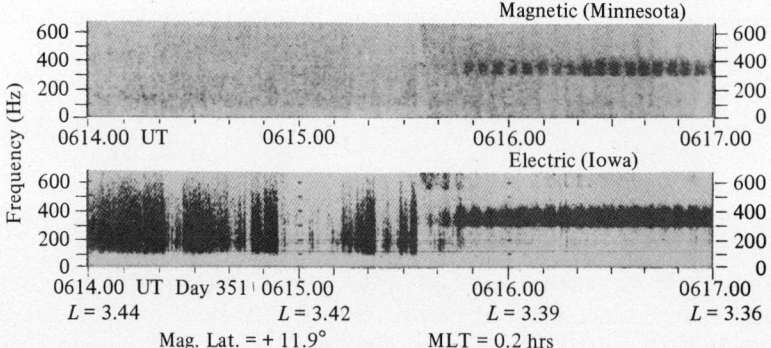

Fig. 17.4. Spectrogram of longitudinal ion-acoustic waves, recorded near the plasmapause ($R \simeq 3.3\ R_0$) with an electric antenna; Explorer 45, 17 Dec. 1971. Hiss-type emissions (depicted on the right) were also recorded on both the magnetic and the electric antennas (Anderson & Gurnett, 1973).

17.1. Ion-acoustic waves and ion-cyclotron waves

observed ion-acoustic waves varied as follows: $E^2 = 2 \times 10^{-10} \, V^2 \, m^{-2} \, Hz^{-1}$ to $10^{-8} \, V^2 \, m^{-2} \, Hz^{-1}$; the maximum of E^2 sometimes corresponded to a frequency $f \simeq 100 \, Hz$. These waves are apparently excited due to the electrostatic instability of the equatorial proton ring.

In the right-hand part of Fig. 17.4 a band of hiss-type *electromagnetic* (transverse) waves was recorded at $f \simeq 300$ to $420 \, Hz$. The modulation of this band (light bands) is associated with the appearance of intensity minima during the rotation of Explorer 45 about its axis. The direction of propagation of these waves is seen to be perpendicular to the Earth's magnetic field. These waves were recorded at a distance of the order of R_0 from the plasmapause, inside the magnetosphere. VLF hiss waves of a similar type have been observed under various conditions and will be discussed below. They could be the result of longitudinal waves being transformed into transverse waves. An analogous band of *electrostatic* waves was observed during a magnetic storm on 4–5 August 1972, on Explorer 45 when it *crossed the plasmapause* (Taylor & Anderson, 1977).

When crossing the plasmapause at geocentric distances $L \simeq 3$, on Hawkeye 1 bands of noise-type electrostatic oscillations were recorded at $f \simeq 10$ to $100 \, Hz$ (Kintner & Gurnett, 1977). The electric field **E** of these oscillations was for the most part perpendicular to the gradient of the electron density of the plasma and perpendicular to the vector of the magnetic field H_0. The oscillations were generated for quite high values of dN/dR, i.e., if the thickness of the plasmapause region did not exceed $0.1 R_0$. This spectrum of longitudinal oscillations was apparently generated due to drift instability of the plasma with respect to the ion-acoustic waves. The intensity of these oscillations varied approximately within the limits $E^2 \simeq 10^{-12} \, V^2 \, m^{-2} \, Hz^{-1}$ to $10^{-10} \, V^2 \, m^{-2} \, Hz^{-1}$.

Ion-acoustic waves excited in the region of the *shock front* and above it were observed on OGO 5 (Fredricks *et al.*, 1970; Scarf *et al.*, 1971a) and on IMP 6 (Rodriguez & Gurnett, 1975, 1976). The spectrum of such oscillations that was recorded on IMP 6 when crossing the shock front (Fig. 17.5) has a broad maximum of the electric field $E^2 \simeq 10^{-9} \, V^2 \, m^{-2} \, Hz^{-1}$ most often in the frequency range $f \simeq 200$ to $800 \, Hz$; it corresponds to a band of ion-acoustic waves. The ratio of the electric and magnetic energy densities of these waves $\varepsilon_0 E^2 / \mu_0 H^2 \simeq 10^2$ to 10^4, $\mathbf{E} \parallel \mathbf{H}_0$. This affirms their electrostatic nature. The broad emission band is associated with a Doppler shift of the frequencies of the excited waves and possibly with nonlinear effects. In the *transition zone* (below the shock-front region) waves with a similar property were observed on IMP 6. However, their intensities were two to four orders of magnitude less than

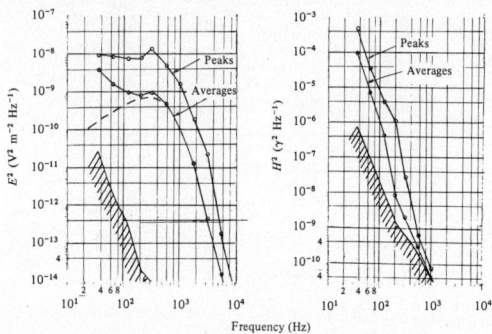

Fig. 17.5. Spectrum of ion-acoustic waves, in the region of the shock front: $R \simeq 18\,R_0$; IMP 6, 20 Apr. 1971 (Rodriguez & Gurnett, 1975).

those of the waves shown in Fig. 17.5. It would seem that the waves in the transition region and at the shock front are generated by the same mechanism. The intensity of the above-examined ion-acoustic waves correlates well with the properties of the solar wind. Above the shock-front region the field amplitude is a maximum when the temperature ratio T_e/T_i of the solar wind is high and the proton temperature T_i is low (Rodriguez & Gurnett, 1976). Still, no clear-cut correlation has been found between the emission intensity of ion-acoustic waves and the magnetohydrodynamic parameters of the shock front, a fact which seems to point to a kinetic nature of this emission.

In the *solar wind*, ion-acoustic waves have been observed both near the Earth at geocentric distances $R \simeq 20R_0$ to $40R_0$ and in the interplanetary medium at distances $R_\odot \simeq 0.3$ to 1 AU (astronomical unit) from the sun. The spectra of these waves, and also their properties, are the same in these different places. However, the waves near the Earth are of 'terrestrial origin'. In most cases they can be attributed to the effect of streams of suprathermal protons with $E_p \simeq 1$ to $10\,\text{keV}$ issuing from the shock-front region, which give rise to two-stream instability of the plasma. These proton streams no longer have any effect at great distances from the Earth. The ion-acoustic waves there are of 'interplanetary origin'. They are presumably generated due to the anisotropic distribution of the electrons in the solar wind, this being caused in turn by streams of hot electrons coming from the Sun. The outcome is electron heat-flux instability (Forslund, 1970).

Ion-acoustic waves *above the shock-front region* were apparently first detected on OGO 5 (Scarf *et al.*, 1970b), while in the solar wind they were observed on Pioneer 8 and 9 (Scarf *et al.*, 1968b, 1971b; Siscoe *et al.*,

17.1. Ion-acoustic waves and ion-cyclotron waves

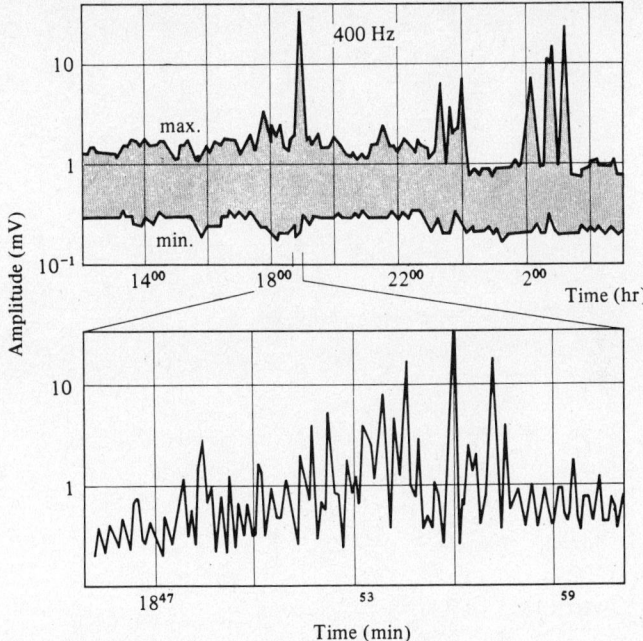

Fig. 17.6. Recordings of antenna potential amplitude for fast ion-acoustic waves observed in the solar wind at a distance $R \simeq 10^6$ km from the Earth; Pioneer 8, 21 Dec. 1967 (Scarf et al., 1968b).

1971). Detailed studies of ion-acoustic waves in the solar wind have been carried out only recently, however, on the basis of observational data from IMP 6 and 8 (circumterrestrial orbits) and Helios 1 and 2 (heliocentric orbits) (Gurnett & Anderson, 1977; Gurnett & Frank, 1978b; see also Gurnett et al., 1978a).

Some results of measuring the amplitude of ion-acoustic waves in the solar wind on Pioneer 8 (heliocentric orbit) at $f = 400$ Hz $< F_0 = \Omega_0/2\pi = 530$ Hz at a distance of $\simeq 10^6$ km from the Earth are shown in Fig. 17.6. The field strength of these waves apparently varied within the limits $E \simeq 0.2$ to 30 mV m^{-1} during these experiments. This corresponds to an energy density $\frac{1}{2}\varepsilon_0 E^2 \simeq 10^{-17}$ to 4×10^{-15} J m^{-3}. These energy-density values exceed those observed in a number of cases in the outer ionosphere (see the table in Chapter 20).

The spectrum of ion-acoustic waves obtained from the data of Helios 2 (Fig. 17.7) illustrates the principal properties of the waves that were observed in these experiments. Their spectral intensity has a broad maximum at frequencies $f \simeq 1$ to 10 kHz higher than F_0, the proton

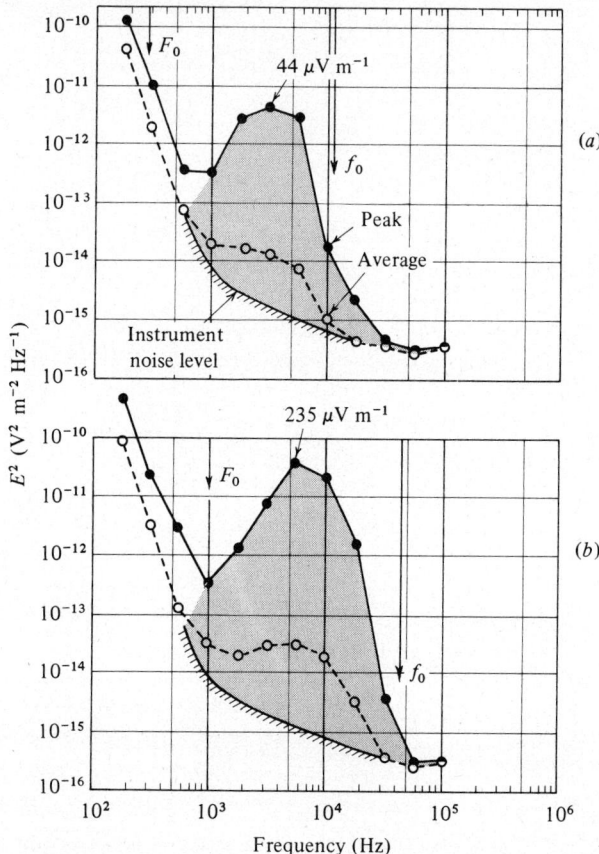

Fig. 17.7. Spectra of ion-acoustic waves observed in the solar wind: (a) 14 Dec. 1974, $R_\odot = 0.98$ AU; (b) 25 Dec. 1975, $R_\odot = 0.47$ AU; Helios 2 (Gurnett & Anderson, 1977).

Langmuir frequency. Since ion-acoustic waves can be excited at frequencies $F_H \lesssim f \lesssim F_0$, a marked increase in their frequency range occurs due to the Doppler shift of the frequencies of these waves associated with the great speed of the solar wind ($V \simeq 400$ to $500 \mathrm{\,km\,s}^{-1}$). These waves are indeed electrostatic (a magnetic field of these oscillations has not been detected), and it has been shown to an accuracy of 15° that their electric field **E** is perpendicular to $\mathbf{H}_{0\odot}$ (where $\mathbf{H}_{0\odot}$ is the constant magnetic field of the solar wind). Another important characteristic of the spectra of these waves is that the maximum (peak) values of the field E_{max} are considerably greater than the average values E. This is because the emission, as shown by the spectrograms, is of the pulse type, consisting of a number of bursts with

17.2. Noise-like waves, plasmaspheric hiss

durations $\Delta t \simeq$ several tenths of a second. The bursts appear simultaneously (accurate to the time resolution) at all the spectral frequencies. The maximum values of the electric field $E_{max} \simeq 50$ to $300\,\mu\mathrm{V}\,\mathrm{m}^{-1}$, and occasionally $E_{max} \gtrsim 500\,\mu\mathrm{V}\,\mathrm{m}^{-1}$; the frequency-average value of the field $\mathbf{E} \simeq 10\,\mu\mathrm{V}\,\mathrm{m}^{-1}$. The results of experiments in the vicinity of the Sun showed that the emission of ion-acoustic waves is a sporadic phenomenon, observed several times a month at all distances from the Sun. The duration of their heightened intensity varies from a few hours to several days. The frequency of the maximum values of the spectrum intensity, the spectral width, and the values of E^2 all increase with increasing proximity to the Sun. It has been found that the frequency of maximum intensity of these waves varies as $(R_0)^{-1}$. An analysis of experimental findings showed this to be short-wave emission. The minimum wavelengths at which the emission spectrum is cut off due to increased Landau damping are determined from the relation $\lambda_{min} \simeq 2\pi D_e$. Estimates show that $\lambda_{min} \simeq 30$ to $70\,\mathrm{m}$. At $f \simeq 1\,\mathrm{kHz}$, on the other hand, the wavelengths $\lambda \gtrsim 100\,\mathrm{m}$.

17.2. Noise-like waves generated in the frequency range $\Omega_H < \omega < \omega_L$. Plasmaspheric hiss (PH)

In this range of frequencies, which corresponds more closely to the definition of VLF waves (see §15.2, above), a number of types of emissions have been recorded. The most common of these is a broad-band emission, known in the literature as *plasmaspheric hiss* (PH). It has been studied most fully in the experiments on OGO 5 (Thorne et al., 1973).

A typical spectrum of PH waves is shown in Fig. 17.8a. In some cases the emission spectrum has two additional maxima and a more intense tail in the lower part of the frequency range (Fig. 17.8b). Such spectra are observed during crossover on the dayside into the region of high geomagnetic latitudes. This kind of emission was recorded during each pass of OGO 5 through the outer ionosphere and was cut off abruptly at its boundary, the plasmapause; thus, plasmaspheric hiss constitutes a constant background in the outer ionosphere. In each individual series of observations the main properties of the PH wave spectrum remain practically unchanged, no matter what the position of the satellite. However, from instance to instance there are marked changes in the PH frequency range, in the lower cutoff frequency f_{min} of its spectrum, and in its frequency of maximum intensity f_{max}. This property of PH indicates that it is generated in a limited region of the plasma and that subsequently it spreads to fill the entire outer ionosphere (plasmasphere) more or less

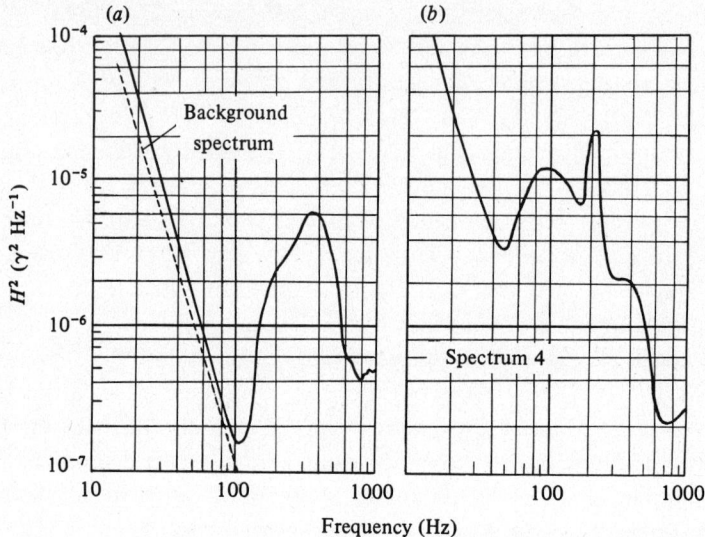

Fig. 17.8. Spectra of VLF hiss waves in the plasmasphere below the plasmapause: (a) usual spectra: (b) rarely observed spectra at higher latitudes; OGO 5, 4 Apr. 1968 (Thorne et al., 1973).

uniformly. For such a situation to occur, the propagation trajectories of the transverse electromagnetic waves must be able to deviate markedly from the direction of the magnetic force lines around which they are generated (see Fig. 17.9, 8.15, and 8.16). This is possible if the frequency of these waves is of the order of, or less than, the lower-hybrid frequencies f_L in the plasma regions through which they pass. Thorne et al., 1973, conclude that the emission in question is generated in the equatorial region near the plasmapause.

The spectrum of PH waves is cut off sharply at the bottom at f_{min}; it is more diffuse at the top at frequencies $f \gtrsim 10^3$ Hz, and its maximum intensity corresponds to f_{max}. The values of f_{min} and f_{max}, as well as the intensity H of the magnetic field of these waves, vary from case to case within the following limits:

$$\left. \begin{array}{l} f_{min} \simeq 30 \text{ to } 250 \text{ Hz}, f_{max} \simeq 100 \text{ to } 350 \text{ Hz} \\ H_{max} \simeq 5 \times 10^{-3} \text{ to } 10 \times 10^{-3} \gamma, H^2 \simeq 10^{-7} \text{ to } 10^{-5} \gamma^2 \text{ Hz}^{-1}. \end{array} \right\} \quad (17.3)$$

Occasionally $H_{max} \simeq 100$ mγ.

The OGO 5 experiments turned up an empirical link between the maximum spectral amplitude H_{max} and the plasma density N at the measurement point, namely:

$$H_{max}/(f_{max})^{1/2} = 1.7 \ln(N/N_0) \, \text{m}\gamma \, \text{Hz}^{-1/2}, \quad (17.4)$$

17.2. Noise-like waves, plasmaspheric hiss

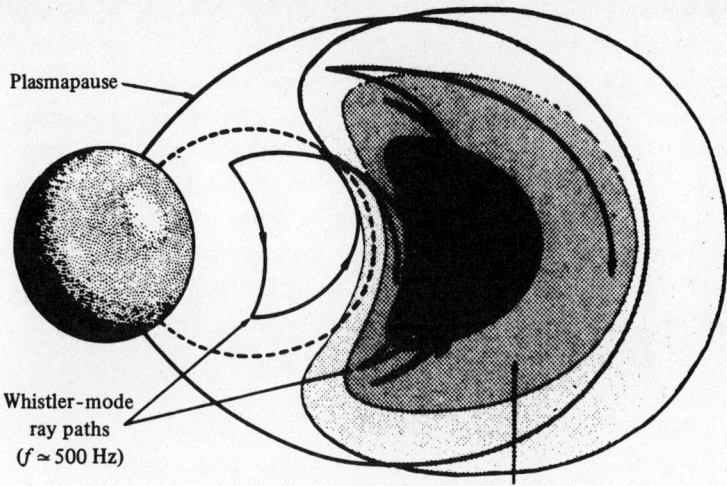

Fig. 17.9. Paths of propagation of VLF hiss waves in the plasmasphere in the frequency range $\omega \lesssim \omega_L$ (Thorne et al., 1973).

where $\ln N_0$ is determined from the linearized dependence of the results of measurements (17.4) at the point where $H_{\max} = 0$ (the origin of the coordinates) (Chan et al., 1974; Chan & Holzer, 1976). In the series of experiments processed with the aid of (17.4), the values $H_{\max}/(f_{\max})^{1/2} = 1$ to $2\,\mathrm{m}\gamma\,\mathrm{Hz}^{-1}$, $f_{\max} \simeq 0.02$ to $0.03 f_H$, $N \simeq 50$ to $500\,\mathrm{cm}^{-3}$ were observed in the plasma region $L \simeq 4$ to 6. From these data, it follows in particular that the frequency of PH emission corresponded to a frequency range near the lower-hybrid frequency $f_L \simeq (m/M)^{1/2} f_H \simeq 0.023 f_H$, in which the whistler-mode properties are already influenced substantially by the presence of ions.

Plasmaspheric hiss emission does not exhibit clearly expressed properties of a state of polarization – it is random. Its maximum intensity is observed in the region $4 < L < 6$ during periods of low magnetic activity (for $k_p \gg 1$). On going from $L \simeq 6$ to $L \simeq 3$, $f_{\max}$ decreases by a factor of two on the average; the value of $f_{\max}$ increases with an increase in the magnetic-activity index K_p.

It is thought that PH emission is generated due to a 'wave–particle' gyroresonant interaction in the region of the geomagnetic equator, with the participation of low-energy electrons ($E_e \simeq 10\,\mathrm{keV}$), which diffuse into the region of the plasmapause from the outer radiation belt.

VLF plasmaspheric hiss waves were earlier detected on OGO 3 (Russell et al., 1969, 1970). In these experiments two types of noise-like oscillations were recorded in the range $f \simeq 10$ to 800 Hz; stable oscillations (lasting several minutes) throughout the whole frequency range, and oscillation bursts (lasting several seconds) whose amplitude varied with the frequency. Stable noise spectra of oscillations were recorded in the plasmasphere for $L \simeq 6$ on the dayside of the magnetosphere in the vicinity of magnetic latitude $\phi \simeq 45°$. Oscillations of variable amplitude were recorded more often during the morning hours and closer to midday, near the geomagnetic equator for $L \simeq 7$. In the vicinity of the geomagnetic equator the spectra of these waves varied in these experiments within the limits $2F_H < f < f_L/2$; they were cut off sharply at $f \simeq 10$ Hz and apparently had maximum values $H^2 \simeq 10^{-6}$ to $10^{-5} \gamma^2 \, \text{Hz}^{-1}$ at $f \simeq 50$ to 200 Hz.

Longitudinal VLF waves of the *noise* type are also generated in the *polar ionosphere* along high-latitude force lines, starting from heights of a few thousand kilometres and extending to great distances from the Earth in the magnetosphere. This emission can be observed at any time of day. It is most intense during heightened auroral activity, and its spectrum is cut off at the electron gyrofrequency. This emission is very intense at frequencies $\Omega_H < \omega < \omega_L$ (Gurnett & Frank, 1977). Fig. 17.10 presents a typical spectrum of this emission, constructed from results of measurements on IMP 6 and Hawkeye 1 at distances $R \simeq 4R_0$ to $4.5R_0$ from the centre

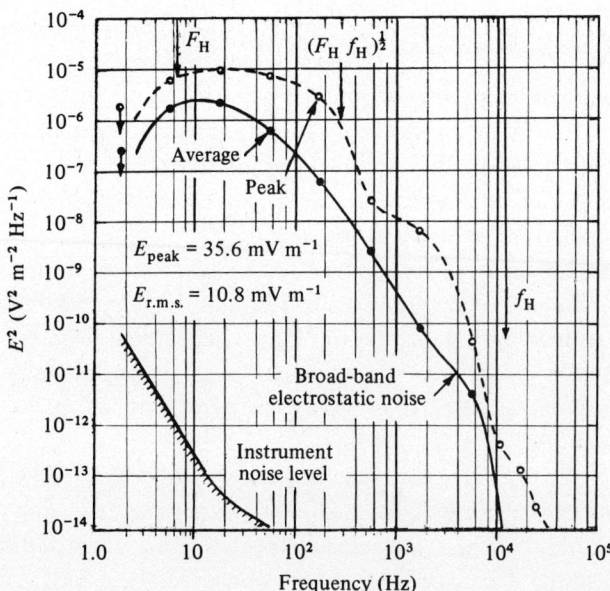

Fig. 17.10. Spectrum of longitudinal VLF waves ($\Omega_H < \omega < \omega_L$) in the polar magnetosphere; $R \simeq 4.5R_0$, Hawkeye 1, 16 Oct. 1974 (Gurnett & Frank, 1977).

17.2. Noise-like waves, plasmaspheric hiss

of the Earth, at a magnetic latitude $\phi \simeq 40°$ for transition from $L = 6$ to 9. An analysis of the results of these experiments showed that the maximum intensity of these waves occurs at frequencies $f \simeq 10$ to 50 Hz, the field strength attaining values $E \simeq 30\,\text{mV}\,\text{m}^{-1}$. The ratio of the electric to the magnetic energy density is about 40; this confirms their quasi-electrostatic nature. For other types of waves, such an intense emission is observed only in the region of the *shock front*! The properties of these waves are closely related to the polar-hiss emission generated in the polar region (Fig. 17.11) and they are associated with streams of particles flowing along

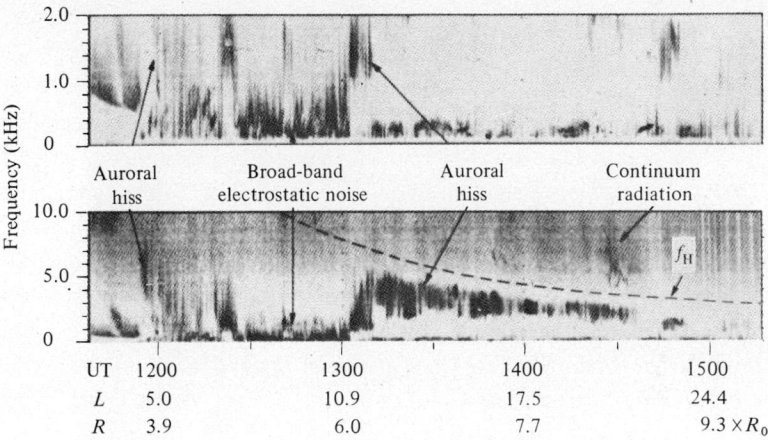

Fig. 17.11. Spectrogram of longitudinal VLF waves ($\omega < \omega_L$) and of 'polar hiss' waves; Hawkeye 1, 12 July 1974 (Gurnett & Frank, 1977).

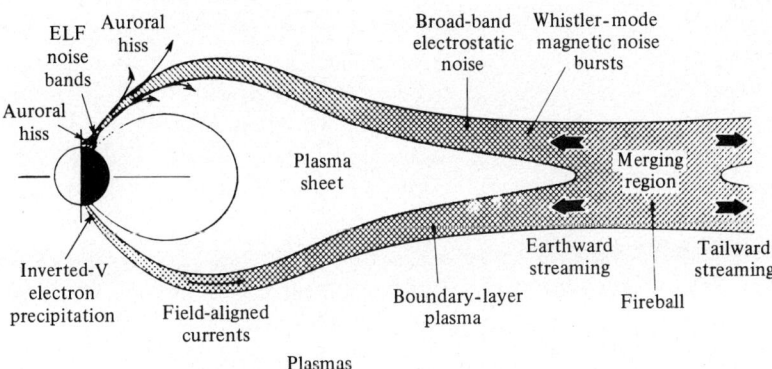

Fig. 17.12. Schematic representation of the distribution of different types of waves in the nightside magnetosphere and in the region of formation of the geomagnetic tail.

the force lines of the magnetic field; these particle fluxes exist in the high-latitude polar magnetosphere (Fig. 17.12). A detailed study of these waves therefore gives us deeper insight into the processes taking place in the polar plasma and in the Earth's geomagnetic tail. The mechanism of their generation is thought to be the same as the generation mechanisms of ion-acoustic waves ($\omega < \Omega_0$) or of oscillations at ion gyroresonance harmonic frequencies $\omega \simeq s\Omega_H$. The electrostatic instability in this case is due to an interaction of the 'wave–hot plasma particle' type. It may be, however, that these waves arise owing to a 'current-driven' instability mechanism (Gurnett & Frank, 1977).

17.3. Waves generated near the lower-hybrid frequency ω_L

At the end of Part 1 we considered the resonance excitation of oscillations in the ionosphere at the lower-hybrid frequency in the vicinity of a satellite, an excitation which was stimulated by whistling atmospherics (see Fig. 14.1). This effect was first observed on Alouette 1 (Barrington & Belrose, 1963; Barrington et al., 1963). Later the same sort of emission was observed in various experiments, and the experimental results were used to determine the effective mass of the ions M_{eff} as well as other parameters of the ionosphere (Barrington et al., 1965).

On Alouette 1 broad-band hiss emission, cut off at the lower-hybrid frequency, was also detected (Brice & Smith, 1965); this emission was recorded mainly on an electrical antenna, i.e., a packet of longitudinal waves, with **E** parallel to **K**, was involved. Fig. 17.13 shows the spectrogram of these waves. The emission is seen to be cut off at a frequency ω_L, the

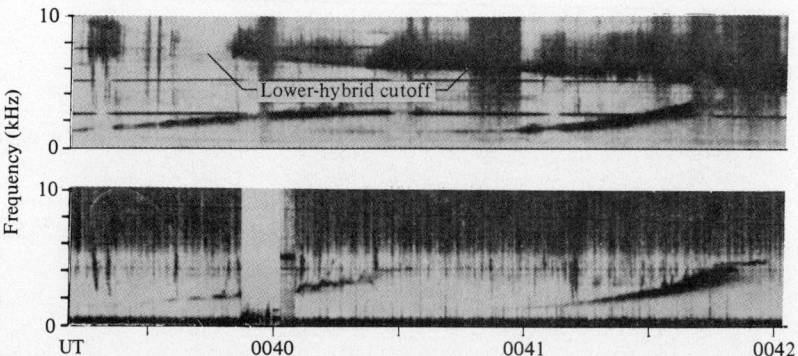

Fig. 17.13. Spectrogram illustrating the cutoff of wide-band VLF hiss waves at the lower-hybrid frequency; Alouette 1, 23 Oct. 1963 (Brice & Smith, 1965).

17.3. Waves generated near ω_L

value of which decreased with time from 10 to 5 kHz, since the satellite moved toward lower latitudes at an almost constant altitude ($z \simeq 1000$ km), while the intensity of the Earth's magnetic field $\mathbf{H}_0$, and thus ω_L as well, decreased. It may well be that the waves in question correspond to the resonance branch of the LF waves excited in a cold plasma (see Fig. 4.7). The lower part of Fig. 17.13 shows data from simultaneous observations at the Earth's surface, below the satellite, where hiss emission was recorded as well; it was, however, not cut off at the lower-hybrid frequency, and presumably had some other origin.

In the *polar ionosphere*, in the height interval $z \simeq 400$ to 1100 km, bands of waves cut off at the bottom at the *lower-hybrid frequency* were observed on OGO 6 (Laaspere & Taylor, 1970). Fig. 17.14a shows a spectrogram with a broad band of such waves in the frequency range $f \simeq 5$ to 10 kHz. The spectrum is cut off abruptly at the bottom. Laaspere & Taylor, 1970, and Laaspere & Johnson, 1973, called this spectrum a 'curtain'. This same spectrogram (Fig. 17.14a) shows a noise band of the polar-hiss type (see

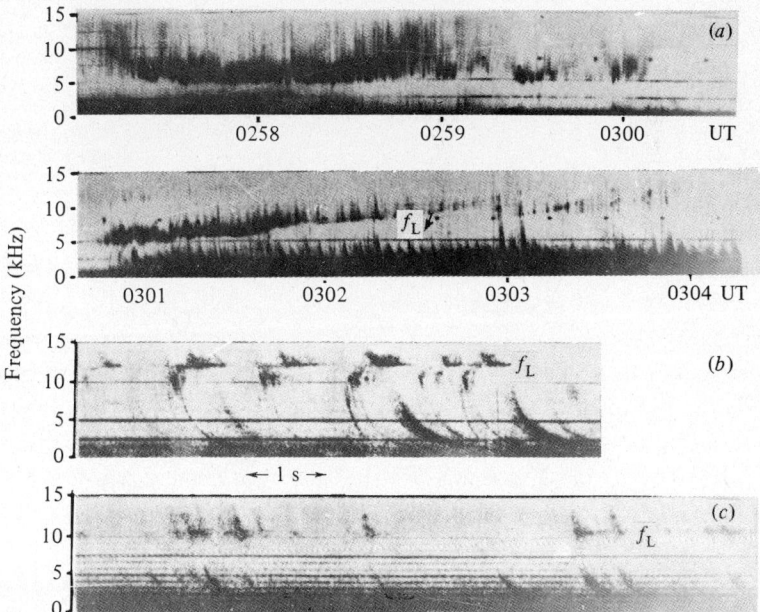

Fig. 17.14. Spectrograms of waves generated in the vicinity of the lower-hybrid frequency ω_L; OGO 6 (Laaspere & Johnson, 1973): (a) 'Curtain effect' band, cut off abruptly at ω_L; $z \simeq 900$ km, $\phi = 65°$, 26 Feb. 1971. (b) Wave packets stimulated by whistling atmospherics, cut off abruptly at ω_L; $z = 950$ km, $\phi = 48°$, 22 Dec. 1969. (c) Wave packets stimulated by whistling atmospherics, with a diffuse boundary in the vicinity of ω_L; $z = 1090$ km, $\phi = 45°$, 5 Oct. 1969.

§17.4, below) at frequencies $f \leqslant 4\,\text{kHz}$. It was observed in this case at relatively low geomagnetic latitudes $\phi < 59°$), apparently because of the heightened magnetic activity ($K_p \simeq 5$ or 6) during the observation period. Interestingly enough, at 0303 UT the cutoff frequency f_L exhibited a jump; the authors cited above did not analyse its cause. The gradual increase in the cutoff frequency f_L from 0301 to 0304 UT was apparently associated with a rise in the lower-hybrid frequency $f_L = \Omega_L/2\pi$, since OGO 6 moved into a region where the effective mass of the ions was lower (with an increase in the height z and decrease in ϕ).

Figure 17.14b shows recordings of oscillations in the vicinity of the lower-hybrid frequency, these being stimulated by whistling atmospherics. Like the waves portrayed in Fig. 17.14a, they are cut off abruptly at the lower-hybrid frequency. However, the excitation of such waves, but with a diffuse 'wedge-shaped' lower boundary, was also observed (Fig. 17.14c). It was assumed by Laaspere & Taylor, 1970, and Laaspere & Johnson, 1973, that the energy of whistling atmospherics or of oscillations of the polar-hiss type served as a source of the energy needed to generate the waves being considered. This mechanism differs from the generation mechanism of the narrow-band wave packets to be described below (see Chapter 18), which are stimulated by telegraph signals, when, in the opinion of Gurevich & Shlyuger, 1975, an external source (signals from a radio transmitter) acts just as a 'trigger' device, the actual source of the wave energy being the streams of energetic particles.

Lower-hybrid resonance waves were recorded in the form of *narrow-band* emissions, for example with the aid of an electric antenna aboard OGO 2 at heights $z = 413$ to $1512\,\text{km}$ (Laaspere et al., 1969). In the experiments on OGO 4 (Laaspere & Taylor, 1970), the ion composition was determined with mass spectrometers at the same time as VLF waves were observed. This made it possible to compare the frequency ω_L of the lower-hybrid resonance, which was known from the VLF oscillation spectra, with the results of direct measurements of the plasma parameters (see (1.42)). On the whole, there was a good fit between the values obtained for ω_L using the two methods. However, in some cases the data did not agree, and this was assumed by Laaspere & Taylor, 1970, to have been because the recordings were not always of lower-hybrid waves excited in the immediate vicinity of the satellite.

Very good agreement with the corresponding values for ω_L was obtained during an analysis of data from measurements made on OGO 6 in the polar zone over Alaska (Morgan et al., 1977). In these experiments the calibration of the ion mass spectrometer was improved, and the location L of the satellite was determined more precisely. To sum up, at heights

17.3. Waves generated near ω_L

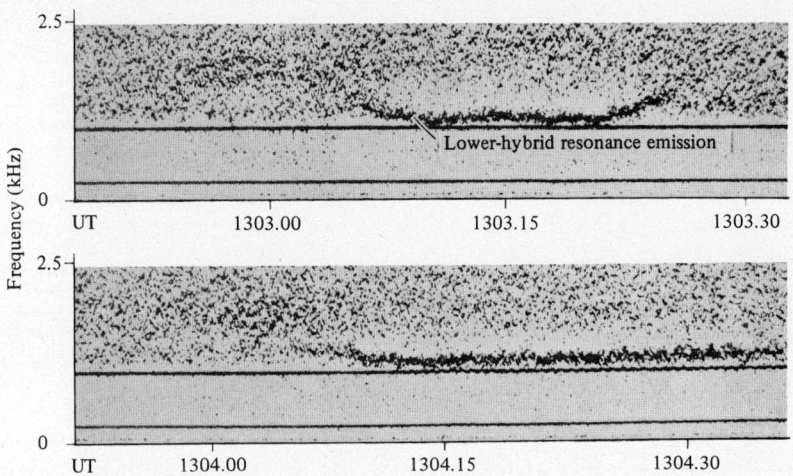

Fig. 17.15. Narrow-band wave packets generated at the lower-hybrid frequency in the outer ionosphere; $z = 16 \times 10^3$ km, OGO 5, 15 May 1969 (Scarf et al., 1971b).

$z \simeq 700$ and 935 km in the interval of L from 2 to 5, values of ω_L (see (4.69)) were obtained from the measured densities of H^+, He^+, N^+, and O^+ ions, and these were found to agree with the resonance frequency ω_L determined from the spectrograms with an accuracy of $\pm 10\%$.

On OGO 5, at greater heights than those explored by OGO 2 and OGO 4, narrow-band emission at the lower-hybrid resonance frequency was observed (Scarf et al., 1971b). Fig. 17.15 shows the corresponding spectrograms; these were obtained from observations at a distance $R \simeq 2.55 R_0$ from the Earth's centre. It was emphasized by Scarf et al., 1971b, that these oscillations were considerable only for reception with an electric antenna; they were not recorded if a magnetic antenna was used. The magnetic field H_0 and the ion density N^+ were measured simultaneously on OGO 5, making it possible to compare the density $N_i \simeq N_e$ obtained from the value of ω_L (the plasma consisted essentially of electrons and protons) with the N^+ value measured directly. For the case shown in Fig. 17.15 values of $N_i \simeq 59$ to $76 \, \text{cm}^{-3}$ were obtained from ω_L (with a maximum spread of 37 to $88 \, \text{cm}^{-3}$), while direct measurements gave $N^+ \simeq 66$ to $88 \, \text{cm}^{-3}$. The agreement between these sets of data is thus very good. Note, too, that emission at the lower-hybrid resonance frequencies was also observed on Injun 5 and OVO 3. For some other experimental and theoretical results pertaining to the vicinity of the lower-hybrid frequency, see also Beghin & Debrie, 1972; Hamelin & Beghin, 1976.

Intense plasma emission in the lower part of the LF range of the resonance branch $\omega_L(\theta)$ (see Fig. 4.7, Volume 1) was observed during one of the most powerful magnetic storms of the solar cycle on 1 November 1968 on OGO 5, while crossing the boundary of the polar cusp at a geocentric distance $R \simeq 3.2R_0$ (or $L = 4.7$) (Scarf et al., 1972). The emission of the hot component of the plasma ($E_e >, \gg 50\,\text{eV}$) was especially strong; it was generated at frequencies $f > f_L$ and F_0. During periods when the characteristic frequencies of the hot plasma component $f_L \simeq F_0$ were of the order of 1 kHz, some especially intense waves at $f = 3\,\text{kHz}$ were recorded in the cusp, their maximum field intensity being $E_{max} \simeq 33\,\text{mV m}^{-1}$. This corresponded to an energy ratio in a narrow frequency band $\varepsilon_0 E^2/2NkT \simeq 3 \times 10^4$, a value which is rare even for the vicinity of a shock front!

Waves at the frequencies $f \simeq f_L, F_0$, corresponding to cold plasma oscillations ($E_e < 10\,\text{eV}$), were less intense. For instance, at frequencies $f_L < f = 1.3\,\text{kHz} < F_0$, a sudden burst of oscillations of the hot plasma component ($E_e >, \gg 50\,\text{eV}$) was observed under conditions when $f_L \simeq F_0 \simeq 400\,\text{Hz}$ and $f_L \simeq F_0 \simeq 500$ to 800 Hz. A similar kind of emission was recorded at geocentric distances $R \simeq 4.5 R_0$ to $5R_0$ during repeated crossings of the boundary of the polar cusp under conditions where $f_L \simeq 370\,\text{Hz}$ and $F_0 \simeq 360$ to 560 Hz. During this period the intense broad-band oscillations corresponded to frequencies $f = 560\,\text{Hz}$ and 1.3 kHz. They took the form of packets of electrostatic waves, cut off close to the lower-hybrid frequency. In some cases this emission was also cut off at the Langmuir frequency of the ions Ω_0. Scarf et al., 1972, concluded, however, that the plasma wave generation in these experiments was associated with the lower-hybrid resonance frequency rather than with ion sound. When the boundary of the polar cusp was crossed, the situation changed somewhat: only narrow bands of oscillations at the lower-hybrid frequency ω_L were recorded.

17.4. Auroral hiss (AH). Saucer-shaped and V-shaped emissions

In the outer ionosphere various kinds of emissions are observed, and these often lie in the VLF range ($\omega \lesssim \omega_L$). In the altitude interval $z \simeq 700$ to 2500 km, systematic studies of hiss emission were carried out in the experiments on Injun 5 (Mosier & Gurnett, 1969a, 1972; Mosier, 1970; Gurnett et al., 1971; Gurnett & Frank, 1972a, 1973). The main kinds of wave spectra recorded on Injun 5 will be presented below. In these experiments observations were made using both magnetic and electric antennas, and the direction of arrival of the radiation was determined.

17.4. Auroral hiss. Saucer-shaped and V-shaped emissions

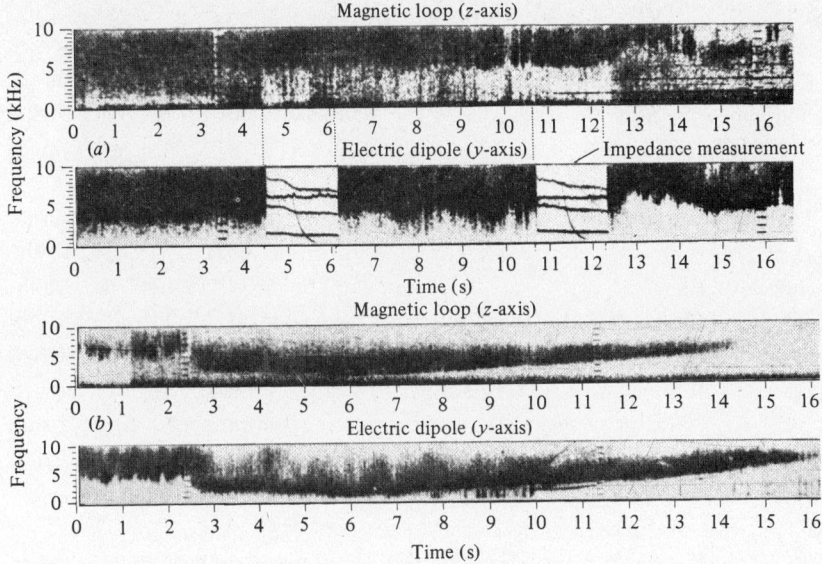

Fig. 17.16. Spectrograms of wide-band packets of VLF waves typical of the polar zone (Injun 5): (*a*) $z = 1172$ km, 24 Jan. 1969; (*b*) $z = 1691$ km, 24 Jan. 1969 (minimum at $\phi = 70°$).

In the polar zone broad-band wave spectra known as *auroral hiss* (AH) were detected (Fig. 17.16), the frequency ranges of which varied markedly from case to case. The lower cutoff frequency of these wave packets was found to change with the magnetic latitude, being a minimum at $\phi \simeq 70°$. At middle latitudes, on the other hand, outside the auroral zone, narrow-band hiss spectra such as those shown in Fig. 17.17 were usually observed during these experiments.

The auroral hiss has a very wide frequency range: $f = 1$ kHz to 10 kHz,

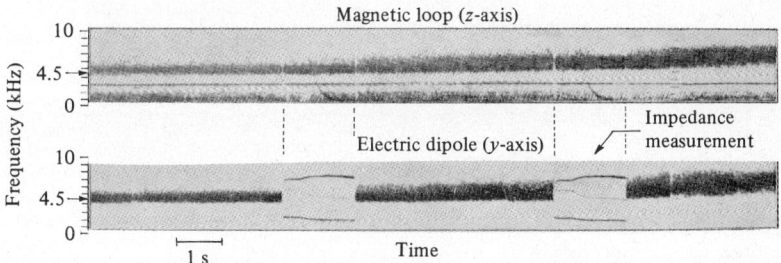

Fig. 17.17. Packet of narrow-band VLF waves typical of mid-latitudes (Injun 5, 8 May 1969).

lying below the local value of the electron gyrofrequency f_H. Consequently, AH waves correspond to the whistler mode $n_2(\omega)$ (see Fig. 4.3). They are very intense; in a wide band their electric field often exceeds $1\,\mathrm{mV\,m^{-1}}$. The emission appears primarily in the vicinity of the auroral oval, in the latitude interval $\phi \simeq 80$ to $70°$, being recorded in a narrow interval of polar latitudes not exceeding $\Delta\phi \simeq 5$ to $10°$, mainly at night.

In these experiments it was shown that auroral-hiss waves are excited in the outer ionosphere by low-energy electrons $E_e \simeq 100\,\mathrm{eV}$ to $1\,\mathrm{keV}$, the fluxes of which $J_e \simeq 10^4$ to $10^7\,\mathrm{el\,cm^{-2}(s\,sr\,eV)^{-1}}$ (Gurnett & Frank, 1972a). The generation mechanism of the auroral hiss is apparently coherent; basically the emission is of the Cerenkov type, but the waves appear to be amplified in some way. Presumably the emission is caused by electrostatic instability of the plasma; during the transformation process the transverse waves are enhanced by the electron streams (see Gurnett & Frank, 1972a, and James, 1973).

Auroral-hiss waves were observed in the VLF range $F_H < f \simeq 0.3$ to $10\,\mathrm{kHz} < f_L$; during the daytime they sometimes exhibited a definite pulsed structure, varying rapidly with time over the entire frequency range. The time intervals between these pulses were about $1\,\mathrm{s}$ (Fig. 17.18). In the evening the traces made by these waves on the spectrograms proved to be similar to the letter V (V-shaped emissions). The traces of this emission appear in green on the original colour spectrograms, reproduced in black and white in Fig. 17.19 (three wave packets). VLF hiss waves generally propagate downward along the geomagnetic field; they arrive from a

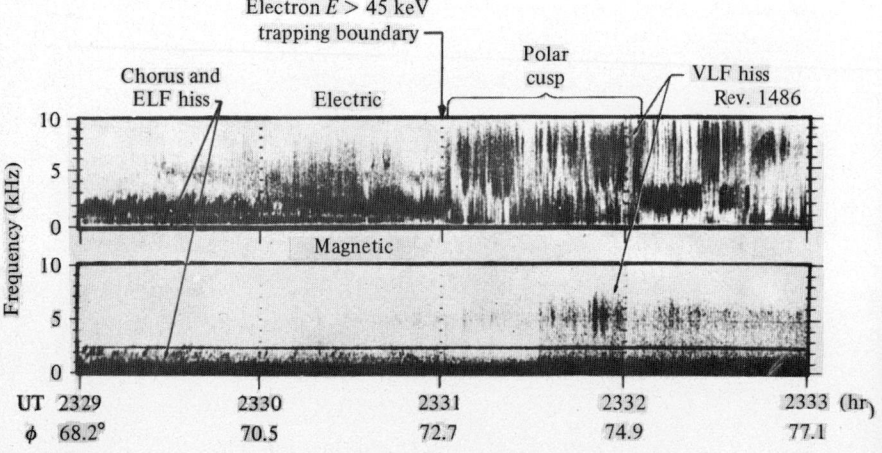

Fig. 17.18. Spectrogram of auroral-hiss waves, Injun 5.

17.4. Auroral hiss. Saucer-shaped and V-shaped emissions

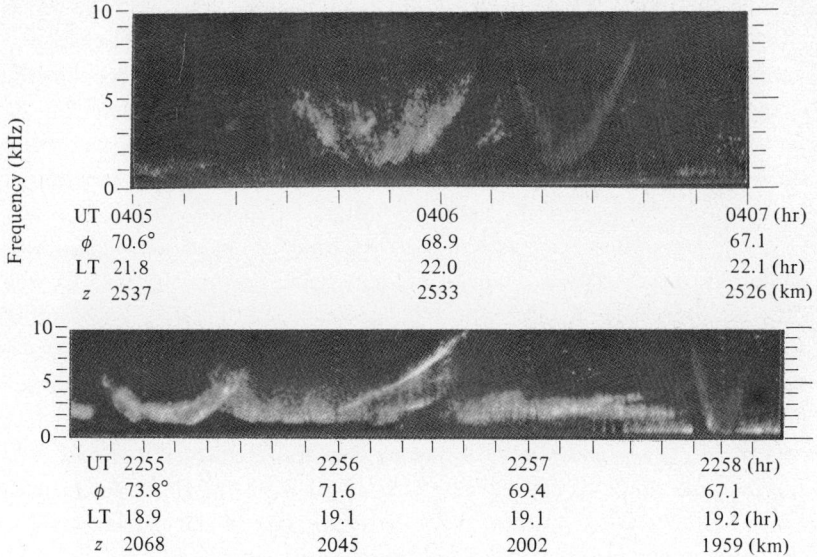

Fig. 17.19. Spectrograms of V-shaped emissions and bands of hiss waves (Injun 5). See Gurnett & Frank, 1972a, for the original versions in colour.

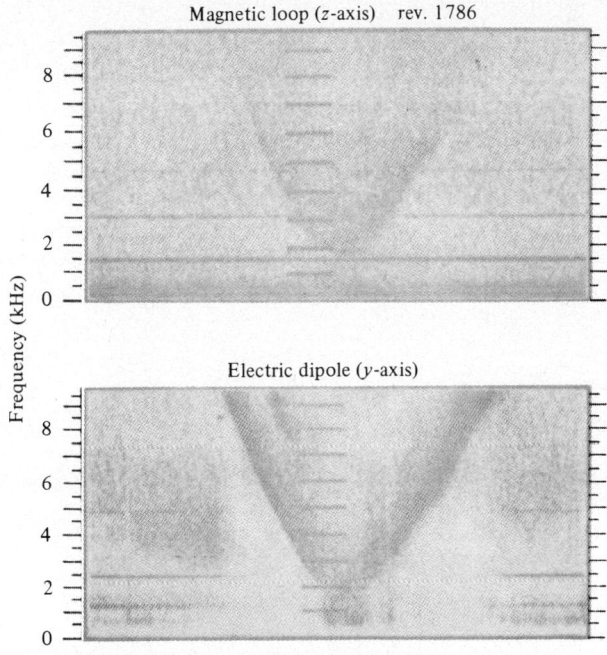

Fig. 17.20. Spectrograms of VLF saucer-shaped emissions; Injun 5, $z = 2530$ km, 2 Jan, 1969 (Mosier & Gurnett, 1969a).

source situated over the observation point. The field strength of the waves recorded on the spectrogram in Fig. 17.19 $E \simeq 4\,\text{mV}\,\text{m}^{-1}$, while their flux varied within the limits 10^{-12} to $10^{-11}\,\text{W}\,\text{m}^{-2}\,\text{Hz}^{-1}$. At the same time the electron flux (energies $E_e \simeq 10\,\text{eV}$ to $1\,\text{keV}$) reached values $T_e \simeq 10^9\,\text{cm}^{-2}(\text{s}\,\text{sr})^{-1}$. The lifetimes of the packets of V-shaped waves during the evening were of the order of 1 min; they were recorded as a function of latitude over a range of the order of 500 km.

In the above-described experiments on Injun 5, emissions of parabolic shape, known as saucer-shaped emissions (SSE), were observed as well (Fig. 17.20) (Mosier & Gurnett, 1969a; see also Smith, 1969). This kind of wave packet always approaches a satellite from below, being generated under the observation point. On the original colour spectrogram of this emission (Gurnett et al., 1971), which is reproduced in black and white in Fig. 17.21, the red colour indicates waves arriving from below. The symmetrical shape of this wave packet is explained by Mosier & Gurnett, 1969a, as an effect connected with the properties of the propagation of electromagnetic waves in the frequency range in which this signal is observed. Part of the wave packet is cut off beyond the surface of the resonance cone, the axis of which lies along the magnetic-field vector $\mathbf{H}_0$, while the minimum signal frequency is equal to the lower-hybrid frequency close to the source. Two wave packets of SSE were also recorded on the colour spectrograms of Fig. 17.19 (see above), at the same time as the auroral hiss and V-shaped waves were observed (green colour on the original spectrogram) which propagated downward relative to the observation point. The maximum field intensity of the waves recorded in

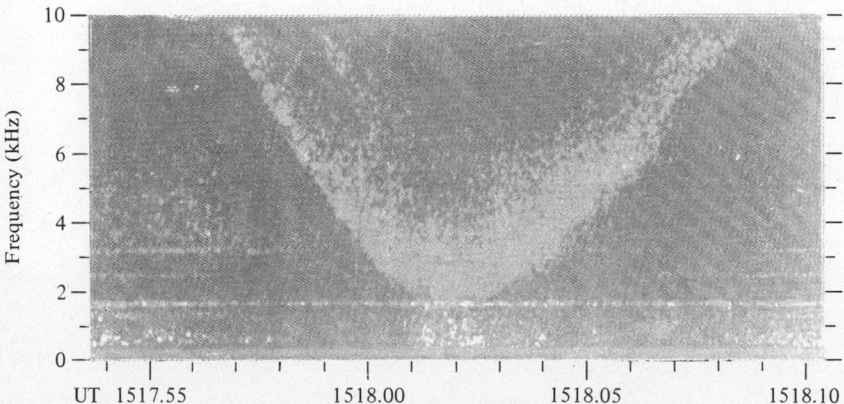

Fig. 17.21. Spectrogram of VLF saucer-shaped emissions, Injun 5, $z = 2539$ km, 2 Jan. 1969. (See Gurnett et al., 1971, for the original version in colour.)

17.4. Auroral hiss. Saucer-shaped and V-shaped emissions

Fig. 17.19 was $E \simeq 1.7\,\text{mV}\,\text{m}^{-1}$, while the flux was approximately $4 \times 10^{-12}\,\text{W}\,\text{m}^{-2}\,\text{Hz}^{-1}$. One of the special features of saucer-shaped emission is its short duration Δt, which is usually only $\Delta t \simeq 10\,\text{s}$.

An analysis of the data obtained with Injun 5 indicated that the source of auroral hiss was situated at a height $z \simeq 5 \times 10^3$ to 10×10^3 km, while the source of the SSE was at a height $z \simeq 1400$ km (Lasch, 1969; Gershman, 1953).

More recently, saucer-shaped emission was investigated on the basis of data obtained with ISIS 2 and Alouette 2 (James, 1976). These results indicated that the source of these waves apparently has linear dimensions of $\rho_x \simeq 0.5$ km in the horizontal and $\rho_z \simeq 10$ km in the vertical directions. It is located in the ionosphere at heights $z \simeq 1000$ km and it emits a wave spectrum in the range 0.1 to 20 kHz. The data of this study and of those cited above imply that SSE is an enhanced coherent Čerenkov radiation of the plasma. It is stimulated by beams of suprathermal electrons with energies not exceeding 5 eV, which promote instability of the thermal plasma electrons (see §15.3)

Auroral hiss was detected in the polar zone over a very wide frequency range on OGO 6 as well, at heights $z \simeq 500$ to 1000 km. The spectra of these waves varied considerably as a function of the time of day and of the location of the observation point. This is associated with a variation in the energy spectra and in the angular distribution of the electrons responsible

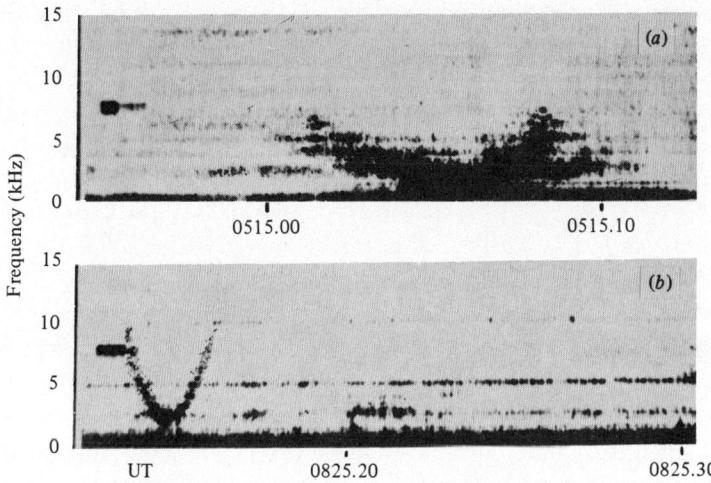

Fig. 17.22. Spectrograms of V-shaped emissions (a) and of saucer-shaped emissions (b), recorded in the polar ionosphere on OGO 6, 15 Dec. 1969, $z = 1000$ km, $\phi = 70°$ (Laaspere & Johnson, 1973).

for this emission. In these experiments it was demonstrated, as in the Injun 5 experiments, that the *polar* hiss is generated by low-energy electrons of energies $E_e \simeq 0.01$ to 1 keV in the range of geomagnetic latitudes $\phi \simeq 70°$ to $78°$ (Laaspere & Johnson, 1973).

The spectrograms in Fig. 17.22 show hiss, V-shaped, and saucer-shaped emissions, the source of which was located below the observation point; these were recorded on OGO 6. Conditions favourable for generating this emission existed in these experiments in a number of cases for a few hours. The minimum height at which SSE was recorded on OGO 6 was of the order of 930 km. This means that the source was located at this height and above it.

The behaviour of hiss emission was investigated during a series of magnetic storms in 1967 (25–27 May, 5–7 June, 19–21 Sept.) on the basis of observations at $f \simeq 3.2$, 9.6, and 16.0 kHz on Ariel 3 and simultaneously at the Earth's surface (Hayakawa *et al.*, 1975a, b; 1977). It was found in these experiments that in the mid-latitudes hiss appears soon after the onset of the main phase of the storm in the morning sector of the magnetosphere. In the evening sector it is observed only during the storm phase known as the 'recovery phase'. The evolution of the hiss emission is attributed to the development of gyroresonant instability, which appears in the electron ring current set up during this period. On ISIS 2, narrow-band hiss with a central frequency $f \simeq 3.1$ to 3.6 kHz was also recorded during the main phase of the storm on 9 Aug. 1972 in the outer ionosphere at a height $z = 1390$ km, between $L \simeq 3$ and 3.2 ($\phi \simeq 54°$ to $56°$). It was similar to the hiss that is excited beyond the plasmapause during periods of low geomagnetic activity (Ondoh *et al.*, 1974). On Alouette 2 narrow-band emission at $f \simeq 5$, 8, and 16 kHz was detected during a magnetic storm on 26 Sept. 1971. In these experiments hiss was recorded during quiet periods (Ondoh & Murukami, 1975).

Bands of chorus oscillations, which are known to consist of successive oscillation bursts of various forms, are observed in the ionosphere, but not too often. The very intense radiation of this type which was recorded on OGO 6 in the polar zone during a period of heightened activity is shown in Fig. 17.23 (Laaspere & Johnson, 1973). These waves were observed with OGO 6 at heights $z = 400$ to 1000 km in a frequency range above the lower-hybrid frequency. They should thus, strictly speaking, be called LF waves. Note that the spectrograms in Fig. 17.23, on which the chorus bands were recorded through different filters, actually represent continuous emission over the entire frequency range $f \simeq 6$ to 25 kHz.

17.5. Trapping of VLF waves. Non-ducted waves

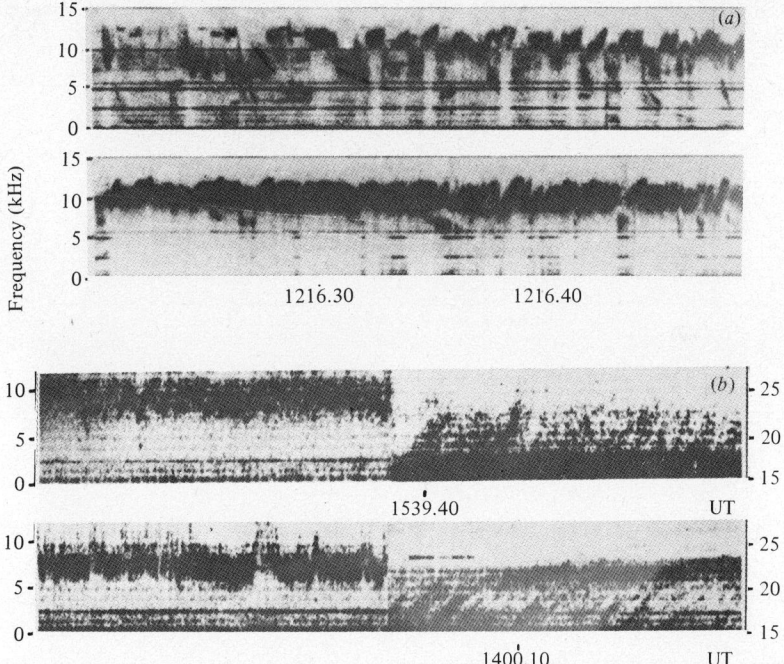

Fig. 17.23. Spectrograms of chorus waves of different types, recorded in the polar ionosphere on OGO 6 at heights $z = 420$ km (a) and 830 km (b), on 29 Sept. 1969 and 18 Aug. 1970 (Laaspere & Johnson, 1973).

17.5. Trapping of VLF waves in the ionosphere and in the magnetosphere. Non-ducted waves

In the Alouette experiments, after the detection of short fractional-hop whistlers which passed once from the Earth's surface to the observation point, a second set of such signals was recorded in the ionosphere which had passed through the latter region three times, having been reflected once in the ionosphere above the satellite and once at the bottom of the ionosphere (Barrington & Belrose, 1963). Detailed studies of this phenomenon were first carried out with Aerobee rockets at altitudes $z \simeq 100$ to 200 km and with Alouette at $z \simeq 1000$ km (Carpenter et al., 1964). These studies showed that the repeated signals travel up and down in the ionosphere, being reflected from the bottom of the ionosphere at $z \simeq 100$ km and from above the satellite below the height region in which protons become the main ionic component. Thus it was shown that this phenomenon can be detected only with the aid of satellites, because the

150   *Results of studies of VLF waves*

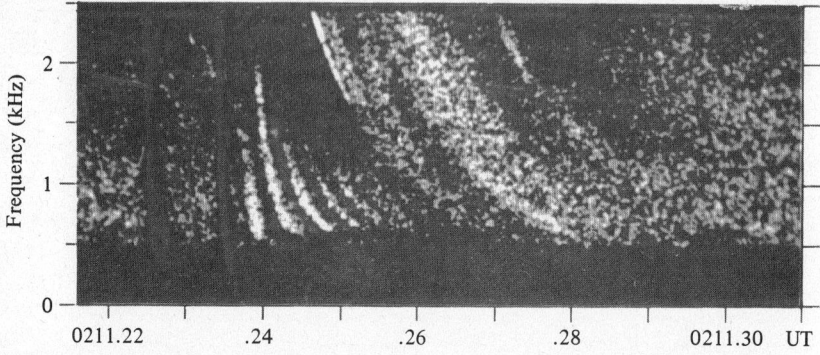

Fig. 17.24. Colour spectrogram of short electron whistlers 'trapped' in the protonosphere (subprotonospheric whistlers); $z = 724$ km, 5 Apr. 1969. (See Gurnett *et al.*, 1971, for the original version in colour.)

wave packet is trapped in the ionosphere beneath the protonosphere. These signals are therefore known as subprotonospheric whistlers. Fig. 17.24 shows a series of alternating short whistlers, signals that have been 'trapped' in the protonosphere, observed on Injun 5 at a height of 724 km. On the original colour spectrogram (Gurnett *et al.*, 1971), red (upward) and green (downward) signals were recorded alternately. After this group of signals, the figure shows a diffuse whistling atmospheric consisting of a mixture of upward and downward components. This is typical of whistlers; they are scattered strongly by inhomogeneities in the ionosphere, creating multiple propagation paths for the wave packets from the source to the satellite.

This trapping of short whistlers in the above-indicated region of the ionosphere can be attributed to the influence of ions on the whistler trajectories, when the spectra lie in a range of ω lower than the lower-hybrid frequency ω_L of the corresponding plasma region. Under such conditions waves can be propagated at any angle, even normal to the direction of the magnetic field $\mathbf{H}_0$ (Hines, 1957). Consequently, under certain conditions they are 'reflected' from the boundary of the region in which they are being propagated, and are not able to leave it (Smith, 1964; Kimura, 1966; Thorne & Kennel, 1967). The trajectories of non-ducted waves appearing due to this effect, which were discussed previously in Chapter 8, explain the following phenomena observed in the near-Earth plasma:

subprotonospheric whistlers (Barrington & Belrose, 1963; Carpenter *et al.*, 1964);

transverse whistlers (Carpenter & Dunckel, 1965; Kimura *et al.*, 1965);

magnetospherically reflected whistlers, and their alter ego, *v* whistlers

17.5. Trapping of VLF waves, Non-ducted waves

(Smith & Angerami, 1968; Edgar, 1976a);

walking-trace whistlers (Walter & Angerami, 1969);

whistler precursors (Helliwell, 1965; Laaspere & Wang, 1968; Dowden, 1972; Reeve & Rycroft, 1976);

various other kinds of VLF waves generated in the near-Earth plasma, and at the plasmapause in particular (see Aikyo & Ondoh, 1971; Aikyo et al., 1972).

A complete treatment of the trajectories of these waves would necessitate taking into account a number of the properties of the plasma through which they propagate, and in particular the effect of horizontal gradients of the ionosphere. The large angles between the wave vector and the magnetic field required for wave reflection from the bottom of the protonosphere ($z \simeq 10^3$ km) to occur can apparently be attributed to the effect of horizontal gradients of the electron density (these must be quite large). Moreover, the refractive index for transverse propagation must decrease with height (Smith, 1964). A number of traits of these waves, evident, in particular, in Fig. 17.24, have not yet been explained. For example, why is there a gradual decrease in the height of reflection of signals above the satellite, i.e., a reduction of the delay time between direct and reflected signals, as well as a systematic change in the upper and lower cutoff frequencies of these signals? To appreciate these points, the reader is invited to examine the colour spectrograms in the publications cited; the direct signals appear in red, and the reflected in green.

The propagation of waves transverse (i.e. perpendicular) to the Earth's magnetic-field vector explains the properties of the so-called *transverse whistlers*, first detected on Alouette 1 (Carpenter & Dunckel, 1965). At frequencies $f \simeq 1$ to 8 kHz, which in these experiments were below the lower-hybrid frequency in the ionosphere, the dependence of frequency on time for this type of whistler was more gently sloping than for whistlers observed at the Earth's surface. The additional delay time Δt reached 0.22 as the magnetic latitude increased from 30° to 44°. This effect can be explained, however, if we assume that on part of its path the whistler followed different trajectories, sometimes travelling transverse to the magnetic field (Kimura et al., 1965). On the other hand, a quantitative comparison with experiment will be possible only if the transverse wave propagation takes place at least over distances of at least several hundred kilometres. To arrive at the corresponding model of the ionosphere, Kimura et al., 1965, used the condition of purely transverse propagation (see Hoffman, 1960), as expressed by the formula

$$n_{\perp g}^2 = (A\cos^6 \phi_0)F(\cos \phi_0/R^{1/2})\,\omega^4 \qquad (17.5)$$

where the group refractive index for transverse propagation is

$$n_{\perp g}^2 = (\omega_0^2/\omega_H^2)M_{eff}. \tag{17.6}$$

The latter formula follows from the expression for the transverse refractive index:

$$n_\perp^2 = [\omega_0^2/\omega_H^2(1 - \omega^2/\omega_L^2)]M_{eff}. \tag{17.7}$$

In (17.5), ϕ_0 is the geomagnetic latitude (the Earth's magnetic field is assumed to be a dipole field), R is the radial distance from the Earth's centre, A is a constant, and F is an arbitrary smooth function.

Finally, (17.7) implies that the distribution of the electron density in the ionosphere is described by the formula

$$N_e = \frac{A}{4\pi e^2}\left(\frac{m}{M_{eff}}\right)(\cos^6\phi_0)F\left(\frac{\cos\phi_0}{R^{1/2}}\right)\omega^4. \tag{17.8}$$

Estimates made by Kimura et al., 1965, indicated that the actual conditions in the ionosphere lead to a good quantitative fit with the experimental values of Δt.

Magnetospherically reflected (MR) whistlers (Fig. 17.25), first detected with OGO 1 (perigee 280 km, apogee 149 400 km, Smith & Angerami, 1968), are trapped in the higher region of the outer ionosphere at heights $z > 1500$ to 2000 km, which is often considered to be a part of the

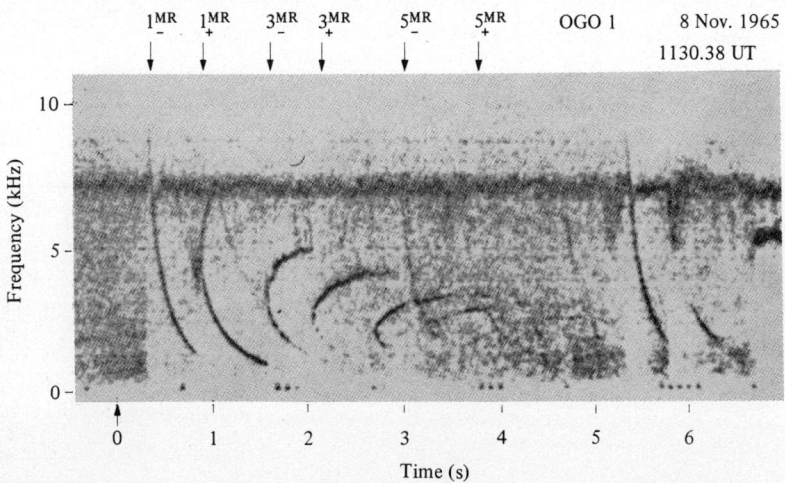

Fig. 17.25. Magnetospherically reflected whistlers, 'trapped' in a region of the ionosphere with its upper boundary at heights of around 10 000 km or more. Signals moving upward are denoted by a minus sign, signals moving downward by a plus sign; OGO 1, 8 Nov. 1965 (Smith & Angerami, 1968).

17.5. Trapping of VLF waves. Non-ducted waves

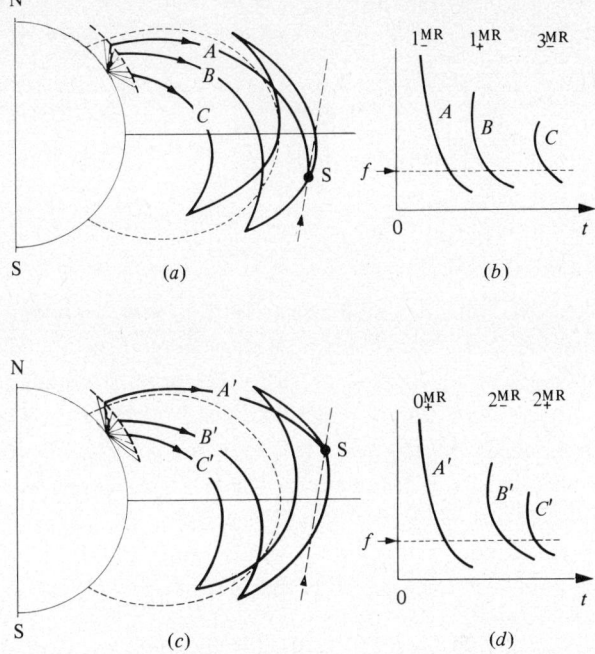

Fig. 17.26. Possible propagation trajectories of MR whistlers.

magnetosphere. Actually, however, this type of whistler is trapped in the outer ionosphere (the plasmasphere).

The propagation of MR whistlers in the outer ionosphere, as mentioned previously, is described adequately by the trajectories depicted in Figs. 8.14 and 8.15 of Volume 1, and Fig. 17.26. There, for clarity, the same notation is used as in Fig. 17.25. The nature of the trajectories of these whistlers changes when they cross from the Southern Hemisphere to the Northern Hemisphere. The various idiosyncrasies of MR whistlers are difficult to account for, however, since they depend on a number of the parameters of the plasma region through which the waves are being propagated. Recently, detailed theoretical calculations of their trajectories have revealed the nature of their frequency cutoffs (Edgar, 1976a, b).

The waves known as v whistlers (nu whistlers), first detected with OGO 1, are also interesting. Two such whistlers appear on the spectrogram in Fig. 17.27, recorded at a height $z \simeq 10\,000$ km. These signals get their name from their similarity to the Greek letter v. It is characteristic of these that they are cut off at some minimum frequency ω_{min}. Waves with frequencies $\omega > \omega_{min}$ are apparently cut off above the satellite. On the other hand, a

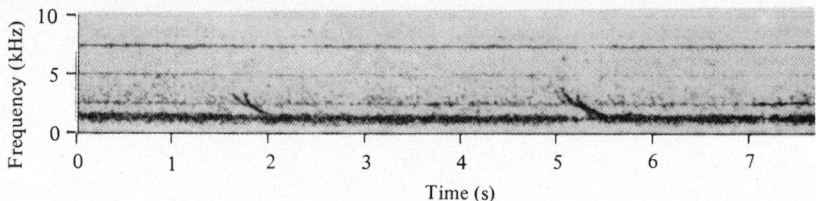

Fig. 17.27. Spectrogram of v whistlers; OGO 1, $z = 10\,000$ km, 17 Oct. 1964 (Smith & Angerami, 1968).

v whistler has two branches, since waves with $\omega > \omega_{min}$ are cut off in the ionosphere below the observation point (satellite). The latter is possible if the lower-hybrid frequency ω_L close to the satellite is higher than the frequency ω_{min} at which the two signals are cut off. Studies of the causes of v-whistler cutoff, by the way, led to some interesting conclusions concerning the properties of the energy spectra of electrons with energies $E_e \simeq 10\,\text{keV}$ (Thorne, 1968).

In later experiments with OGO 5 (Scarf et al., 1971b), just as in the OGO 1 experiments, together with the described MR and v whistlers, other kinds of signals were detected as well, which were not guided by the Earth's magnetic field, being trapped in limited regions of the plasmasphere. More recently, detailed studies of a number of properties of unguided waves at $f = 16.8\,\text{kHz}$ were carried out with the French satellite FR 1 (Cerisier & James, 1970; Cerisier, 1973, 1974). Experiments over a wide frequency range ($f \simeq 1$ to $8\,\text{kHz}$) with the Japanese rocket K 9M 26 also made it possible to investigate a number of properties of these waves (Hayakawa, 1974).

17.6. Broad-band emission of VLF waves. 'Lion's roar' LF waves

We will describe here briefly an emission of the near-Earth plasma of the mixed type, in the sense that it pertains both to the VLF range and to the LF range, and at the same time even to ELF oscillations at the ion gyrofrequencies.

In the vicinity of the geomagnetic equator ($\phi \simeq \pm 10°$), in the *plasmasphere* at geocentric distances $R \simeq 2.5 R_0$ to $5.0 R_0$, *broad-band* noise emission of an unusual kind in the range $f \simeq 10$ to 10^2 Hz was often picked up by OGO 3 (Russell et al., 1970), IMP 6, and Hawkeye 1 (Gurnett, 1976). After it first appeared, this emission persisted for several weeks, being observable at any time of day. The maximum field strengths were usually $H \simeq 20\,\text{m}\gamma$ and $E \simeq 200\,\mu\text{V m}^{-1}$ (Gurnett 1976). The spectrum of

17.6. Broad-band emission of VLF waves. 'Lion's roar' waves

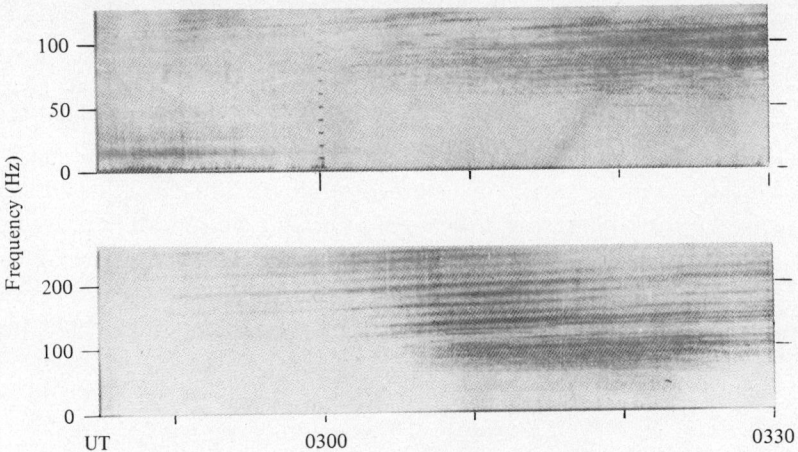

Fig. 17.28. Spectrograms of a broad-band emission with a line structure, observed in the vicinity of the geomagnetic equator in the plasmasphere; IMP 6, 1972 (Gurnett, 1976).

this emission is unusual in that it has a linear structure, with a complicated superposition of narrow oscillation bands having widths of a few hertz or less.

An analysis of spectrograms of these waves (Fig. 17.28) showed that in a number of instances the emission bands were spaced in frequency at intervals $\Delta f \simeq F_H(H^+)$ equal to the proton gyrofrequency. During the measurement period corresponding to the spectrogram in Fig. 17.28, the interval $\Delta f \simeq F_H \simeq 13.1$ to 15.7 Hz. However, in addition to the bands with $\Delta f \simeq 16$ Hz, during different time intervals bands with $\Delta f \simeq 2$ Hz and 4 to 5 Hz were also recorded. An analysis of the experimental data led Gurnett, 1976, to conclude that this emission stems from a gyroresonant interaction with the plasma in the region of equatorial ring currents of energetic protons, α particles, and other heavy ions present in the magnetosphere.

In the *magnetosheath* region of the near-Earth plasma, on OGO 1 at distances $R \simeq 12 R_0$ to $17 R_0$ broad-band emission that apparently consisted of a packet of transverse waves was detected with a loop antenna over a wider frequency range with $f \simeq 3$ to 300 Hz (Smith et al., 1967). Under the conditions of these experiments the lower boundary of the frequency range was of the order of $F_H(H^+)$, while the upper was much higher than the lower-hybrid frequency f_L, already being of the order of the electron gyrofrequency f_H. Consequently, in these experiments a continuous spectrum of VLF and LF waves was observed. Their intensities varied with frequency as f^{-3} over the following limits: $H^2 \simeq 5 \times 10^{-8}$ $\gamma^2\,\text{Hz}^{-1}$ to $10^{-1}\,\gamma^2\,\text{Hz}^{-1}$.

In the magnetosheath at distances from the Earth $R \simeq 10R_0$ to $15R_0$ another type of emission was detected with OGO 5. This took the form of intense discrete wave packets with $f \simeq 100$ to 200 Hz, observed primarily as bursts during magnetically disturbed periods (Smith *et al.*, 1969). Later such ELF wave packets were detected at heights $z \lesssim 300$ km in the polar ionosphere with Injun 5 (Gurnett & Frank, 1972b). According to established tradition, they were called by Smith *et al.*, 1969, ELF noise of the *lion's roar* type, which is what they sound like. Fig. 17.29 shows

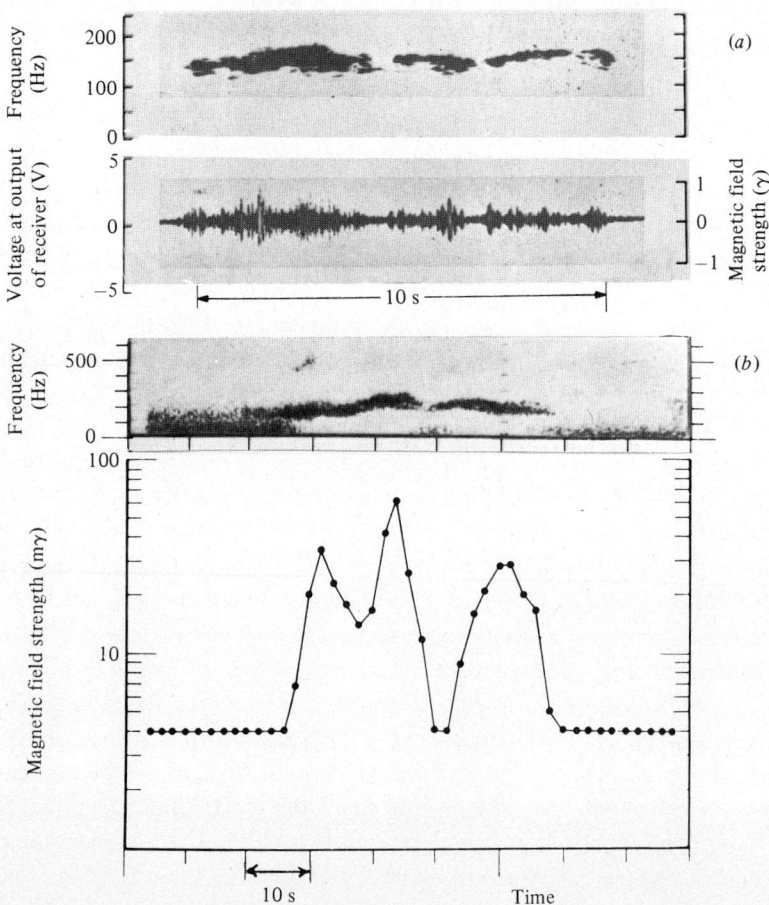

Fig. 17.29. Narrow-band ELF hiss (lion's roar): (*a*) OGO 5, magnetosheath, $R \simeq 12R_0$ to $15R_0$, 12 March 1968 (Rodriguez & Gurnett, 1971); (*b*) Injun 5, outer ionosphere, $z < 3000$ km, polar zone, 9 Dec. 1968 (Gurnett & Frank, 1972b).

17.6. Broad-band emission of VLF waves. 'Lion's roar' waves

recordings of this emission obtained with OGO 5 and Injun 5.

On OGO 5, lion's roar waves were picked up with a magnetic antenna (loop) in the range $f \simeq 90$ to $160\,\text{Hz}$; the central frequency of their spectrum $f_e \simeq 120\,\text{Hz}$. Later detailed studies of the properties of this emission (see Smith & Tsurutani, 1976) showed that the field amplitude H, the frequency f_c, the duration of the emission Δt, and the time interval Δt_0 between bursts all varied mainly within the following limits:

$$H \simeq 40 \text{ to } 160\,\text{m}\gamma,\ H = 85\,\text{m}\gamma;\ f_c \simeq f_H/2$$

$$\Delta t \simeq 4 \text{ to } 6\,\text{s},\ \overline{\Delta t} \simeq 1.6\,\text{s};\ \Delta t_0 \simeq 2 \text{ to } 14\,\text{s},\ \overline{\Delta t_0} \simeq 5\,\text{s}.$$

Lion's roar emission is actually made up, according to Rodriguez & Gurnett, 1971, of packets of transverse electromagnetic waves in the whistler mode. Since at the considered distances from the Earth the values of f_c are much higher than the proton gyrofrequency $F_H(H^+)$ and also higher than the lower-hybrid frequency f_L, these are electron VLF and LF waves. They correspond to the frequency range of the resonance branch $\omega_2(\theta)$ of a cold plasma (see Fig. 4.7, Volume 1) or of the branch $\omega_1(\theta)$ of a warm plasma (see Fig. 5.4, Volume 1). According to the data of Smith & Tsurutani, 1976, the propagation direction of these waves in the magnetosheath makes an angle $\theta < 30°$ with the magnetic field $\mathbf{H}_0$.

In the experiments with Injun 5, corresponding emission bands in the range $f \simeq 100$ to $300\,\text{Hz}$ with a width $\Delta f \simeq 100\,\text{Hz}$ were recorded in a narrow latitude region a few degrees in extent in the polar zone at geomagnetic latitudes from 70 to 80°. The lifetimes of these waves varied from a few seconds to some tens of seconds, depending on the event. In these experiments fields E and H were measured simultaneously. The values found were $E \simeq 3$ to $10\,\text{mV\,m}^{-1}$ and $H \simeq 10$ to $30\,\text{m}\gamma$. Gurnett & Frank, 1972b, comparing their data with those of Rodriguez & Gurnett, 1971, concluded that they had recorded the lion's roar wave packets described by Smith et al., 1969, which were generated in the magnetosheath and trapped there in open tubes of force of the Earth's magnetic field $\mathbf{H}_0$. These waves are propagated along $\mathbf{H}_0$, being guided by field-aligned irregularities, and they reach the polar zone at heights $z \lesssim 3000\,\text{km}$. It is assumed, too, that the propagation of the wave packets is purely longitudinal, i.e., it deviates very little from the direction of the force line of $H_0(R, \phi)$.

Consequently, even though the wave frequency f along the propagation trajectory gradually becomes lower than the ion gyrofrequency in the ionosphere, the waves are not cut off. Despite the fact that in these experiments both field components (E and H) were measured, it cannot be proven decisively on the basis of the obtained data that these components satisfy the relation $H = En/\mu_0 c$ describing transverse electromagnetic waves, since sufficiently accurate data on the electron density N and the magnetic field H_0 are lacking. It would be of great interest to investigate further the described type of wave packets and to analyse their spectra, especially under conditions of their simultaneous observation in the transition zone of the near-Earth plasma i.e. the magnetosheath and in the polar zone.

18

Results of studies of LF waves

LF waves correspond to the frequency range $\omega_L \ll$ or $< \omega \lesssim \omega_H$, and they also include the branch of slow electron-acoustic waves in a warm plasma (see Chapter 5, Volume 1). In this chapter we will discuss the results of studies of wave processes in the frequency range $\omega > \omega_L$, the properties of which are determined mainly by electron oscillations. These are whistler-mode waves $n_2(\omega)$, *guided* by the magnetic field $\mathbf{H}_0$.

18.1. Whistling atmospherics

A popular type of whistler-mode waves are the long whistling atmospherics, or electron whistlers, that are observed in magnetically conjugate regions on the Earth's surface or in the ionosphere; they originate from the electromagnetic waves emitted by a lightning flash. A spectrogram of a whistling atmospheric and its $\omega(t)$ characteristic, obtained on Vanguard 3, are shown in Fig. 18.1. In these experiments on Vanguard 3 it was noted that during certain periods 20 whistlers per minute were recorded (Gurnett & O'Brien, 1964). During the same period, instruments

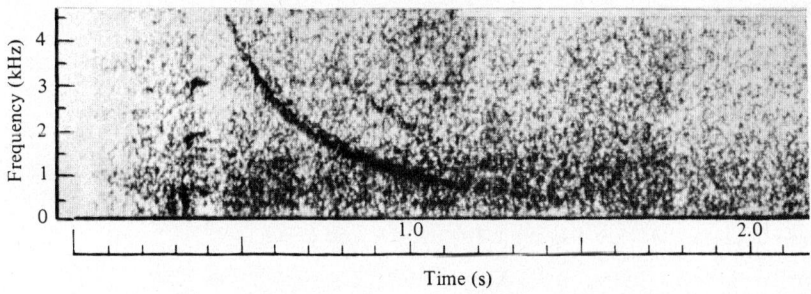

Fig. 18.1. Spectrogram of an electron whistler, or whistling atmospheric; Vanguard 3, $z = 2889$ km, $\phi = 18°$S, $\lambda = 4°$E, 1959 (Gurnett & O'Brien, 1964).

with the same sensitivity at the Earth's surface recorded only three whistlers in the course of 15 hours!

A detailed discussion of whistling atmospherics will not be given here. See Barkhausen, 1919, 1930; Eckersley, 1928, 1929, 1935; Storey, 1953, 1957, 1958, 1962; Helliwell, 1965; Gendrin, 1971; Walker, 1976; Al'pert, 1980. Nevertheless, it is appropriate to mention briefly the following important facts that have been discovered during studies of whistling atmospherics.

(1) Whistling atmospherics led to the discovery and comprehensive study of the 'knee' at the boundary of the outer ionosphere; specifically, the *plasmapause* region was discovered, where the electron density drops in a number of cases about tenfold to a hundredfold in the distance interval $\Delta R \simeq 0.15 R_0$ at distances from the Earth's centre in the equatorial plane $R \simeq 3.0 R_0$ to $5 R_0$ (Carpenter, 1963; Angerami & Carpenter, 1966). Subsequently it was shown that the plasmapause boundary and the electron-density drop are subject to still greater changes (see for instance, Carpenter *et al.*, 1969a).

(2) An analysis of the results of observations of whistling atmospherics established that the process of their guiding is governed to a considerable extent by field-aligned irregularities that arise in the near-Earth plasma along the force lines of the Earth's magnetic field $\mathbf{H}_0$. As a consequence of this, electron whistlers are cut off primarily at the frequency $\omega_c \simeq \omega_{H0}/2$, where ω_{H0} is the electron gyrofrequency in the equatorial plane at the apogee of the whistler trajectory (Carpenter, 1968). This effect was predicted theoretically in studies dealing with the trapping of waves in inhomogeneities extended along $\mathbf{H}_0$ (Smith, Helliwell, & Yabroff, 1960; Smith, 1961). A number of studies have been made of the cutoff of whistling atmospherics at the electron gyroresonance frequency ω_{H0}, which is described by the spatial coefficient of cyclotron attenuation of an electron wave

$$\kappa_{He} = (\pi^{1/2}/2)(c/v_e)[(\omega_H - \omega)/\omega] \exp(-z_{He}^2), \qquad (18.1)$$

where

$$z_{He} = [(\omega_H - \omega)/\omega](c/v_e n_{20}) \gg 1 \qquad (18.2)$$

(see Chapter 5). For the analysis of whistling atmospherics at the apogee of their trajectories, use has also been made of a kinetic correction to the refractive index describing whistler-mode waves. Specifically, taking into account the thermal motion of electrons,

$$n^2 = n_{20}^2 (1 + \delta) = \omega_0^2 (1 + \delta)/\omega(\omega_H \cos\theta - \omega), \qquad (18.3)$$

18.1. Whistling atmospherics

where, for $\theta = 0$ and granted the condition (18.2), we have

$$\delta = \tfrac{1}{2}(v_e/c)^2[\omega\omega_0^2/(\omega_H - \omega)^3]. \qquad (18.4)$$

It turned out that in some cases formulas (18.1)–(18.4) all the same depict quite accurately the results of observations of whistling atmospherics. Results of studies concerning the cyclotron cutoff of whistling atmospherics are to be found, for example, in Scarf, 1962; Liemohn & Scarf, 1962, 1963, 1964; Liemohn, 1965a, b.

(3) An analysis of the findings from the OGO 3 experiments (Angerami, 1970) revealed the following properties of the irregularities that guide whistling atmospherics:

(a) the radial thickness (shell thickness) Δ of the field-aligned ducts varies from $0.035R_0$ to $0.070R_0$, i.e. $\Delta \simeq 200$ to 400 km;

(b) the ducts are spaced radially by $0.017R_0$ to $0.18R_0$ from one another, i.e. 100 to 1000 km apart;

(c) the width ρ of the ducts in the East–West direction at the geomagnetic equator is of the order of $0.3R_0$, i.e. $\rho \simeq 2000$ km;

(d) the irregularities involve enhancements rather than reductions of the electron density; $\delta N_e = N_e - N_0 > 0$, where N_0 is the unperturbed value of the electron density. In other words, they are much more likely to be 'peaks' than 'troughs'.

The above data refer to observations made between about $L = 4.7$ and $L = 4.1$, close to the geomagnetic equator, i.e., $R \simeq (4.1-4.7)R_0$.

(4) During observations of whistling atmospherics at the Earth's surface, the spectrograms revealed narrow bands of waves excited in the near-Earth plasma and stimulated by the whistling atmospherics (Helliwell, 1963). A study of these narrow-band wave packets showed that they are recorded primarily in two frequency ranges and have the following properties (Helliwell, 1965; Carpenter, 1968):

(a) one type of wave is excited in the vicinity of the frequency $\omega_c \simeq \omega_{H0}/2$ of whistling-atmospheric cutoff, where ω_{H0} is the gyrofrequency at the apogee of the magnetic field line along which the whistling atmospheric propagates. The frequency of this wave packet increases very slowly with time; its width is of the order of 50 Hz. The duration of the emission may be as long as 20 s in some cases.

(b) Another type of narrow-band signal is excited near the tail of the whistling atmospherics, at a frequency $\omega \simeq \omega_{H0}/6$. The frequency of this wave packet increases rapidly and markedly with time (rising tone) and has a width of the order of 50 Hz. Sometimes these signals have a more complex form; in other words, they also have a branch of falling frequency (falling tone, hooks).

18.2. Emissions generated at the boundary of the plasmasphere by means of radio waves (artificially stimulated emissions, ASE). Intensification of radio waves in the whistler mode

Shortly after the discovery of narrow-band packets of waves generated, as has been shown, by whistling atmospherics at the apogee of their trajectory above the magnetic equator, it was found that analogous phenomena are observed during the reception of signals from longwave radiotelegraph transmitters. 'Triggered' emissions of this type, emanating from an external source of quasi-monochromatic electromagnetic waves (Fig. 18.2), were detected for the first time during the reception of telegraph signals from the NAA transmitter ($f = 14.7$ kHz, transmitted power $W_0 = 1000$ kW) and the NPG transmitter ($f = 18.6$ kHz, $W_0 = 200$ kW) (Helliwell et al., 1964; see also Helliwell, 1965). Later, during observations of signals from the Omega transmitter ($f = 10.2$ kHz) it was shown that these artificially stimulated emissions (ASE) are also excited for lower transmitter powers ($W_0 \simeq 10^{-1}$ kW), when sufficiently long dashes of Morse code are sent (Kimura, 1968). In earlier experiments it had been found that triggered emission is observed more frequently when the dash duration $\Delta t \gtrsim 150$ ms. Moreover, it is more varied in character than for the transmission of dots ($\Delta t \lesssim 50$ ms). During the reception of dashes, narrow bands of triggered ASE wave packets were recorded, the frequency of which increased or decreased with time (risers and fallers; rising and falling tones) or had the form of hooks (also called branching spectra). During the reception of dots, on the other hand, very weak ASE only of the faller type were recorded (Lasch, 1969). The early experiments had also ascertained that triggered emission is observed more often when the transmitter frequency $\omega_0 \simeq \omega_{H0}/2$, specifically when it differs from the minimum value of $\omega_{H0}/2$ along the propagation trajectory of the whistler-mode radio waves, (i.e. the value at its highest point) by only a few percent

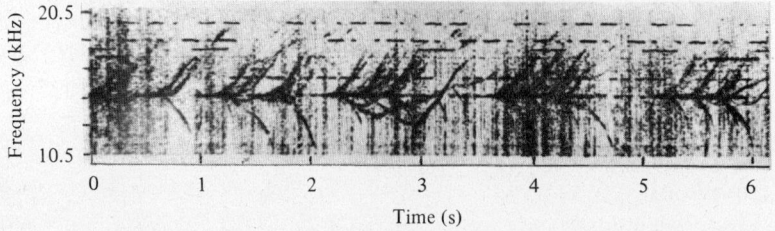

Fig. 18.2. Spectrogram of artificially stimulated emissions (ASE). Oscillations (rising tones and falling tones) stimulated by radiotelegraph signals transmitted on 19 Oct. 1962 at 14.7 kHz (Helliwell, 1965).

18.2. Artificially stimulated emissions

(Carpenter, 1968; Carpenter et al., 1969b). During this same period a possible theoretical explanation of triggered emission was considered for the first time (see, for instance, Helliwell, 1967; Kimura, 1967; Matsumoto, 1972).

However, it was only during the last few years that extremely systematic and comprehensive studies were performed of a highly interesting group of phenomena that are observed on the outskirts of the plasmasphere and in the plasmapause during the interaction there between the whistler-mode radio waves and particles, leading to the excitation of ASE. The relevant experiments were set up with the aid of the Siple radio station in Antarctica, 76°S, 84°W (Helliwell & Katsufrakis, 1974). This station spans the frequency ranges $f_0 = 1.5$ to 16 kHz. It is equipped with a horizontal antenna (balanced dipole) 21.3 km long. The power fed to the antenna $W_0 \simeq 100$ kW, while the power emitted at various frequencies is of the order of 0.1 to 1 kW. The site of the transmitter corresponds to $L \simeq 4.1$. The observation point is situated near the magnetically conjugate point, in Canada: Roberval, Quebec, 48°N, 73°W. A number of publications that appeared after these experiments had begun contain highly significant and interesting new experimental findings (Helliwell & Katsufrakis, 1974; Stiles & Helliwell, 1975; Carpenter & Miller, 1976; Helliwell et al., 1975). The observed effects are presumed to be due to a nonlinear type of gyroresonant interaction between the whistler-mode radio waves and electrons, which is accompanied by a 'feedback' mechanism (Helliwell & Crystal, 1973; Helliwell, 1974). A number of features of LF emission generated in the region of the geomagnetic equator had before this given rise to the assumption that a mechanism of this type is operating in the magnetosphere (Brice, 1963; Helliwell, 1967), and various possibilities were explored for explaining the causes of generation and the properties of triggered emission from the magnetosphere. The results of these studies can be found in, for example: Sudan & Ott, 1971; Dysthe, 1971; Ashour-Abdalla, 1972; Nunn, 1974; Karpman et al., 1974; Istomin et al., 1976; Roux & Pellat, 1976, 1978. However, to date no clearly defined and exhaustive theory has been found for these interesting and complex phenomena. Such a theory will require, above all, further in-depth experimental studies under various conditions.

Let us take a brief look at some of the data emanating from the above-mentioned series of experiments. The most important of the results are the dependences of the intensity of the whistler-mode radio waves recorded at the reception point and the triggered emission excited at the apogee of the wave-propagation trajectory on the duration of the transmitted radio-wave packets–radio pulses (dots, dashes) (see Figs. 18.3, 18.4, 18.5).

The plasma very rapidly responds to *the effect of each wave packet* (Fig. 18.3). A single pulse, the duration of which varied by an amount $\Delta t = 50$ to 400 ms, is accompanied by the same effects.

As the duration Δt of the emitted radio signals increases, the intensity of the wave packets received at the magnetically conjugate point *increases*

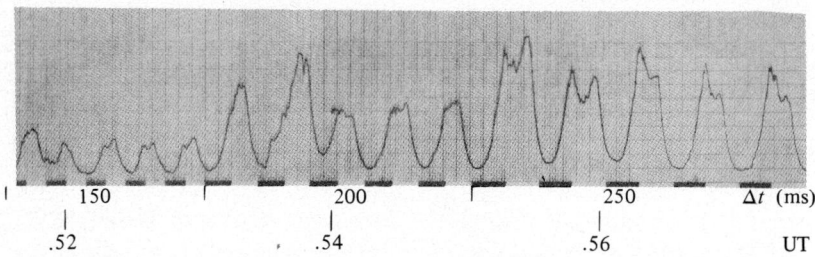

Fig. 18.3. Envelopes of single wave packets recorded at the magnetically conjugate point (Canada) following the emission of radio pulses (bold horizontal line segments, Morse-code dashes) at Siple Station (Antarctica). The time scales at the reception and transmission points have been matched, taking into account the time of propagation along the magnetic field line ($\tau \simeq 2.02$ s). The figure shows the increase in the amplitude of the received signals with increasing pulse width $\Delta t = 150$, 200 and 250 ms, $f = 5.5$ kHz, band width of receiver $\Delta f = 130$ Hz, 23 June 1973 (Helliwell & Katsufrakis, 1974).

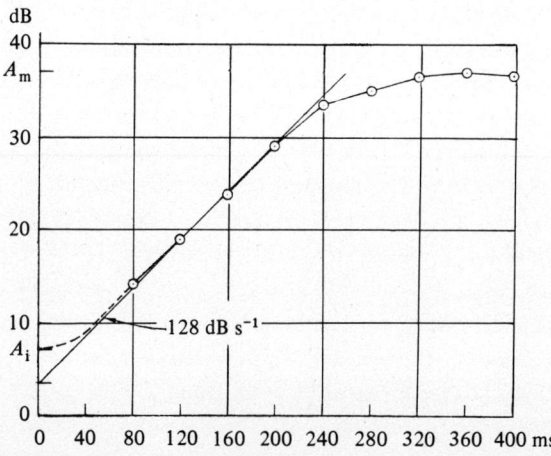

Fig. 18.4. Average amplitude of wave packets received at the magnetically conjugate point (Canada) as a function of the pulse width of transmitted 'dash' telegraph signals (Antarctica, $f = 5.5$ kHz, 23 June 1973). (Helliwell & Katsufrakis, 1974.)

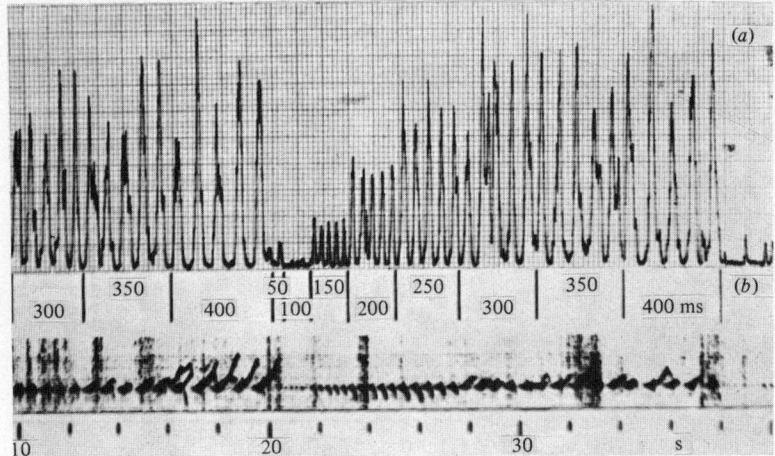

Fig. 18.5. (a) Recordings of the amplitude of wave packets received at the magnetically conjugate point (Canada) during an observation session when at Siple Station (Antarctica) successive transmissions were made of 'dash' wave packets of width Δt = 50, 100, 150, 200, 250, 300, 350, and 400 ms. (b) Spectrogram of received ASE signals generated at the geomagnetic equator – rising and falling tones, $f = 5.5\,\text{kHz}$, 23 June 1973 (Helliwell & Katsufrakis, 1974).

exponentially with time. For $\Delta t = 50$ to 100 ms the waves received at the magnetically conjugate point are very weak. They become quite intense, however, for $\Delta t = 150$ ms. The enhancement of the whistler mode takes place on the average at a rate of $\simeq 100\,\text{dB s}^{-1}$ (i.e., by a factor of 10^{10} per second) for a variation of the pulse duration in the range $\Delta t \simeq 80$ to 250 ms. At $\Delta t \simeq 200$ to 250 ms the enhancement reaches saturation and does not increase any further (Fig. 18.4). Under varying conditions the rate of signal enhancement was found to be from 30 to $200\,\text{dB s}^{-1}$. As is seen from Fig. 18.4, the total increase in wave intensity as a result of enhancement, from the initial amplitude A_i to the maximum amplitude A_m, is of the order of 30 dB (i.e., three orders of magnitude in energy).

Both the whistler mode of radio waves and triggered emission were observed at the reception point only in the range $f_0 \simeq 2.5$ to 7.6 kHz, even though there were transmissions at many fixed frequencies in the range $f_0 = 1.5$ to 7.6 kHz and at frequencies $f_0 = 14$ and 16 kHz (see Helliwell & Katsufrakis, 1974).

What is still not clear is whether the obtained enhancement of the received waves is the maximum possible enhancement that can take place in the corresponding regions of the near-Earth plasma. It would be of great interest to clarify this point.

Triggered emission begins to manifest itself clearly at $\Delta t \simeq 150$ ms. The first signals to be observed are fallers, and then at $\Delta t \simeq 200$ to 250 ms their shape changes, until at $\Delta t \gtrsim 300$ ms what we have are by now pronounced rising tones of great intensity (Fig. 18.5). When two transmission frequencies were used alternately, strong excitation of triggered signals was usually observed only at one of them. It often happened that a triggered emission excited at one of these frequencies interrupted the emission at the other frequency. Triggered signals of identical shape were observed to be shifted in time with respect to one another, which suggests that they arrived at the observation point along different trajectories.

According to the data of Carpenter & Miller, 1976, the waves observed at a reception point were guided along force lines the apogee heights of which varied between $R \simeq 3.5 R_0$ and $4.5\ R_0$; the electron density at the apogee varied between $N = 10^2$ and 10^3 cm^{-3}. Usually this region lay below the plasmapause boundary in the plasmasphere. Occasionally, however, during periods of moderate magnetic activity the wave trajectory traversed a region of large gradients dN/dR, and the apogee lay beyond the plasmapause boundary ($N \simeq 1$ cm^{-3}). The ratio of the radio-wave frequency f_0 to the gyrofrequency f_{H0} at the equator varied in these experiments from 0.2 to 0.45. The most common value was $f_0 \simeq 0.4 f_{H0}$.

In the experiments described, triggered emission stimulated in the plasmasphere by radio waves emitted in Antarctica was found to be regulated by the emission of high harmonics of a local electrical power grid situated in the vicinity of the reception point (Roberval) in Canada. Here it is harmonics of high order s that have an effect, namely $s \simeq 50$ to 70, i.e., at frequencies $f_p \simeq 2.5$ to 5.0 kHz $= 40 F_0$ to $80 F_0$, where $F_0 = 60$ Hz is the frequency of the power grid (Helliwell & Katsufrakis, 1974).

In a number of cases the sign of the slope of the triggered emission df/dt changed when the equality $f = f_p$ was valid. This correspondence was perceived from an examination of (f, t) spectrograms, on which parallel lines of the harmonic emission of the electrical power grid were clearly recorded. Further experiments revealed an enhancement of the emitted harmonics in the magnetic-field-aligned duct, owing to their interaction with electrons whose energy must have been of the order of several keV. These waves were recorded both in the vicinity of the reception point (Roberval) and at the magnetically conjugate point at Siple Station (Helliwell et al., 1975). These harmonic-emission lines (f_p = const.) are already of *magnetospheric origin*; they are wider ($\Delta f_p \simeq 30$ Hz) and are subject to slow changes in amplitude and frequency. What makes them so interesting is: (1) that they are observed only on odd harmonics of the

18.2. Artificially stimulated emissions

a.c. grid, (2) that the minimum interval between the lines $\Delta f_p \simeq 120\,\text{Hz}$, and (3) that their frequency is always higher than the frequency of the corresponding power-grid harmonic $(f - f_p) \simeq 20$ to $30\,\text{Hz}$. Spectrograms obtained simultaneously at both sites show that the frequencies of the corresponding emission lines of magnetospheric origin agree well, within the limits of accuracy of the frequency measurements ($\pm 5\,\text{Hz}$).

The above data on the interaction between power-grid harmonics and the magnetosphere are extremely interesting, since they indicate that emission from a power grid on Earth can apparently play a significant role in processes observed in the plasmasphere and magnetosphere. In this connection it is appropriate to mention the results of experiments carried out on Ariel 3 and 4 (Lefeuvre & Bullough, 1973; Bullough et al., 1975; Bullough et al., 1976; Kaiser et al., 1977; Tatnall et al., 1978). In this series of experiments the world-wide distribution of whistler-mode emissions at an altitude $z \simeq 500\,\text{km}$ was studied over a wide range of frequencies. It was discovered on Ariel 3 at $f \simeq 3.2$ and $9.6\,\text{kHz}$ that there exist zones of intense radiation over North America and in the magnetically conjugate region in the Southern Hemisphere in the latitude range $2 < L < 3$. An analysis of these data showed that this radiation can be attributed to the effect of harmonic emissions from power grids, as well as partly to the effects of lightning discharges. The above scientists estimated that the *enhancement of the harmonics* of a power grid in the plasmasphere was usually of the order of 20 to 30 dB and sometimes even amounted to 50 dB, i.e., it varied from 10^2 to 10^5. These results led researchers to conclude that the pitch-angle diffusion of electrons with energies $E_e > 100\,\text{keV}$ in the radiation belts ($2 < L < 3$) is regulated to a marked extent by the harmonic emissions of power lines (Tatnall et al., 1978). Let us also note at this point that a detailed analysis of wave spectra generated in the plasmasphere in the range $f = 0.5$ to $10\,\text{kHz}$ and recorded at the Earth's surface in the USA also invited the conclusion that during magnetic storms the strongest waves are associated with the emission of odd harmonics by a power station (Park, 1977).

To conclude this section let us point out that triggered-emission spectra have been studied in detail with the aid of an instrument with a resolving powers of $\Delta t \simeq 30\,\text{ms}$ in time and $\Delta f \simeq 45\,\text{Hz}$ in frequency (Stiles & Helliwell, 1975). In these experiments observations were performed for the following radio transmitters:

NAA, Maine;	$f_0 = 14.7\,\text{kHz}$,	$W_0 = 10^3\,\text{kW}$,
Omega, New York;	$f_0 = 10.2\,\text{kHz}$,	$W_0 = 0.1\,\text{kW}$,
Siple Station, Antarctica;	$f_0 = 5.5\,\text{kHz}$,	$W_0 = 0.4\,\text{kW}$.

An analysis of large number of signals of the riser ($d\omega/dt > 0$) and faller ($d\omega/dt < 0$) types showed that triggered wave packets always begin to be generated at the transmitter frequency about $\Delta t \simeq 70$ to 160 ms after the appearance of dashes of whistler-mode radio waves. During the initial period of transmission the wave-packet frequency always increases, no matter what the frequency and intensity of the generating wave. For a change in dash duration, the amplitude of the triggered emission ceases to increase at $\Delta t \simeq 260$ ms.

An important consequence of the effect of electromagnetic waves injected into the plasmasphere is a variation in the energy and pitch angles of the electrons that take part in this process. This gives rise to a number of secondary effects, particularly in the ionosphere, as a result of precipitation of these particles into its lower-lying regions (see Dungey, 1963). The effect of these precipitated electrons was observed, for example, during the reception in Antarctica (Eights Station) of radio waves from the NSS transmitter ($f = 22.3$ kHz, Annapolis, Maryland), which were propagated in the Earth-ionosphere waveguide. The intensity of the electromagnetic field of the radio waves at this station increased during periods of heightened activity of whistling atmospherics (Helliwell et al., 1973). This apparently occurred due to an increase in the conductivity of the D region (the upper wall of the waveguide) along the path of propagation of the received radio waves, on account of the effect of electrons which 'poured out' of the region of the geomagnetic equator during periods of intense and frequent lightning flashes. Effective precipitation of electrons is known to take place through a gyroresonant interaction with the whistler mode, if the electron energy $E_e \simeq 30$ to 300 keV. As a result of this interaction, electrons trapped by the wave are scattered in pitch angle, as had been predicted theoretically (Kennel & Petschek, 1966; see also Lyons et al., 1972). More indirect evidence of the role of energetic electrons precipitated into the D region of the ionosphere consists of the following. There have been observations of X-ray intensity bursts (bremsstrahlung), recorded by balloons in Antarctica (Eights Station), each time that an emission with a central frequency $f \simeq 2.5$ kHz, coming from the near-Earth plasma, was enhanced (Rosenberg et al., 1971).

A final note: on OGO 6 at altitudes $z \simeq 1000$ km, radiation bands stimulated by telegraph signals were recorded; the signals were transmitted by the GBR station at $f_0 = 16$ kHz and the Omega station at $f_0 = 10$ and 15 kHz. A spectrogram of such an emission, stimulated by the GBR transmitter, is shown in Fig. 18.6 (Laaspere & Johnson, 1973). These wave packets were observed during a period of heightened magnetic activity

18.3. Hiss and chorus low-frequency waves

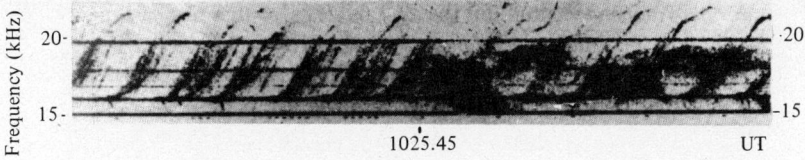

Fig. 18.6. Spectrogram of ASE wave packets stimulated by telegraph signals from the GBR station, and observed by OGO 6, on 10 Nov. 1969, in a region of the ionosphere magnetically conjugate with respect to the transmitter (Laaspere & Johnson, 1973).

($K_p = 4$ to 5) around midday local time in a region of the ionosphere magnetically conjugate to the site of the transmitting station.

18.3. Hiss and chorus low-frequency waves

Low-frequency (LF) waves of various types, the frequencies of which are less than the electron gyrofrequency ω_{Hs} at the satellite, were observed during a series of experiments on OGO 1 and OGO 3 (Burtis & Helliwell, 1969; Dunckel & Helliwell, 1969; Dunckel et al., 1970). Measurements were performed only on a magnetic antenna (loop). Plasma emission was recorded mainly in the vicinity of the geomagnetic equator at $f \simeq f_{H0}/2$ (where $f_{H0} = \omega_{H0}/2\pi$ is the gyrofrequency at the equator) in the range $f = 0.3$ to $3\,\text{kHz}$ in the *plasmatrough* between the plasmapause and the magnetopause, at distances $R \simeq 4R_0$ to $10R_0$. The intensity of the emission and the probability of its appearance diminished with increasing distance

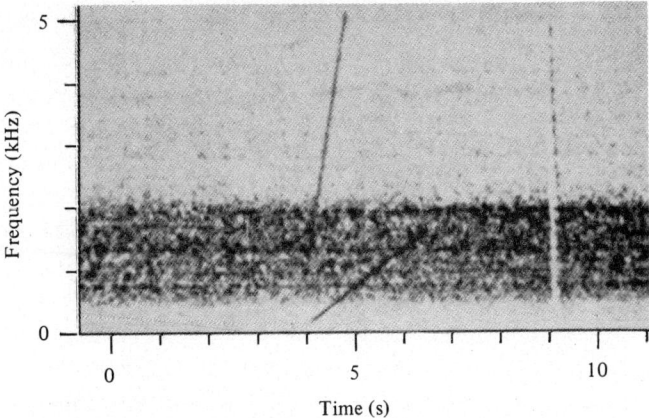

Fig. 18.7. Spectrogram of a band of LF hiss waves, OGO 1, $z \simeq 10^4\,\text{km}$, 4 Dec. 1964 (Dunckel & Helliwell, 1969).

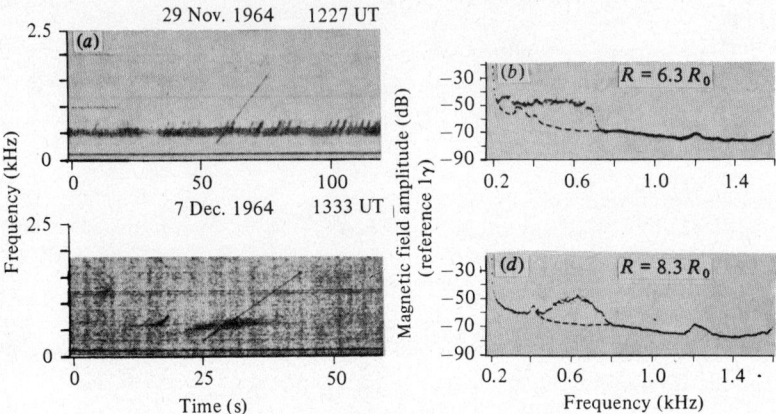

Fig. 18.8. Spectrograms of discrete bands of chorus emission, OGO 1 (Dunckel & Helliwell, 1969).

from the Earth. In no case were LF oscillations of the described type detected when the front of the shock wave was crossed. Examples of the emission spectrograms are shown in Figs. 18.7 and 18.8.

Chorus-type emission was usually cut off sharply at the upper frequency (Fig. 18.7). The nature of the behavior of the chorus oscillations changed with increasing distance from the Earth; the appearance of emission bands became more sporadic. For instance, Fig. 18.8 shows that at a distance from the Earth's centre $R \simeq 8.3 R_0$ only one emission band was recorded during the 50 s observing period. The same figure also shows the field amplitude H of the observed oscillations as a function of frequency. The duration of such emission bands was of the order of 10 s, and they were usually repeated at intervals of several minutes. Detailed studies of these bands showed that the central value of their frequency increased at lower geomagnetic latitudes, and that it correlates well with the variation in the minimum gyroresonance frequency of electrons along a force line of the the Earth's magnetic field, i.e., with the value f_{H0} at its apogee, at the geomagnetic equator (Burtis & Helliwell, 1969). The dependences of f_{H0} and the frequency f of the recorded emission on the location of OGO 1 (time) are shown in Fig. 18.9. This figure also gives the dependence of the local (close to the satellite) value of the gyrofrequency f_{Hs}. A similar comparison showed that the frequency of chorus-type emission in general varies within the limits $f = 0.2$ to $0.5 f_{H0}$. However, a subsequent analysis of the experimental data, together with the calculations performed by Burtis & Helliwell, 1969, led to the conclusion that in fact the emission is generated in

18.3. Hiss and chorus low-frequency waves

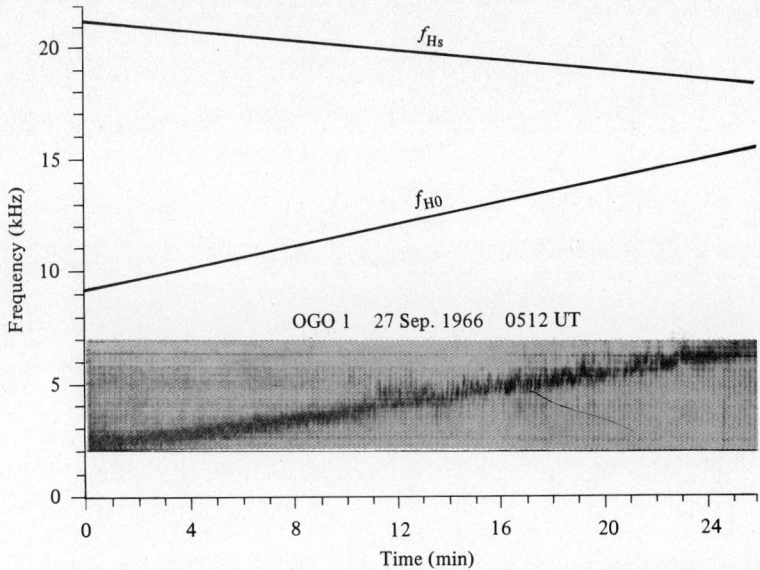

Fig. 18.9. Spectrogram of a narrow band of LF waves (mixture of hiss and chorus) and variations of the electron gyrofrequencies f_{H0} at the apogee of the force lines of the Earth's magnetic field and f_{Hs} in the vicinity of the satellite; OGO 1, $z = 18 \times 10^3$ to 23×10^3 km, 27 Sep. 1966 (Burtis & Helliwell, 1969).

the equatorial plane at a frequency $f \simeq f_{H0}/2$. Yet since the paths of propagation of these wave packets to the observation point deviate from the force line of the magnetic field $H_0\,(R, \phi)$, their trajectories are more extended than the force lines $H_0\,(R, \phi)$ and the local values of f_{Hs} differ from f_{H0}. The possibility that such trajectories might exist for whistler-mode waves had already been shown theoretically (see, for example, Thorne & Kennel, 1967). Therefore, when interpreting the experimental data, if we associate them with lines of the Earth's magnetic field, we get values of the ratio f/f_{H0} which are too low, especially for high geomagnetic latitudes.

Similarly, in these experiments it was established that the upper cutoff frequency of the packets of received waves varies with a change in the position of the satellite and also is proportional to f_{H0}. This follows, for example, from Fig. 18.10, which presents a spectrogram of a hiss emission band and its details for three short time intervals, for values of $L \simeq 6.0$ to 8.5. As L increased, f_{H0} decreased, and the upper cutoff frequency of the emission decreased proportionally.

More detailed studies of the properties of chorus emission were performed on OGO 5 (Burton & Holzer, 1974; Tsurutani & Smith, 1974).

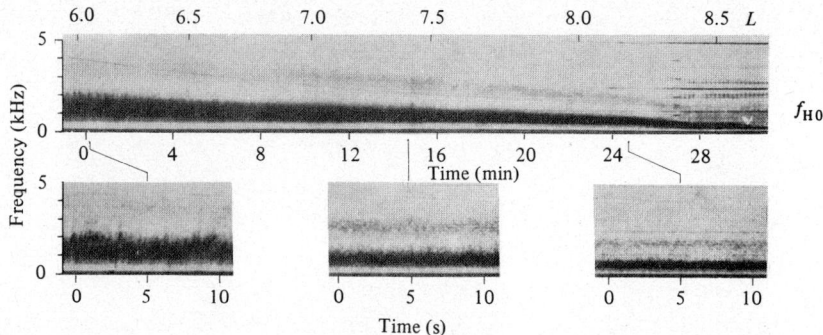

Fig. 18.10. Spectrogram of LF hiss waves; OGO 1, $L = 6$ to 8.5, 1 Oct. 1964 (Burtis & Helliwell, 1969).

In the experiments described by Burton & Holzer, 1974, measurements were made of the direction of the wave normal of these signals, and at the same time the energy spectra of the electrons, their pitch-angle distribution, and the magnetic field and electron density of the plasma were determined. The principal results of these experiments were the following.

As we have already seen from the preceding discussion, LF chorus emission is generated within the plasmatrough in the vicinity of the geomagnetic equator more often on the dayside of the equatorial plane in the geomagnetic-latitude range $|\phi| \lesssim 25°$, and on the nightside in the range $|\phi| \lesssim 2°$. The mechanism of generation of these waves is a Doppler-shifted gyroresonant emission from electrons with energies within the limits $E_e \simeq 5$ to 150 keV. The frequencies of the observed waves in the equatorial plane vary from $f \simeq 0.15 f_{H0}$ to $0.25 f_{H0}$. An essential condition for their generation is the presence of a maximum in the pitch-angle distribution of the electrons in the direction perpendicular to the vector of the Earth's magnetic field $\mathbf{H}_0$. The parameter characterizing the anisotropic pitch-angle distribution must exceed its critical value (see §6.2.3, Volume 1; Kennel & Petschek, 1966). It has been shown that the generation process is accompanied by the precipitation of electrons with energies $E_e > 20$ keV into the lower ionosphere. Near the source the distribution of the wave normals of the excited waves is limited by a cone with an angular spread $\theta \simeq \pm 20°$ around the direction of $\mathbf{H}_0$. Chorus waves are propagated during different periods both by being guided along the magnetic field line $\mathbf{H}_0$ and in arbitrary directions in the angular range $\theta \simeq 20°$ to $90°$. An interesting feature of this emission is the difference between its properties on the dayside and on the nightside of the Earth.

18.3. Hiss and chorus low-frequency waves

On the dayside rising tones ($d\omega/dt > 0$) are always observed, while on the nightside both rising tones and falling tones ($d\omega/dt < 0$) are in evidence. However, on this particular pass of OGO 5 through the magnetosphere, only one type of signal was observed. On the dayside the appearance of chorus emission is more sporadic in nature. These waves are observed during periods of moderate magnetic activity, being repeated irregularly and more seldom. On the nightside this emission is recorded only during periods of heightened magnetic activity.

It seems that the first attempts to study experimentally the fine structure of different types of emissions observed in the near-Earth plasma, using a spectrum analyser with a high time resolution, were made by Coroniti et al., 1971. They gave the results of an analysis of spectrograms of LF chorus, recorded at a distance from the Earth's centre $R \simeq 30 \times 10^3$ km near the equatorial zone (Fig. 18.11). The sequential nature of the recordings in this figure at time intervals $\Delta t = 12.5$ ms describes the structure and its evolution in time. The chorus spectrum, conclude Coroniti et al., consists of narrow-band signals of width $\Delta f \simeq 20$ to 30 Hz and of frequency-modulated waves. No doubt, further studies of the fine structure of the spectrograms will permit us to probe more deeply into the nature of the diverse type of waves observed in the near-Earth plasma.

Certain peculiarities distinguish the chorus emission during the buildup of magnetic storms, when it is observed for only about three hours after midnight local time and in the early hours of the morning. This emission is known as *postmidnight chorus*. The results of a study of it on OGO 5, in the vicinity of the geomagnetic equator (Tsurutani & Smith, 1974), are briefly as follows. This kind of emission was observed in the region $L \simeq 5$

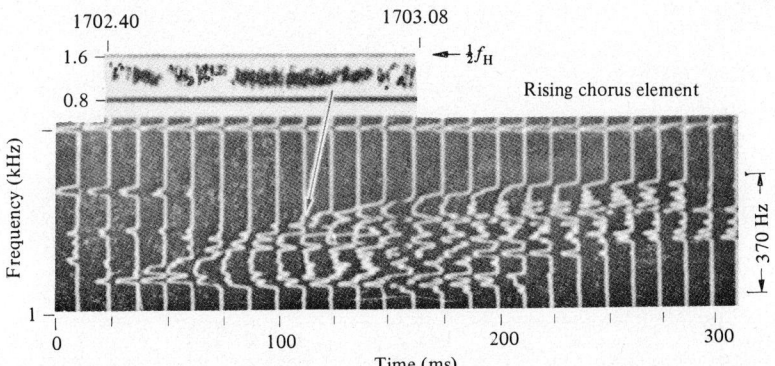

Fig. 18.11. Results of a frequency–time analysis of the fine structure of an LF chorus wave packet; OGO 5, 8 Jan. 1968 (Coroniti et al., 1971).

to 9, primarily at distances from the Earth $R \simeq 6.5R_0$ to $8R_0$ only in the geomagnetic-latitude interval $\phi \simeq \pm 15°$. The chorus wave packets more often than not had a band structure, or a falling-tone structure, as well as a modulated quasi-periodic form primarily with a period $T_0 \simeq 5$ to 15 s; Values of $T_0 \simeq 3$ s were also observed. The time average of the intensity of these waves was $H^2 \simeq 10^{-8} \gamma^2$ to $10^{-6} \gamma^2$. Their frequency ranges were $f \simeq 0.25 f_{H0}$ to $0.75 f_{H0} = 10$ to 10^3 Hz, where f_{H0} is the electron gyrofrequency at the geomagnetic equator; the width of the emission bands was $\Delta f \simeq 50$ to 200 Hz. If the chorus emission encompassed the frequency $f \simeq 0.5 f_{H0}$, then it had two frequency bands with maxima $f_{max} < 0.5 f_{H0}$ and $f_{max} > 0.5 f_{H0}$ and with a minimum $f_{min} \simeq 0.5 f_{H0}$, at which these waves were apparently strongly absorbed. It appears that postmidnight-chorus emission is generated due to gyroresonance, with the participation of electrons having energies $E_e \simeq 40$ keV, trapped in the magnetosphere during severe magnetic storms. In such a case, after midnight an additional role is played by the curvature of the trajectories of the fluxes of injected electrons in the western direction and by their drift under the influence of the spatial-distribution gradient. Other characteristics of this type of emission are discussed by Tsurutani & Smith, 1974.

In the *shock-front* region on IMP 6 bands of transverse *electron-whistler* LF waves were observed in the range $f \simeq 20$ to 200 Hz $< f_H$ (Rodriguez & Gurnett, 1975, 1976). In these experiments both the electrical component E and the magnetic component H of the field were measured. The amplitude of the magnetic component H decreased sharply near $f \simeq 100$ to 200 Hz, where the ratio of the magnetic to the electric energy density was $\mu_0 H^2 / \varepsilon_0 E^2 = 10^{-3}$ to 10^{-4}. The dependences $E^2 \simeq \omega^{-2 \pm 0.5}$ and

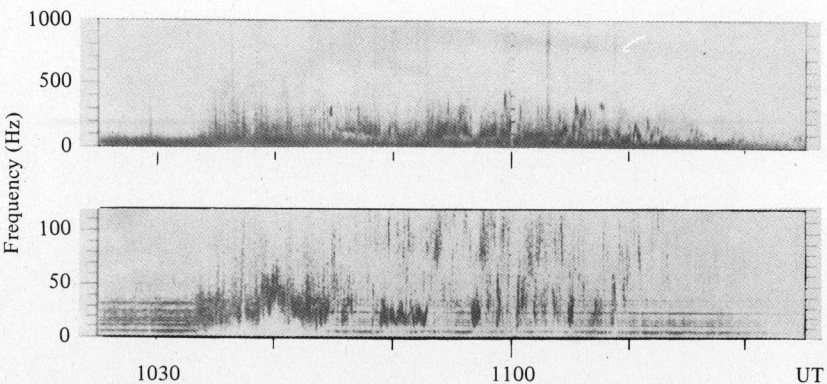

Fig. 18.12. Spectrograms of LF electrostatic waves; IMP 8, 5 April 1974, in the geomagnetic tail close to the neutral sheet, $R = 37R_0$ (Gurnett et al., 1976).

18.3. Hiss and chorus low-frequency waves

$H^2 \simeq \omega^{-4\pm0.5}$ were obtained; these describe accurately the properties of electron-whistler transverse waves, since their components are linked by the relation $H^2/E^2 \sim n^2 \sim \omega^{-2}$, when $\omega \lesssim \omega_H$. In these experiments $f_H \simeq 350$ Hz in the shock-front region. At $f \simeq 36$ Hz the mean square intensity of these waves was $E^2 \simeq 3 \times 10^{-9}$ V m^{-1} Hz^{-1}. These transverse waves are assumed to have arisen as a result of the transformation of electrostatic waves generated in the region of high temperature gradients.

Longitudinal electrostatic LF waves were recorded in the *geomagnetic tail* on IMP 8 over the entire range $\omega_L < \omega < \omega_H$ (Gurnett et al., 1976). LF-emission spectrograms obtained in these experiments are shown in Fig. 18.12. The spectra of these waves are cut off sharply at $f \simeq 10$ Hz $\simeq f_L$. The extent to which their intensity varied is seen from Fig. 18.13, which presents the entire spectrum of the electrostatic waves that were recorded in the geomagnetic tail. The band of LF electrostatic waves shown in Fig. 18.12 consist of bursts of duration $\Delta t \simeq 1$ s to 1 min., the burst frequency as a function of time often having a V shape (the value of f at first drops and then rises). When the boundary of the geomagnetic tail's neutral sheet was

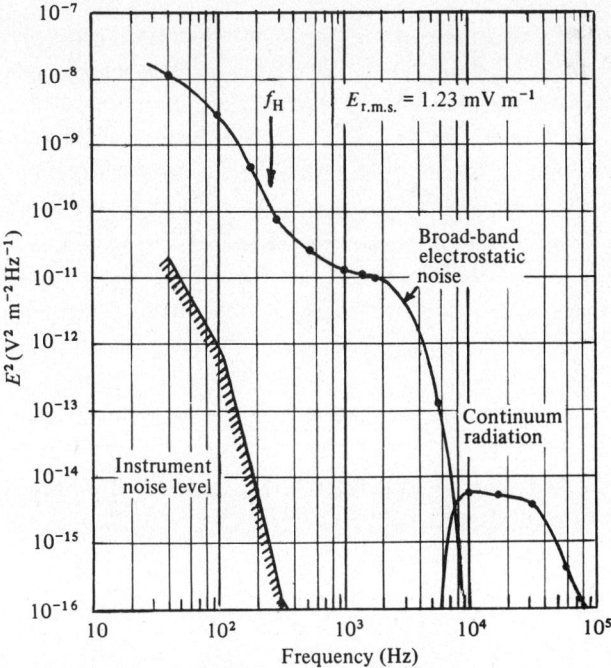

Fig. 18.13. Spectrum of LF electrostatic waves; IMP 8, 18 April 1974, in the geomagnetic tail, $R \simeq 34R_0$ (Gurnett et al., 1976).

crossed, the intensity of these waves increased. The duration of their generation periods was of the order of half an hour or more, a typical value of the field strength $E \simeq 1\,\mathrm{mV\,m^{-1}}$. As a rule, emission was observed near the boundary of the neutral sheet, as well as at distances of $10R_0$ to $15R_0$ from it in the plasma sheet (see Fig. 2.9, Volume 1), and it was virtually absent inside the tail, where the electron density N is high, and also in regions where, on the contrary the density is low; $N \simeq 10^{-2}\,\mathrm{cm^{-3}}$. Generation occurred most often in regions of high magnetic-field gradients, where the proton fluxes are strongly anisotropic, i.e., their velocities V_p are directed toward the Sun and in the opposite direction. In these regions $V_\mathrm{p} \simeq 10^3\,\mathrm{km\,s^{-1}}$ and the particles are strongly accelerated. This kind of emission is apparently brought about by ion-cyclotron or current-driven instabilities, appearing under the influence of the fluxes of these protons, $E_\mathrm{p} \simeq 1\,\mathrm{keV}$ (see Gurnett *et al.*, 1976 and Scarf *et al.*, 1973a). The intensity of this emission and the probability of its being generated are correlated with polar-aurora activity and polar hiss. Its frequency f is sometimes higher than the electron gyrofrequency, i.e. $f > f_\mathrm{H}$. This is apparently due to the Doppler shift of the frequency, which is considerable because of the high particle velocities.

Close to the neutral sheet in the *geomagnetic tail*, on IMP 8 in the frequency range $f_\mathrm{L} \ll f \simeq 40$ to $300\,\mathrm{Hz} < f_\mathrm{H}$, *transverse* electron-whistler waves were also observed. The structure of these emission bands is different from the structure of electrostatic waves. The former consist almost entirely of monochromatic bursts of duration $\Delta t \simeq 1$ to 10s, during which the frequency changes rapidly. The intensity of these waves is considerably less than that of electrostatic waves. For example, Gordon & LaLonde, 1961, present a case where the field intensities H of the longitudinal and transverse waves in the same series of observations were, respectively, $H \simeq 126\,\mathrm{m\gamma}$ and $H \simeq 4.6\,\mathrm{m\gamma}$. The authors suggest that the generation

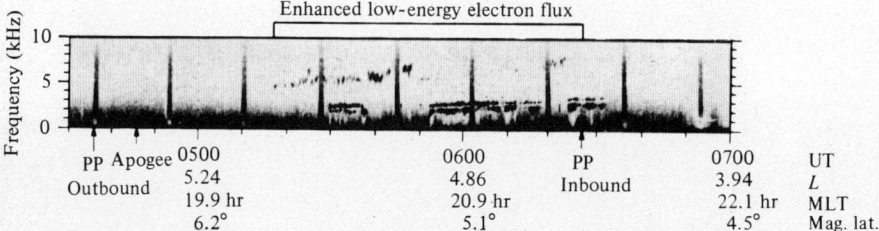

Fig. 18.14. Spectrograms of LF chorus electrostatic waves; Explorer 45, 25 Jan. 1972. A period of enhanced low-energy electron flux is shown; PP signifies the position of the plasmapause, and MLT means Magnetic Local Time.

18.3. Hiss and chorus low-frequency waves

mechanisms of the two types of waves are different. It is mooted that the transverse waves are related to a current-driven instability, which leads to the generation of oscillations with **K** parallel to $\mathbf{H}_0$.

On Explorer 45 recordings were made of electrostatic *LF chorus emission* in the vicinity of the frequency $f_H/2$ on the nightside of the magnetosphere above the plasmapause, during periods of heightened magnetic activity (Anderson & Maeda, 1977). Spectrograms on which similar emission bands were recorded at $f = 2$ and $3 \text{ kHz} \simeq f_H/2$ are shown in Fig. 18.14. The appearance of this emission always coincides with an abrupt increase in the electron flux $E_e \simeq 1$ to 10 keV beyond the plasmasphere for $L > 4$, this increase attaining values of $J_e \simeq 10^8 \text{ el cm}^{-2} (\text{s sr keV})^{-1}$. The emission disappears when $J_e \lesssim 10^6 \text{ el cm}^{-2} (\text{s sr keV})^{-1}$. When the plasmapause is crossed, this type of LF emission is cut off sharply.

19

Results of studies of HF waves ($\omega \gtrsim \omega_H$, $\omega \gtrsim \omega_0$)

In the frequency range where electron oscillations play the main role, wave processes of all possible types are observed, just as in other frequency ranges; these processes have been detected in all the regions of the natural plasma in which *in-situ* measurements have been carried out. Here, in addition to relatively wide bands of noise-like oscillations, resonance oscillations with quite narrow bands are generated. Recordings have been made of emissions of local origin, excited in the vicinity of the observation point, as well as of waves arriving from remote sources. The same types of wave processes were detected in the outer ionosphere, at heights of several hundred to a thousand kilometres, and also at distances of hundreds of thousands or of a million or more kilometres from the Earth. The mechanisms of excitation of the various kinds of oscillations can also be assumed to be very diverse. In many cases beam instability plays a role, the plasma waves being generated by streams of energetic particles either arriving from the Sun or produced in the vicinity of the Earth's bow shock. However, the anisotropic distribution of the electron velocities is frequently a major factor, and in particular the instability arising when the distribution of electron velocities has a maximum in a direction normal to the vector of the static magnetic field H_0. Many experimental data can also be explained in terms of a nonuniform spatial distribution of particles, or by the plasma structure.

Other types of mechanisms were involved in some experiments in the outer ionosphere, where the plasma excitation occurred under the influence of packets of radio waves (radio pulses), transmitted by a satellite in the vicinity of their source. Nonlinear mechanisms of the wave–particle or of the wave–wave type may also account for a number of effects observed in the magnetosphere and in other plasma regions. For instance, the generation of oscillations at a frequency $2\omega_0$ can be explained by such a nonlinear interaction. Emission at the Langmuir frequency ω_0 can be assumed to be a result of beam instability, and then the emission at $2\omega_0$

can be attributed to the scattering of this radiation by the oscillating electrons.

On the whole, it can be stated quite definitely at present that diverse nonlinear processes are important factors in the HF range, as well as at frequencies where the motion of ions has considerable influence. Some results pertaining to HF waves that illustrate this situation were presented in §7.1, Volume 1, where it was seen that in a number of cases the experimental data gathered in the ionosphere are in good agreement with the theoretical results. However, this is unfortunately not typical for the near-Earth and interplanetary plasma. It is only in exceptional cases that sufficiently satisfactory and, to some extent, unambiguous explanations of the observed facts have been obtained, especially from the quantitative point of view. The difficulties involved here are complex in nature. They are related to the complexity of the theoretical problems and even, as was mentioned above, to a fundamental confusion in the very formulation of the problems. However, an equally important factor, and one which is often decisive, is the lack of the data required for a comparison of theory with experiment.

19.1. Resonances in the outer ionosphere (plasmasphere)

High-frequency resonance waves in the outer ionosphere were apparently first observed with rocket-borne ionosondes (Knecht et al., 1961). Narrow packets of radio waves (radio pulses) emitted directly in the plasma from a rocket or satellite excite natural plasma oscillations, which last for times $\Delta t \simeq 5 \times 10^{-3}$ to 10×10^{-3} s after switching off the transmitter or shifting its frequency away from the resonance frequency of the plasma.

Resonance plasma oscillations of various types were observed in a large series of interesting experiments carried out with Alouette 1 and 2, Explorer 20, and ISIS 1 and 2 in the outer ionosphere at heights $z \simeq 800$ to 3000 km (Proceedings of the IEEE, 1969). Some of these correspond to the principal resonances expected from the theory of linear plasma oscillations. Other types of resonances require a special explanation, and their appearance is undoubtedly the result of a nonlinear interaction between waves and particles, and in particular a result of the parametric decay instability of the plasma. Resonance oscillations were recorded whose genesis attests to the complex, unusual nature of these processes. The most typical of these experimental results will be decribed briefly below.

Essentially, the experiments with ionosondes carried out on satellites can be summed up as follows. Ionosondes (see also §7.1, Volume 1) are devices that are able to transmit, over a wide frequency range, narrow-band wave packets ($\Delta\omega \ll \omega$) in the vicinity of the carrier frequency ω, and to receive them simultaneously. By continually running through the entire frequency range quite rapidly, it is possible to use these devices to obtain 'ionograms', on which are recorded wave packets reflected from different regions of the ionosphere. When the satellite is in orbit above the level where the electron density of the ionosphere is maximum, the radio pulses are reflected from the ionospheric region between this level and that of the orbit, specifically from the altitude where $n^2(\omega) \to 0$; also they are scattered by various kinds of irregularities.

Such experiments have established that, when the transmitter frequency is equal to some resonance frequency of the plasma, both reflected waves and packets of natural plasma oscillations, generated at this frequency, are received. On the ionograms the natural plasma oscillations appear in the form of spikes. The lifetimes and shapes of these spikes vary from case to case over wide limits and depend on the type of resonance and on the experimental conditions. Fig. 19.1 shows the corresponding characteristic, obtained on Alouette 2; the resonance spikes are evident on the diagram (Calvert & McAfee, 1969). The vertical coordinate is the so called virtual range $z_v = c\Delta t/2$, where c is the velocity of electromagnetic waves in vacuo, and Δt is the delay time of the reflected waves. The horizontal coordinate shows the frequencies scanned by the ionosonde. At the top are the resonance frequencies $f_0 = \omega_0/2\pi$, f_H, $2f_H$, etc., at which the gyroresonance spikes were recorded in the given case $f = f_H$, $2f_H$, and $3f_H$, for plasma

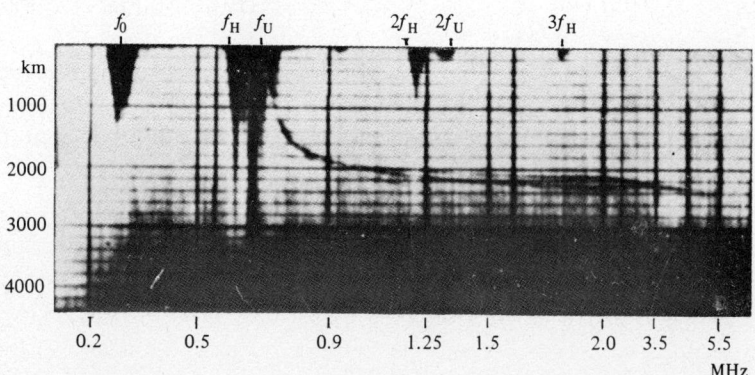

Fig. 19.1. Topside ionogram showing spikes, which correspond to resonance excitation of plasma oscillations at the Langmuir frequency $f_0 = \omega_0/2\pi$, at the gyrofrequency f_H and its harmonics $2f_H$ and $3f_H$, and at the upper-hybrid frequency f_U and its harmonic $2 f_U$; Alouette 2, $z \simeq 1000$ km, 1966 (Calvert & McAfee, 1969).

19.1. Resonances in the outer ionosphere

excitation at the Langmuir f_0 and upper-hybrid f_U and $2f_U$ frequencies.

Numerous observations have indicated that almost always in these experiments resonances were recorded at frequencies $f_0, f_H, f_U, 2f_H, 3f_H,$ and $2f_U$ (see Calvert & Goe, 1963; Lockwood, 1963; Hagg et al., 1969). Gyroresonant oscillations of the plasma at harmonic frequencies are excited quite easily. For example, Alouette recorded gyroresonance harmonics $f = sf_H$ up to $s = 22$ (Lockwood, 1965). Excitation at twice the plasma frequency $2f_0$, was rarely observed, and apparently it was never observed at the frequencies $3f_0, 4f_0...$ and $3f_U, 4f_U,...$

In a number of experiments excitation of diffuse oscillations between the harmonic gyroresonance frequencies sf_H and $(s+1)f_H$ was detected, s being a whole number (Warren & Hagg, 1968; Nelms & Lockwood, 1966). These resonances of a 'diffuse type', the central frequencies of which will be denoted here as $f_{D1}, f_{D2},...,$ are actually broadened wave packets occupying a wide frequency band (Fig. 19.2). Their central frequencies do not always correspond exactly to the half-integral values $(s+\frac{1}{2})f_H$. An analysis of a large number of measurements (Oya, 1970) showed that in instances where $f_0 < 1.8f_H$ (900 instances were considered) there was no

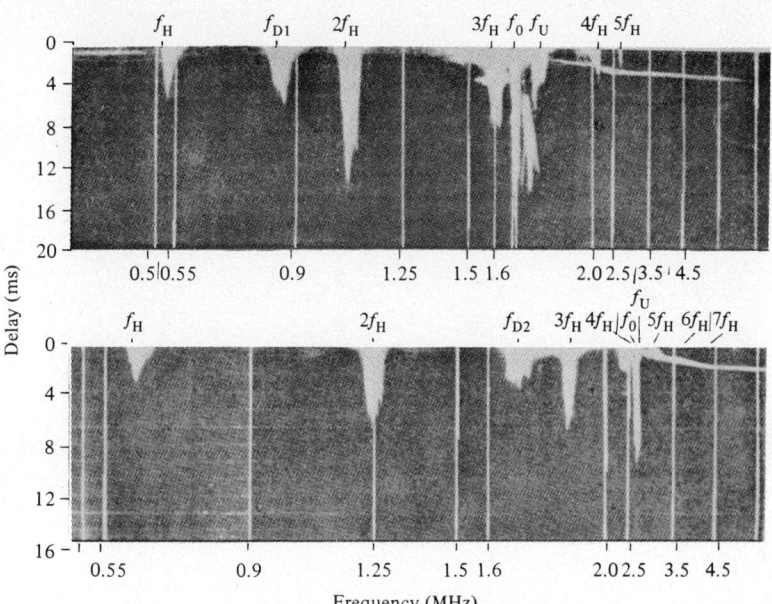

Fig. 19.2. Ionograms showing diffuse resonances f_{Ds} between the electron gyrofrequency sf_H and $(s+1)f_H$, resonances at gyrofrequencies from f_H to $7f_H$, and at the Langmuir and upper-hybrid frequencies f_0 and f_U; Alouette 2, $z = 100$ km, 22 April, 1966 (Nelms & Lockwood, 1966).

resonant excitation of the plasma. When, on the other hand, $f_0 > 1.8 f_H$, such excitation was always observed. The resonances f_{Ds} were recorded only up to $s = 4$. In some cases the resonant oscillations f_{Ds} were divided into two wave packets. On the whole, the resonance frequencies f_{Ds} varied over the following limits:

$$\left.\begin{array}{l}(f_H - 2f_H); \quad f_{D1} \simeq (1.4 - 1.85)f_H; \quad f_{D1} \simeq (1.7 - 1.9)f_H; \\ \qquad\qquad\qquad\qquad\qquad\qquad f_0 \simeq (2.0 - 3.7)f_H; \\ (2f_H - 3f_H); \quad f_{D2} \simeq (2.4 - 2.9)f_H; \quad f_{D2} \simeq (2.6 - 3.0)f_H; \\ \qquad\qquad\qquad\qquad\qquad\qquad f_0 \simeq (3.4 - 4.8)f_H; \\ (3f_H - 4f_H); \quad f_{D3} \simeq (3.6 - 3.9)f_H; \quad f_0 \simeq (4.5 - 5.8)f_H; \\ (4f_H - 5f_H); \quad f_{D4} \simeq (4.5 - 4.9)f_H; \quad f_0 \simeq (5.4 - 6.7)f_H.\end{array}\right\} \quad (19.1)$$

The following formula is recommended for determining the values of f_{Ds}:

$$f_{Ds} \simeq f_H[s + (0.464/s^2)(\omega_0/\omega_H)^2] \qquad (19.2)$$

(Dougherty & Monaghan, 1966).

On Alouette 2 'half-integral' resonances at $f = f_H/2$, $3f_H/2$, and $f_0/2$ were observed (Nelms & Lockwood, 1966; see also Hartz & Barrington,

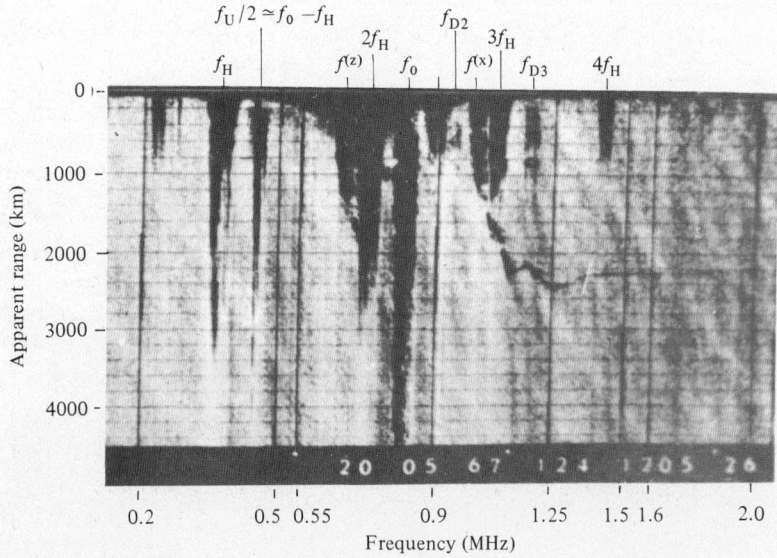

Fig. 19.3. Ionogram illustrating a floating spike $f = \dfrac{f_U}{2} = f_0 - f_H$, the resonances f_{D2} and f_{D3}, and resonances at cutoff frequencies of two branches of the extraordinary wave $f^{(x)}$ and $f^{(z)}$. Resonances at f_H, $2f_H$, $3f_H$, and $4f_H$ were also recorded; Alouette 2, 4 May 1967, $z = 2158$ km, 2.6°S, 67.9°W (Hagg et al., 1969).

19.1. Resonances in the outer ionosphere

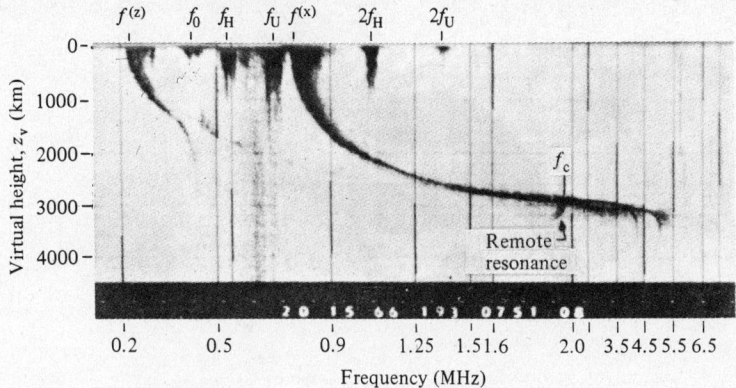

Fig. 19.4. Ionogram illustrating a spike due to resonant plasma excitation on a branch of the extraordinary wave below the satellite at the height $z_0 = 445$ km, at the frequency $f_c = 2f_H$; Alouette 2, 12 July 1966, orbit height $z_s = 2995$ km, 79.5°N, 56.7°E (Hagg et al., 1969).

1969), as well as a resonance at $f = f_U/2$, known as a *floating spike*; the latter was observed only when the condition $f_U/2 \simeq f_0 - f_H$ was satisfied (Fig. 19.3; Hagg et al., 1969). In these experiments both 'sum' $(f_0 + f_H)$ and 'difference' $(f_0 - f_H)$ resonances were recorded. Fig. 19.4 shows an interesting type of plasma excitation below the satellite orbit, along which the above-described resonance spikes were recorded (see Figs. 19.1–19.3). On the branch corresponding to the extraordinary waves $f^{(x)}$, that were reflected from lower-lying regions of the ionosphere, *remote resonance* was observed, i.e., resonant excitation of the plasma at the height z_0 where the extraordinary wave was reflected and where the condition $f^{(x)} = 2f_H$ was satisfied. On the ionogram in Fig. 19.4, such plasma excitation occurred below the satellite orbit at 2500 km, at a height $z_0 \simeq 445$ km.

Resonances in the ionosphere were also studied by sounding it at fixed frequencies. In these experiments the frequency of the moving transmitter can be equal to or quite close to one of the principal resonance frequencies of the plasma region through which the transmitter is passing. This causes the excitation of plasma oscillations, which are recorded on the ionogram. Here, such resonance conditions may sometimes exist for several minutes, since the characteristic frequencies of the ionosphere vary quite slowly along the orbit in comparison with the bandwidth of the transmitted wave packets and of the receiver. Consequently, observations like these can provide detailed recordings of the space–time characteristic of the oscillations generated in the plasma – an expanded sweep of the resonance spikes Fig. 19.5 (Calvert & Van Zandt, 1966) and Fig. 19.6. (Muldrew,

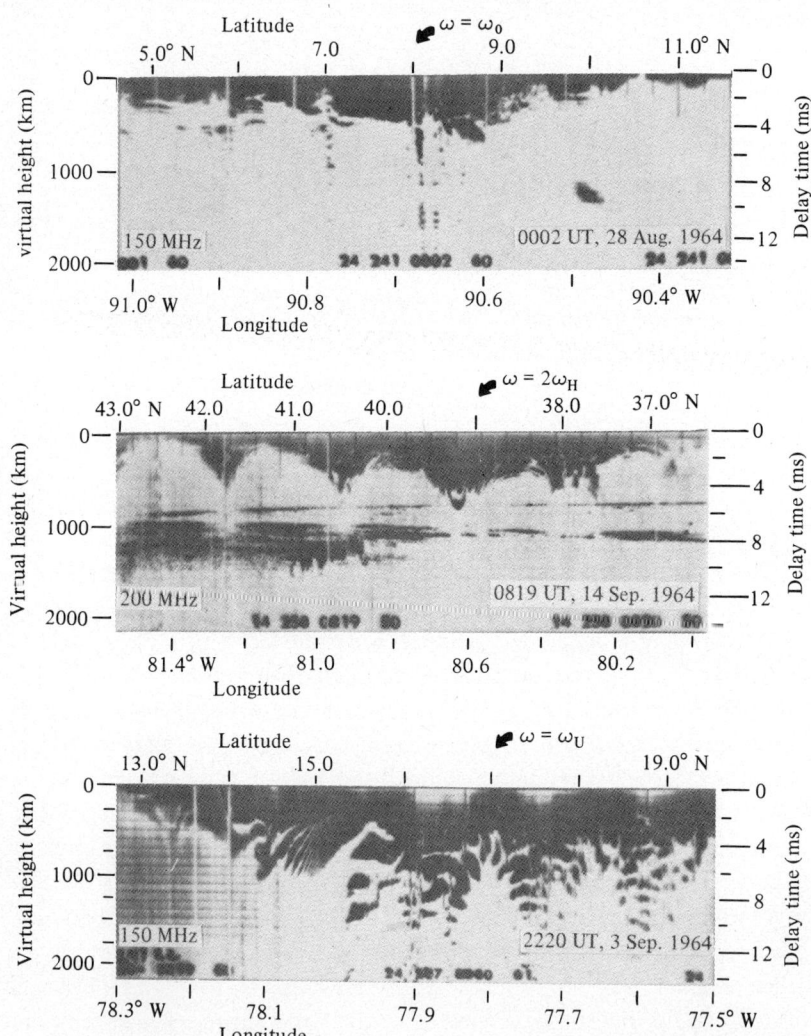

Fig. 19.5. Ionograms at fixed frequencies: Langmuir $f = f_0$, upper-hybrid $f = f_U$, and the second harmonic of the gyrofrequency $f = 2f_H$ (Explorer 20). (Calvert & Van Zandt, 1966.)

1972) show some typical such ionograms. The recordings of the resonances at $f = f_0, 2f_H$, and f_U in Fig. 19.5 are quite complex: the 'edges' of the ionograms have a complicated *fringe pattern*. This ionogram pattern is apparently explained in part by the fact that oscillations excited near the source were reflected from lower-lying plasma layers, which caused interference bands to appear.

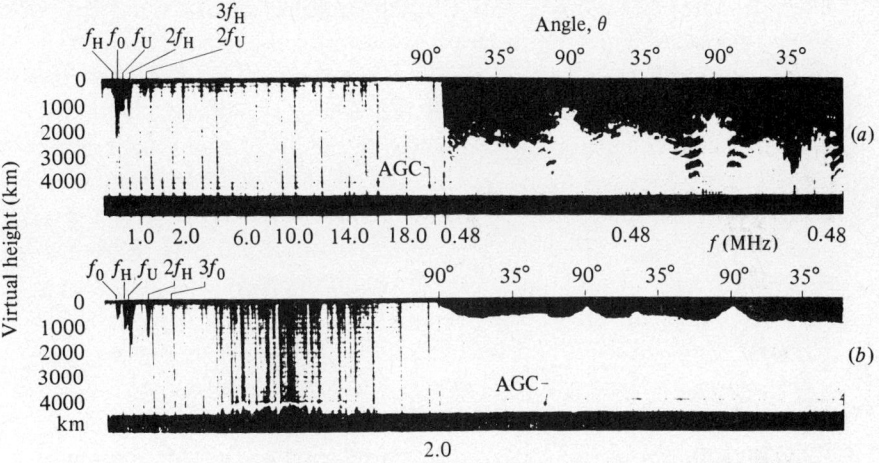

Fig. 19.6. Ionograms taken using the swept-frequency method and also at the fixed frequency $f = f_0 = 0.48$ MHz under conditions where (a) $f_H < f_0$ and (b) $f_H > f_0$ (ISIS 1). The lowermost trace on each record, labelled AGC, is the setting of the automatic gain control for the receiver (Muldrew, 1972).

However, it is also possible that modulation of the 'edges' of the resonance oscillations was caused by oscillations with regard to which the plasma exhibits parametric instability. These oscillations build up quite slowly, which may also complicate the intensity patterns of the fundamental resonance oscillations. Note that the lower ionogram in Fig. 19.6., which was recorded at the resonance frequency $f \simeq f_0 < f_H$, is quite smooth. It has minima (zeros) when the angle θ between the antenna and the vector $\mathbf{H}_0$ is $\pi/2$ (upper scale), which is explained by the fact that the observed oscillations are longitudinal Langmuir waves. For their resonance excitation the wave vector must satisfy the condition $\mathbf{K} \parallel \mathbf{E} \parallel \mathbf{H}_0$. Here, when $f < f_H$, there are no waves reflected from below, which may also explain the shorter duration of the resonance oscillations, and apparently the absence of a fringe pattern on the ionogram as well (see Calvert & Goe, 1963; McAfee, 1968, 1969a,b, 1970).

Note here, however, that the resonances at the upper-hybrid frequency f_U and at the gyrofrequency f_H and its harmonics sf_H are naturally a result of the excitation of longitudinal waves, like the resonance at the Langmuir frequency f_0. Therefore, the intensities of the oscillations at f_H and at the harmonics sf_H, which correspond to Bernstein modes, should be a maximum when the condition $\mathbf{K} \perp \mathbf{H}_0$ is satisfied. On the other hand, the waves excited at the frequencies f_0 and $2f_0$ should have a maximum intensity when $\mathbf{K} \parallel \mathbf{H}_0$ (see §4.5, Volume 1). The results of the experiments described agree on the whole with these theoretically predicted properties of the various kinds of resonances. However, for the resonance at twice the upper-hybrid frequency $2\omega_U$, no preferred direction relative to the vector $\mathbf{H}_0$ was detected. This means that it is more complicated in nature.

Soon after the detection of resonances in the upper ionosphere in the Alouette experiments, a theoretical evaluation of the results was initiated. The corresponding data can be found in, for instance, Fejer & Calvert, 1964; Dougherty & Monaghan, 1966; Crawford et al., 1967; Bitoun et al., 1970; McAfee, 1968, 1969a,b, 1970; Muldrew, 1967, 1972; Muldrew & Estabrooks, 1972, and in the references cited in these.

Intense *noise spectra* of HF oscillations between the plasma and upper-hybrid frequencies (upper-hybrid resonance noise) were observed regularly in the outer regions of the ionosphere with IMP 6 (Mosier et al., 1973). The spectra of these waves correspond to the HF resonance branch of the plasma waves $\omega_1(\theta)$ (see Figs. 4.7 and 4.9, Volume 1). This emission ceased when the boundary of the plasmapause was crossed. It was recorded at heights $z \simeq 3 \times 10^3$ to 15×10^3 km in the range $f \simeq 250$ to 600 kHz. These waves can be assumed to be a result of incoherent resonant Čerenkov radiation of electrons with energies $E_e \simeq 10$ eV to 10 keV (see Mosier et al., 1973, and Taylor & Shawhan, 1974). The propagation of these waves in the outer ionosphere is described by the formulas for a cold plasma given in Chapter 4, Volume 1. Their direction of arrival is mainly perpendicular to the vector of the magnetic field $\mathbf{H}_0$. This agrees with the condition for the most intense excitation of upper-hybrid resonance waves when the wave vector $\mathbf{K} \perp \mathbf{H}_0$. Mosier et al., 1973, used the effect of the cutoff of the noise spectrum at the upper-hybrid frequency to determine ω_0, and thus the local electron density as well.

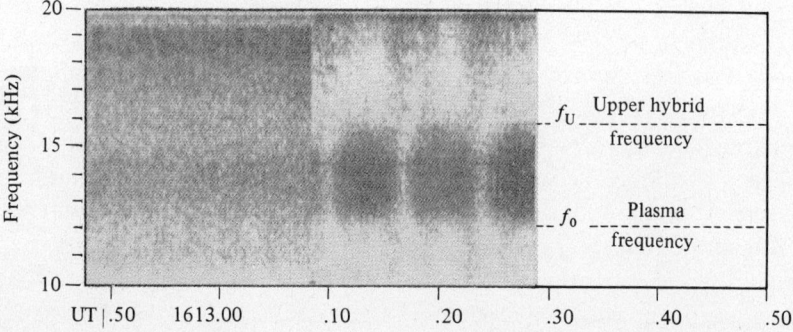

Fig. 19.7. Spectrogram of a band of HF plasma waves, showing the resonance branch $\omega_1(\theta)$; IMP 6, 28 Sep. 1971, $z = 3.67 R_0$ (Shaw & Gurnett, 1972).

19.2. Waves in the magnetosphere, in the interplanetary medium, and in the solar wind

This resonance method is one of the most accurate and one of the simplest means of obtaining the height dependence $N(z)$ of the electron density. Fig. 2.3, Volume 1, gave a profile of $N(z)$ obtained using this method; it extends out to distances from the Earth's centre $R \simeq 9R_0$ in the magnetosphere (Shaw & Gurnett, 1972). Fig. 19.7 gives a spectrogram of longitudinal oscillations recorded near the plasmapause boundary during these experiments.

19.2. Waves in the magnetosphere, in the interplanetary medium, and in the solar wind

In this section we will describe the main HF oscillatory processes observed between the plasmapause and the magnetopause (plasmatrough), in the geomagnetic tail, and in the Sun's vicinity. Some of these processes are typical of all these plasma regions. However, the generation mechanisms of given kind of emission are not always the same.

19.2.1. Half-integral gyroresonances $\omega = (s + \tfrac{1}{2})\omega_H$

In different regions of the magnetosphere and interplanetary plasma, narrow-band wave packets of electrostatic radiation at frequencies $\omega = (s + \tfrac{1}{2})\omega_H$ (where $s = 1, 2, \ldots$) have been detected. These half-integral

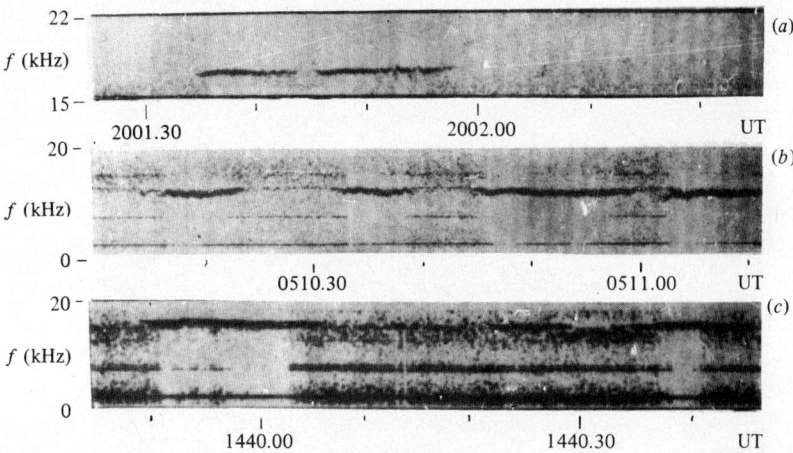

Fig. 19.8. Spectrograms of narrow-band emission at $f = \tfrac{3}{2} f_H$ in the magnetosphere (OGO 5); (a) 17 Oct. 1970, $R = 4.5R_0$, $\phi = -25.4°$; (b), (c) 6 Dec. 1970, $R = 5R_0$, $\phi = -28°$. During some time intervals the oscillation intensity is comparable to the noise intensity (Fredricks & Scarf, 1973).

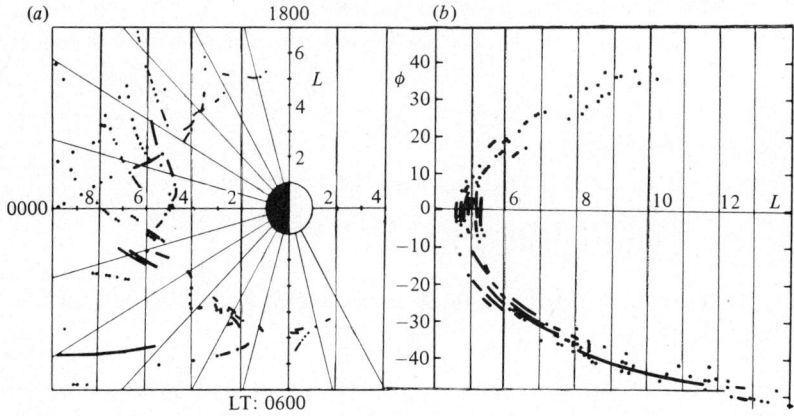

Fig. 19.9. Distribution of $f = \frac{3}{2}f_H$ emission in the magnetosphere with respect to local time, (a), and geomagnetic latitude ϕ and L-value, (b); OGO 5, 17 Oct. 1970 to 10 April 1971 (Fredricks & Scarf, 1973).

gyroresonances were apparently first observed with OGO 5 (Kennel et al., 1970). They were recorded most frequently at $\omega = \frac{3}{2}\omega_H$, where their intensity is highest and where they persist continually for many minutes. Figs. 19.8 and 19.9 (Fredricks & Scarf, 1973) show typical OGO 5 recordings of this emission at $f = \frac{3}{2}f_H$. The results of these experiments, carried out over a period of six months, indicate that this emission is frequently very intense and masks various kinds of noise. Practically each time that OGO 5 crossed the geomagnetic equator, oscillations at $\frac{3}{2}f_H$ were recorded. The region of their generation spans the geomagnetic latitudes $\phi = +40$ to $-50°$ and L-values from 5 to 10 (Fig. 19.9a). However, they are most intense and last longest in the vicinity of the geomagnetic equator ($\phi \simeq 0$), and at other latitudes for $L \simeq 5$ to 7. Emission at $\frac{3}{2}f_H$ was observed during the entire period from 1900 to 0700 LT, when OGO 5 was traversing the corresponding plasma regions (Fig. 19.9b). Series of such oscillations that occurred around midnight (0000 LT) turned out to be the most persistent and most intense. Their field strengths varied over the range $E \simeq 1$ to $10\,\mathrm{mV\,m^{-1}}$.

Fig. 19.10 shows an IMP 6 spectrogram on which at a distance from the Earth's centre $R \simeq 7R_0 \simeq 45 \times 10^3$ km simultaneous recordings were made of narrow-band electrostatic emissions at the two frequencies $f = \frac{3}{2}f_H$ and $\frac{5}{2}f_H$ (Shaw & Gurnett, 1975).

Electrostatic waves which were apparently also associated with the half-integral gyroresonances $f = \frac{3}{2}f_H$ and $\frac{5}{2}f_H$ were detected by IMP 7 and IMP 8 in the *geomagnetic tail*, in its distant surroundings, and in the

19.2. Magnetosphere, interplanetary medium and solar wind

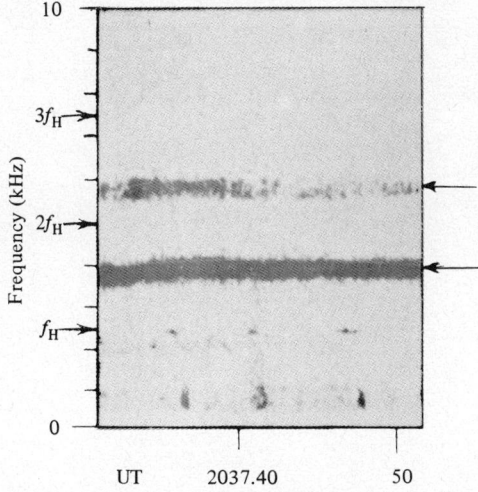

Fig. 19.10. Bands of $f = \frac{3}{2}f_H$ and $f = \frac{5}{2}f_H$ emission; IMP 6, 9 Aug. 1971, $\phi = -18°$, $R = 7R_0$ (Shaw & Gurnett, 1975).

region of transition from the plasma sheet to the neutral sheet (Scarf et al., 1974b; Gurnett et al., 1976).

Various mechanisms that might explain the nature of the half-integral plasma gyroresonances were considered. It can be assumed in particular that excitation of the $(s+\frac{1}{2})\omega_H$ emission takes place when the velocity distribution of the energetic electrons has a not very wide maximum $v_\perp > 0$ in a direction normal to the magnetic-field vector (ring electron distribution; Fredricks, 1971). It was shown that this kind of plasma instability is enhanced considerably in the presence of a cold plasma component (Young et al., 1973; Ashour-Abdalla et al., 1975; Ashour-Abdalla & Kennel, 1978). Emission at the frequencies $(s+\frac{1}{2})\omega_H$ can also be interpreted on the basis of a nonlinear explanation of wave–wave interactions (Oya, 1972, 1975).

During a strong magnetic storm (substorm) intense emission at $f = \frac{3}{2}f_H$ is observed in the magnetosphere. Such a spectrogram, obtained with OGO 5 at distances from the Earth's centre $R = 40 \times 10^3$ to 50×10^3 km, is shown in Fig. 19.11a, together with a graph of the field strength of these waves during the substorm (Scarf & Fredricks, 1971). The sudden rise in the amplitude of these oscillations during the second phase of the substorm, when a value $E \simeq 100\,\text{mV m}^{-1}$ was reached, is connected with abrupt variations of the magnetic field (Scarf et al., 1973b). At this time the electric-field intensity was seen to rise by several orders

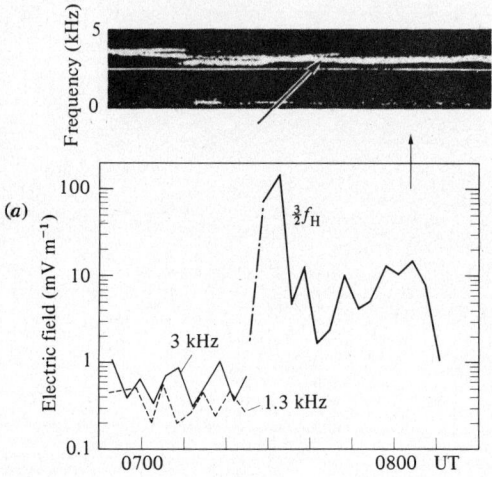

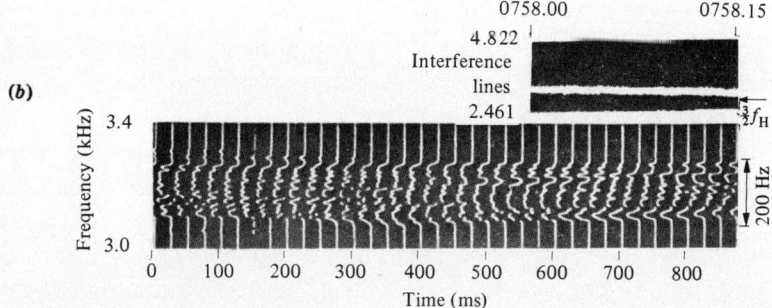

Fig. 19.11. (a) Band of $f = \frac{3}{2}f_H$ HF waves and variation in their amplitude: OGO 5, 15 Aug 1968 [508]. (b) Recording of the fine structure of these waves (Coroniti et al., 1971).

of magnitude. Figure 19.11b shows the fine structure of the $\frac{3}{2}f_H$ emission during a time $\Delta t = 0.42$ s, at 12.5 ms intervals, for $R \simeq 42 \times 10^3$ km and a magnetic latitude $\phi = -4°$ (Coroniti et al., 1971). The frequency–time characteristic of this emission obtained here shows that it apparently consists of several discrete signals in a narrow frequency band $\Delta f \simeq 200$ Hz. Naturally, such studies of the fine structure of the emission of these and other waves observed in the near-Earth plasma are of great interest; methods for experimental investigation of the fine structure of electromagnetic processes are described in, for example, Beghin & Siredey, 1964, and Al'pert et al., 1975.

Let us note here that half-integral gyroresonance oscillations of the electric field $f = (s + \frac{1}{2})f_H$ have also been observed under laboratory conditions. These results, which are very interesting, were obtained during experiments using a large vacuum device (Bernstein et al., 1975). In a vacuum chamber 30 m in diameter and 36 m high the iron chamber walls set up a magnetic field with an intensity $H_0 \simeq 0.33$ Oe. This value of H_0 corresponds to an electron gyrofrequency $f_H \simeq 0.93$ MHz. An electron gun emitted a quasi-monoenergetic electron beam along the $\mathbf{H}_0$ direction, and the beam was incident upon a grid and collector at an angle $\phi = 78°$ to $\mathbf{H}_0$. These were located 20 m away from the electron gun! The beam was focused to a point a few metres away from the grid, and its maximum diameter was about 3 m. A negative potential was applied to the grid. The electron energy was varied over a range $E_e \simeq 50$ eV to 5 keV. For a grid potential $\phi_g \simeq -40$ V, a reflected counterstream appeared in the chamber, the relative energy of which was 10^{-2} to 10^{-1} times the energy of the incident beam; for $\phi_g \simeq -200$ to -300 V, electrical oscillations at $f = \frac{3}{2}f_H, \frac{5}{2}f_H$, and $\frac{7}{2}f_H$ were excited. During these experiments weak oscillations at the Langmuir frequency f_0 were also detected, but without a counterstream; the ratio f_H/f_0 varied approximately from 0.2 to 1.

19.2.2. Waves in the vicinity of the electron Langmuir frequency ω_0 and gyrofrequency ω_H

The wave processes generated at the frequencies ω_0, ω_H, and $s\omega_H$ (where $s = 1, 2, \ldots$), either regulated by or fundamentally connected with electron resonances, may have either narrow-band or diffuse spectra. In some plasma regions such oscillations are regularly observed.

The *constant background of the magnetosphere* is apparently *electrostatic*, consisting of longitudinal plasma oscillations ($\mathbf{K} \| \mathbf{E}$). These are 'regulated' in this case by the plasma frequency ω_0 (Shaw & Gurnett, 1975). These waves are, however, very weak. In the frequency band their field strength $E \simeq 2 \,\mu\text{V m}^{-1}$. Most frequently, the intensity of their emission $E^2 \simeq 10^{-15} \,\text{V}^2 \,\text{m}^{-2} \,\text{Hz}^{-1}$. It was detected on magnetically quiet days at all geomagnetic latitudes ($\phi < 45°$) crossed by the orbits of IMP 6, and it was observed during the seasonal period of the experiments on about $\frac{2}{3}$ of the passes of IMP 6 through the magnetosphere. Emission of this type appears at the boundary of the plasmasphere and is cut off sharply at geocentric distances $R \simeq 10 R_0$ (close to the magnetopause). From there right up to the apogee of IMP 6 ($R \simeq 30 R_0$), these electrostatic waves were then not recorded. They were observed in the plasmasphere at all hours, but mainly from 1800 to 2400 LT.

The plasma emission under review has both narrow-band and diffuse

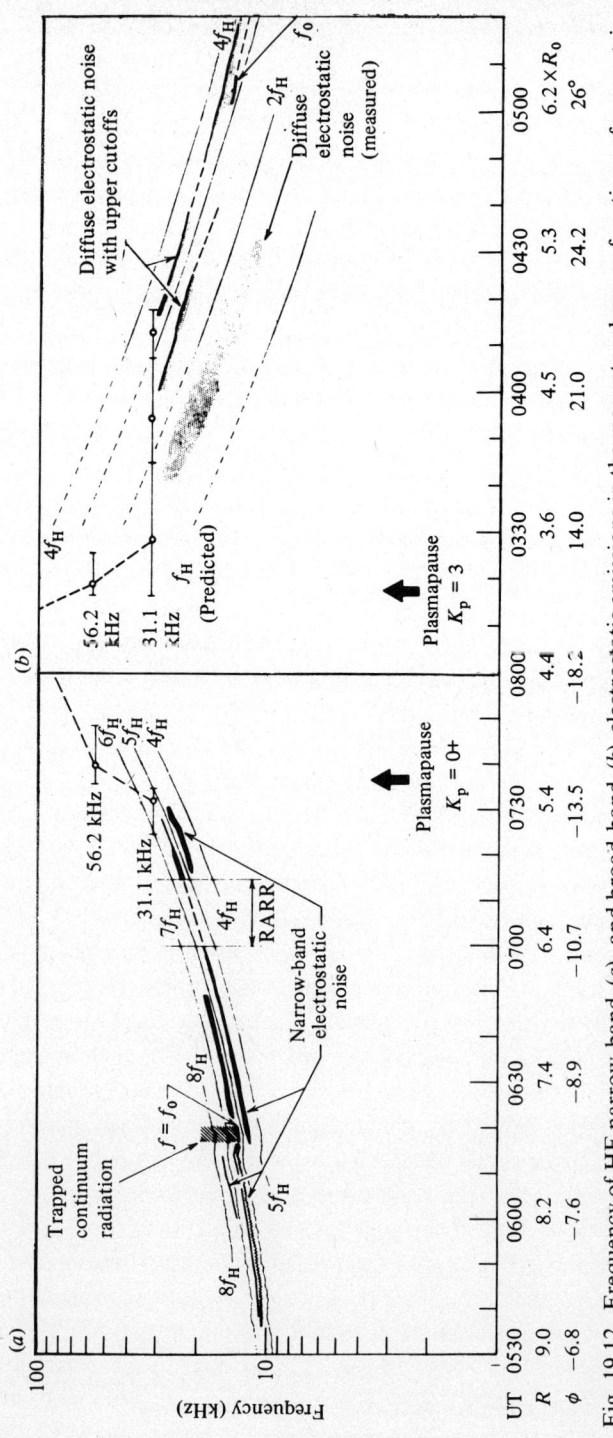

Fig. 19.12. Frequency of HF narrow-band, (a), and broad-band, (b), electrostatic emissions in the magnetosphere, as functions of geocentric distance from the Earth; IMP 6, (a) 27 May 1971, (b) 27 Jan. 1972 (Shaw & Gurnett, 1975).

spectra. The *narrow-band* emission consists of one or several bands of width $\Delta f \simeq 10^2$ Hz, the frequency of which lies between the electron gyrofrequency harmonics $s\omega_H$ and was found to vary with the position of IMP 6 in the magnetosphere, following the function $\omega_H(\phi, R)$. In the series of observations in Fig. 19.12a, for instance, the narrow emission bands are visible between the frequencies $f = 4 f_H$ to f_H. It is typical of these that their maximum frequencies are close to the plasma frequency f_0. Such a relationship between the plasma frequency and the bands of electrostatic waves observed both at the lower boundary of the magnetosphere and within it may prove to be very important as an aid to understanding the generation mechanisms of this emission. It should be noted with regard to this that the electric field **E** of the narrow-band emission has no definite direction relative to the vector $\mathbf{H}_0$.

The *diffuse* type of electrostatic emission in the magnetosphere has a bandwidth $\Delta f \simeq 1$ kHz. Its spectrum is often cut off sharply at the top. The frequency bands of this type of emission lie between the harmonics $s\omega_H$ and $(s+1)\omega_H$; they also follow the variation of the electron gyrofrequency $\omega_H(\phi, R)$ in the magnetosphere. Fig. 19.12b portrays three bands of this diffuse emission, one of which is not cut off sharply at the top. This kind of diffuse plasma emission is apparently similar to the oscillations recorded in the outer ionosphere with Alouette (see Fig. 19.2, formulas (19.1) and (19.2) above). It is observed in lower magnetospheric regions than the narrow-band emission, and usually not simultaneously with it. The electric vector **E** of these waves is perpendicular to the magnetic field $\mathbf{H}_0$.

An analysis of a whole set of data on electrostatic emission of the type being considered led Shaw & Gurnett (1975) to the following development pattern for the various kinds of oscillations in the plasmasphere and magnetosphere. They assume that the HF waves in the frequency region adjacent to the upper-hybrid frequency that are observed in the outer ionosphere (plasmasphere) are transformed at great heights into the magnetospheric emissions under review here (see also Liemohn & Scarf, 1963). Accordingly, processes taking place in the cold plasma below the plasmapause are controlled at the boundary of the plasmasphere by other processes taking place in the hot plasma. In remoter plasma regions the excitation mechanism of this emission is related to the non-Maxwellian electrostatic instability. Here intense narrow-band longitudinal oscillations at $(s + \frac{1}{2})\omega_H$ are an important factor. Such a transformation of one type of oscillation into another is naturally connected with the change in the plasma properties on going from the plasmasphere to the magnetosphere.

For instance, when the boundary of the plasmasphere is crossed, the electron density N of the plasma decreases, but the electron temperature T_e may rise approximately a thousandfold. Shaw & Gurnett, 1975, also assume that the continuous spectrum of nonthermal HF radiation of the Earth, to be described below in §19.2.4, may be connected with the electrostatic plasma oscillations being considered here as well.

Some important properties of this kind of HF emission, which may help to reveal its nature and the details of its structure in the magnetospheric trough, were discerned recently with GEOS 1 at geocentric distances $R \simeq 5R_0$ to $7R_0$ (Christiansen et al., 1978; Gendrin & Jones (Eds.), 1978). During these experiments plasma emissions of natural origin as well as emissions created artificially with GEOS 1 were recorded, the artificial radiation being narrow packets of monochromatic radio waves with f up to $\simeq 76$ kHz. The radio waves had a pulse width $\Delta t \simeq 3\,\mu s$, and the carrier frequency f was varied in steps with intervals $\Delta f = 300$ Hz, in any part of the above range of f, as telecommanded from the Earth. However, plasma emission was recorded for 10 ms after the radio source was turned off. These recording sessions were repeated successively every $\simeq 20$ ms. Tentative data from these experiments revealed the following properties of HF emission in the magnetosphere.

In the frequency range below the plasma frequency ω_0 and the upper-hybrid frequency ω_U (which were very close, since in these experiments $\omega_0 \simeq 6\omega_H$), only emission bands at the electron gyrofrequencies were recorded. In a number of cases gyroresonances were detected up to $\omega = 32\omega_H$. Here longitudinal waves at half-integral gyroresonances $\omega_s = (s + \frac{1}{2})\omega_H$ were also observed in the region $L \simeq 5$ to 6. In the daytime magnetosphere their lifetimes ranged from half an hour to an hour; their field strength $E \simeq 10\,\mu V\,m^{-1}$ for $s = 1$. At night, emission with this intensity was recorded only at $\omega_1 = \frac{3}{2}\omega_H$.

For frequencies $\omega > \omega_0, \omega_U$, Bernstein modes were first detected in their pure form in these experiments and their main properties were studied. In particular, it was found that this emission is excited at certain frequencies ω_q, intermediate between $s\omega_H$ and $(s + 1)\omega_H$, corresponding to values at which the condition $(d\omega/dk)_{\omega = \omega_q} = 0$ is satisfied (see. Fig. 5.6. Volume 1). Here the natural plasma emission usually consisted of bands $\omega = \omega_0, \omega_U$ and the triplet $\omega_{q1}, \omega_{q2}, \omega_{q3}$. Their field strength amounted to a few microvolts per metre in the band of width $\Delta f = 300$ Hz. For active modification of the plasma, on the other hand, the Bernstein modes were then enhanced by a factor of around 10^3 to 10^4, and the 4th and 5th bands of ω_{qs} emission were even observed. However, there were also

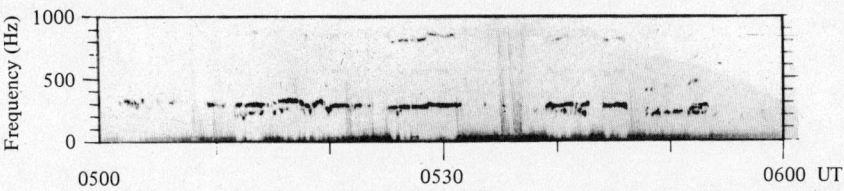

Fig. 19.13. Narrow bands of electrostatic waves close to the electron gyroresonance in the geomagnetic tail; IMP 8, 6 April. 1974, $R = 33\ R_0$ (Gurnett et al., 1976).

instances in which the intensities of the natural and artificial emissions were commensurable. These were cases when the emission intensity at the plasma frequency ω_0 was very high and of the same order as the emission intensity at ω_{q1}. But the plasma-wave intensity was usually lower than at the frequency ω_{q1}. In such cases the higher harmonics at $\omega_{q2}, \omega_{q3}, \ldots$ were suppressed.

In the *geomagnetic tail* at a distance $R \simeq 33R_0$ *narrow-band* spectra of electrostatic plasma oscillations were also observed with IMP 8 at $\omega \simeq 1.21\omega_H$ ($f = 280$ Hz) (Fig. 19.13; Gurnett et al., 1976). Some weaker oscillations close to the second and third harmonics of the electron gyrofrequency were also recorded. These waves, which were quite weak, were detected only rarely in these experiments. Their field intensity did not exceed $300\,\mu\text{V m}^{-1}$. However, Gurnett et al., 1976, accorded these cases great significance, since the oscillations were generated only in very hot regions of the geomagnetic tail, where the electron density is high. They can thus serve as an indicator, revealing regions which are situated around the neutral zone of the geomagnetic tail.

Narrow bands of sporadic emission of electrostatic waves close to the electron gyrofrequency, namely at $f \simeq 1.1 f_H$ to $1.2 f_H \simeq 2$ to $4\,\text{kHz}$, were also observed in the *polar cusp*; these were recorded with Hawkeye 1 at geocentric distances $R \gtrsim 6R_0$. Occasionally weak emission at the second or higher harmonics was detected (Gurnett & Frank, 1978b). The field intensity of these waves at $f \simeq f_H$ usually ranged from $E \simeq 10$ to $100\,\mu\text{V m}^{-1}$. This emission was generated in the magnetosphere in the polar cap, on both sides of the cusp, but it was generally not detected in the magnetospheric plasmatrough. Consequently, the appearance of these waves could serve as an indicator of the boundary between the polar cusp and the plasmatrough.

Fig. 19.14 portrays another kind of quite wide noise band close to the *gyroresonance frequency* f_H. This oscillation spectrum was recorded with Explorer 45 during a *magnetic storm* in the magnetosphere (Taylor &

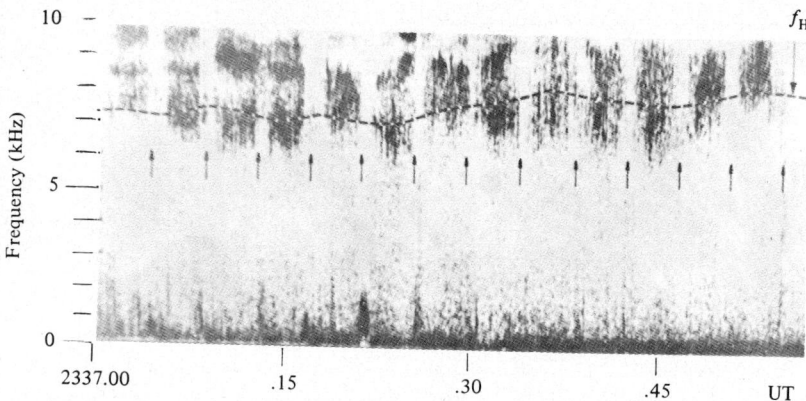

Fig. 19.14. Spectrogram of a wide-band magnetospheric emission observed during a magnetic storm, cut off close to the electron gyrofrequency f_H; Explorer 45, 4–5 Aug. 1972. The white bands (breaks in the spectrogram, indicated by arrows) correspond to instants when the direction of the electric antenna was close to the direction of the vector $\mathbf{H}_0$ (Taylor & Anderson, 1977).

Anderson, 1977). The electric field of these waves is closely perpendicular to $\mathbf{H}_0$; they are transverse electromagnetic waves. The regions of disappearance of the oscillations in Fig. 19.14 (vertical white bands with arrows) correspond to the directions of the electric antenna in which minima of E were observed as Explorer 45 rotated (it was spin-stabilized). Oscillation spectra like these, associated with a frequency ω_H, were recorded during these experiments in the outer regions of the magnetosphere at various frequencies. The lower boundary of their frequencies is always close to ω_H. Taylor & Anderson, 1977, assumed that the recorded oscillations were not necessarily generated close to the observer. Their excitation mechanism was the electrostatic gyroresonance instability, accompanied by a transformation process. Since the plasma in which these waves are excited is moving relative to the satellite, they are observed at Doppler-shifted gyroresonance frequencies. It may also be that these waves are generated in a very extended plasma region, which would explain their wide band and the possible transformation of the generated longitudinal waves into transverse waves. The source of this emission is assumed to be somewhere between the plasmapause and the magnetopause, and it is thought to be trapped in this trough region of the magnetosphere.

Various kinds of emission in the frequency range associated intimately with *Langmuir oscillations* ω_0 of the plasma are generated both in the magnetosphere and beyond it. In the *solar wind* over the shock-front region (bow shock), the generation of HF Langmuir oscillations ω_0 is a

19.2. Magnetosphere, interplanetary medium and solar wind

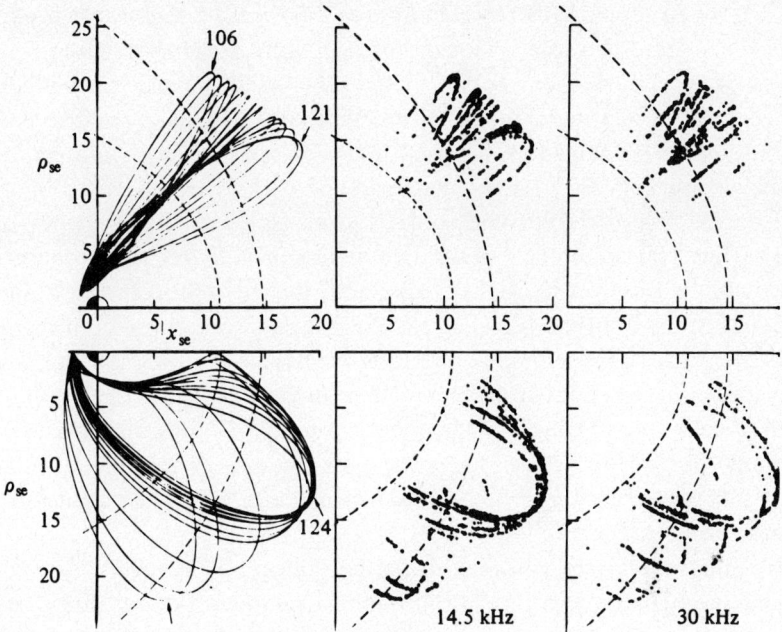

Fig. 19.15. Distribution of HF emission at the Langmuir frequency in the magnetosheath (between the magnetopause and the bow shock); OGO 5, 12 Dec. 1968 to 8 April. 1969. The oscillations at ω_0 (dots) were almost always recorded along the OGO 5 orbit (left-hand part of figure) (Fredricks et al., 1972).

regular occurrence. They appear to be distributed isotropically in space. Fig. 19.15 shows the results of some observations of these waves with OGO 5 from 2 Dec. 1968 to 8 April 1969. It illustrates the situation clearly (Fredricks et al., 1972). The left-hand part of the diagram shows OGO 5 orbits 106–121 and 124–154, in the solar-ecliptic coordinates x_{se} and ρ_{se}. In the upper part of the figure, the apogee of each orbit corresponds to a time after local noon, while in the lower part to morning. The dashed curves indicate the average locations of the magnetopause and of the shock-front region during this period. In the right-hand part of the figure, dots indicate cases where on the channels $f = 14.5$ and $30\,\text{kHz}$ oscillations with a field strength $E > 0.3\,\text{mV}\,\text{m}^{-1}$ (the noise level was $\simeq 0.07\,\text{mV}\,\text{m}^{-1}$) were recorded along the satellite orbit. The field strengths of these waves varied in different cases from $E \simeq 0.1$ to $10\,\text{mV}\,\text{m}^{-1}$. It is seen that the observed HF oscillations with frequencies $\omega \simeq \omega_0$ were recorded uniformly along each orbit, apparently partly in the magnetosheath and always in the shock-front region. Previous studies with OGO 5 (Scarf et al., 1971a; Fredricks et al., 1971) had shown that streams of electrons with energies

$E_e > 700\,\text{eV}$, apparently reflected from the shock-front region, are responsible for the excitation of Langmuir oscillations in the shock front.

Over the shock-front region in the solar wind, intense narrow-band Langmuir oscillations with $f \simeq 10$ to $30\,\text{kHz}$ were also observed frequently with IMP 6 (Rodriguez & Gurnett, 1975). These experiments showed that the electric field **E** of the waves is parallel to $\mathbf{H}_0$, which agrees with the expected properties of longitudinal plasma waves at the frequency ω_0. The mechanism of their generation is assumed to be the two-stream instability. Electron streams are apparently ducted in the solar wind along the force lines of the magnetic field $\mathbf{H}_0$ that cross the shock-front region. In these experiments wide-band electrostatic oscillations $(\mathbf{E} \| \mathbf{H}_0)$ were recorded in the shock-front region, their upper frequency limits being at the local values of the plasma frequency $f_0 \simeq 20$ to $30\,\text{kHz}$ and their lower limits being as low as $f \simeq 1\,\text{kHz}$. These are assumed to be Doppler-shifted plasma waves with $f < f_0$, propagated in the direction opposite to the solar-wind velocity **V**.

In the *interplanetary medium*, under the influence of the variations taking place in the solar wind, some collisionless phenomena of the shock type, analogous to the Earth's bow shock but usually weaker, are also observed. Narrow-band plasma waves with $f \simeq f_0$, apparently excited by shock-type phenomena, were recorded at heliocentric distances from the Sun $R_\odot \simeq 0.47\,\text{AU}$ (Gurnett *et al.*, 1978b), but such phenomena seem to be rare. These electrostatic waves are weak. In the given case at $f = 31.1\,\text{kHz} \simeq f_0$ their field strength was of the order of $E \simeq 35\,\mu\text{V}\,\text{m}^{-1}$. Note, too, with respect to this, that electric and magnetic fields in the interplanetary medium associated with shock-type phenomena were also recorded earlier in other experiments (see Scarf *et al.*, 1974a; Neubauer *et al.*, 1977; Scarf, 1978). On Helios 2 wide-band electrostatic waves in the range $f \simeq 1$ to $30\,\text{kHz}$ with a maximum at $f \simeq 3\,\text{kHz}$ were also observed. The excitation of such electrostatic waves is typical of the Earth's bow shock; they can be assumed to be ion-acoustic waves (see §17.1). In these experiments a monotonically diminishing emission spectrum in the range $f \simeq 10$ to $1000\,\text{Hz}$ was also observed, these waves being recorded on a magnetic antenna. Their field strength varied from $H \simeq 10\gamma$ at $f \simeq 10\,\text{Hz}$ to $H \simeq 10^{-3}\gamma$ at $f \simeq 600$ to $800\,\text{Hz}$. The cutoff of this emission spectrum corresponded to $f \simeq f_H/2$; it naturally pertains to the electron-acoustic mode of the waves $n_2(\omega)$ (see Fig. 4.3, Volume 1).

An interesting case is shown in Fig. 19.16. It portrays the spectrum of electromagnetic waves observed with IMP 8 in the solar wind at a geocentric distance $R \simeq 43.4 R_0$ on the illuminated side of the Earth

19.2. Magnetosphere, interplanetary medium and solar wind

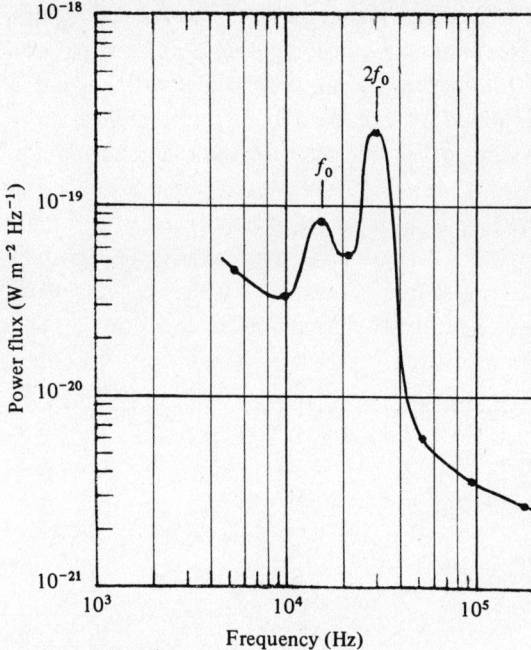

Fig. 19.16. Spectrum of HF emission in the solar wind in the shock-front region, with maxima at the plasma frequency f_0 and at its second harmonic $2f_0$; IMP 8, geocentric distance $R = 4.34R_0$, 18 March 1974 (Gurnett & Frank, 1975).

(Gurnett & Frank, 1975). This spectrum describes the properties of waves arriving from the shock-front region. The emission has two maxima, at the plasma frequency f_0 and at its second harmonic $2f_0$. Such spectra were often observed when IMP 8 passed through the shock-front region. They are stable for long periods of time, of the order of an hour or more. This emission, as is evident from Fig. 19.16, is generally weak. The maximum intensity is always at $f = 2f_0$; the peak at f_0 is sometimes absent. At times these oscillations disappear suddenly, at the same time as the direction of the magnetic field $\mathbf{H}_0$ of the solar wind changes. The intensity of the waves at $f = 2f_0$ is usually not associated with local Langmuir oscillations of the plasma. This is apparently because the source of the emission is far away from the observation point. Measurements of the direction of arrival of these waves indicate that the source has an angular size of $\simeq 20°$. The appearance of these waves is probably connected with the excitation of longitudinal Langmuir oscillations in a region close to the shock front, due to the two-stream instability of the plasma caused by the influx of hot particles. These oscillations are transformed into electromagnetic (transverse) waves as a result of a nonlinear wave–wave interaction. The emission at f_0 apparently appears in this case because of an interaction of the plasma oscillations with an ion-acoustic wave, while the emission at $2f_0$ is caused by an interaction between longitudinal f_0 waves propagating in different directions (Papadopoulos, et al., 1974).

Fig. 19.17 shows some other results of observations of plasma electrostatic waves. These measurements were carried out in the *interplanetary medium* with Helios 1, the heliocentric orbit of which had a perihelion $R_\odot = 0.309$ AU and an aphelion $R_\odot = 0.985$ AU (Gurnett & Anderson, 1977). In these experiments narrow-band emission in the range $f_0 \simeq 20$

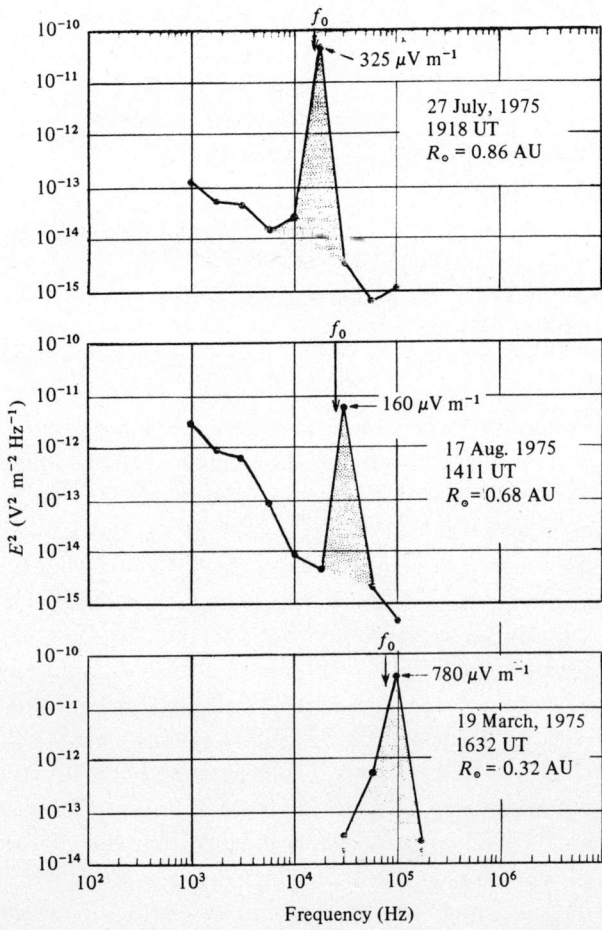

Fig. 19.17. Spectra of HF plasma waves; Helios 1, in a heliocentric orbit at various distances $R_\odot$ from the Sun (Gurnett & Anderson, 1977).

to 100 kHz was observed at different distances from the Sun in the solar wind. The generation periods of these oscillations had durations ranging from a few minutes to about half an hour; such events occurred two or three times a month. The emission was especially intense during periods of high solar-flare activity, when low-energy electrons with $E_e \simeq 1$ to 20 keV were observed in the solar wind. The spectra of plasma oscillations given in Fig. 19.17 illustrate their general behavior: with the approach to the Sun, their frequency increases, $f \sim R_\odot^{-1}$, but the intensity of these waves does not always increase. An analysis of the structure of this radiation shows it to consist of short bursts with durations Δt of some seconds or minutes. Consequently, the mean field strengths E of the plasma oscillations will be considerably lower than their peak value $E_{max} \simeq 100$ to 500 μV m; cases where $E_{max} \simeq 1$ mV m^{-1} were rare. It was shown that, to an accuracy of $\simeq 20°$, the electric field $\mathbf{E}$ was perpendicular to $\mathbf{H}_0$.

Gurnett et al., 1978a, considered the totality of observational data on plasma waves ω_0 in the solar wind, gathered over 4 to 7.5 years at different times with IMP 6 and 8, Helios 1 and 2, and Voyager 1 and 2. They concluded that the appearance of these waves is always associated with type-III solar flares. Their field strength diminishes with the heliocentric distance $R_\odot$ approximately as $R_\odot^{-\frac{3}{5}}$. At distances from the Sun $R_\odot \simeq 0.3$ to 0.4 AU, where AU is the astronomical unit, the maximum values of $E \simeq 7$ mV m^{-1}, while at $R_\odot \simeq 1$ AU, we have $E \simeq 0.2$ mV m^{-1}. The probability of exciting Langmuir oscillations in the solar wind decreases rapidly further away from the Sun.

The resonant excitation of *longitudinal* Langmuir waves ω_0 in the *solar*

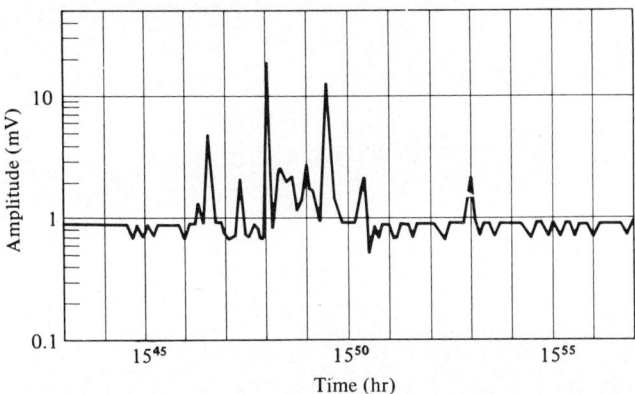

Fig. 19.18. Recording of the amplitude of longitudinal plasma waves at the Langmuir frequency ω_0 of the electrons, in the solar wind at $R \simeq 10^6$ km from the Earth; Pioneer 8, 21 Oct. 1967 (Scarf et al., 1968b).

wind was apparently first recorded with Pioneer 8, the heliocentric orbit of which had an aphelion at 1.09 AU and a perihelion at 1 AU (Scarf et al., 1968b). Fig. 19.18 shows a recording of the amplitude of these waves at $f = 20\,\text{kHz}$, at a distance from the Earth of the order of $R \simeq 10^6\,\text{km}$. Evaluations of the experimental results indicated that the energy density of the Langmuir waves was about $E^2 \simeq 10^{-17}$ to $10^{-15}\,\text{erg cm}^{-3}$. At the same time, the particle-flux kinetic-energy density NMV^2 was $\simeq 10^{-9}\,\text{erg cm}^{-3}$, while the thermal energy of the electrons NkT_e was $10^{-10}\,\text{erg cm}^{-3}$. Consequently, the energy of the Langmuir waves only constituted a small fraction of the total energy present.

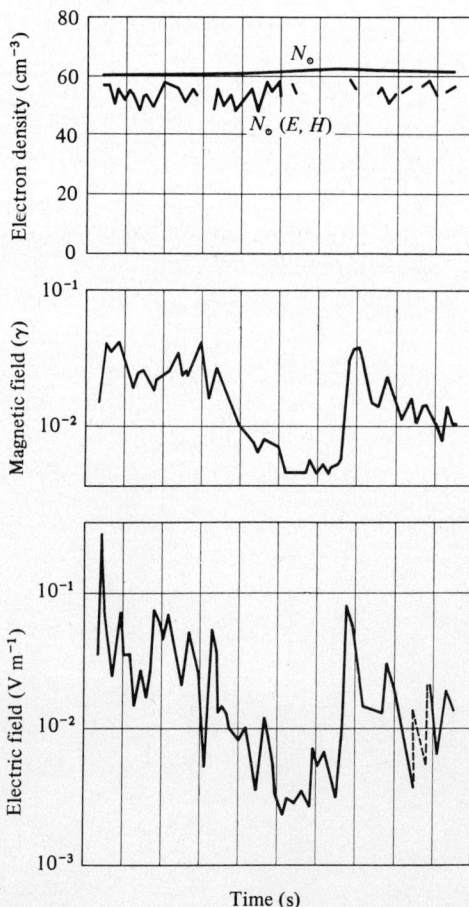

Fig. 19.19. Determination of the electron density $N_\odot$ from the ratio of the electric and magnetic field strengths of transverse waves at the Langmuir frequency ω_0. Data recorded by OGO 5, on 5 April 1968, in the solar wind (Scarf et al., 1970a).

19.2. Magnetosphere, interplanetary medium and solar wind

Transverse electromagnetic waves in the solar wind close to the Langmuir frequency were apparently first detected in the OGO 5 experiments at distances from the Earth $R \simeq 10^5$ km (Scarf *et al.*, 1970a). Fig. 19.19 shows some observational results obtained at the time of a powerful flare at $f = 70$ kHz, when the electron density $N_\odot$ in the solar wind was 8 or 10 times its normal value. The plasma frequency in the solar wind is usually $f_0 \simeq 14$ to 30kHz. The electric and magnetic field components, **E** and **H**, were measured with OGO 5 simultaneously, and probe measurements of the plasma density $N_\odot$ were carried out. Fig. 19.19 portrays the electron densities $N_\odot$ and $N_\odot(E,H)$, respectively measured by the probe and determined from the ratio of E to H for transverse electromagnetic waves; the agreement is seen to be good. In another work (Scarf *et al.*, 1971b) similar determinations of $N(E,H)$ were carried out using E and H values measured during a violent magnetic storm (frequency $f \simeq 70$ kHz, OGO 5, distance from the Earth's centre $R \simeq 4.7 \times 10^3$ km). A direct measurement of the electron density in this experiment gave $N = 63$ cm^{-3}, while the value obtained from the measurements of E and H was $N(E,H) \simeq 61$ cm^{-3}. Thus, in this experiment, just as in the results shown in Fig. 19.19, a very good fit is obtained between the values of N and $N(E,H)$.

The detection of transverse electromagnetic waves (**K**$\perp$**E,H**) in the solar wind at distances from the Sun greater than those at which the above-mentioned electrostatic waves were observed indicates that in the solar wind longitudinal waves are transformed into transverse waves. This transformation of one kind of wave into another apparently takes place fairly close to the Sun. However, the transverse waves may be trapped by field-aligned irregularities in the solar-wind plasma, ducted by them, and then detected close to the Earth.

To conclude this section, let us note that during one of the most violent *magnetic storms* of the present solar-activity cycle, on 1 November 1968, as OGO 5 crossed to boundary of the *polar cusp*, some very intense waves of diverse types, including HF waves, were observed (Scarf *et al.*, 1972, 1973a,b,1974 a,b,c). Some of the results obtained during this period in the VLF–LF ranges were presented in Chapter 17. In the frequency range of interest to us here, at a geocentric distance $R \simeq 3.2R_0$ ($L = 4.7$), at times intense HF waves with an electric-field strength $E \simeq 27$ mV m^{-1} were detected at $f = 70$ kHz, a frequency close to the Langmuir frequency f_0 and to the upper-hybrid frequency f_U of the cold plasma component ($E_e < 10$ eV). At the same time, at $f \simeq 30$ kHz, which corresponded to the plasma frequency f_0 of the energetic electrons ($E_e > 50$ eV) – the hot component of the plasma – no sizable oscillations were recorded. Later, on the other hand, oscillations were excited at $f \simeq 30$ kHz and none were detected at $f \simeq 70$ kHz. Apparently, the two processes of plasma-wave generation (cold and hot plasma) are not related to one another. Gyroresonance excitation of high-intensity oscillations was also observed at $f = 14$ kHz $\simeq f_H$. Scarf *et al.* assumed that the main generation mechanism for the oscillations in question at the boundary of the polar cusp is drift instability of the plasma.

19.2.3. Continuous spectrum of HF transverse electromagnetic waves trapped in the plasmatrough (trapped nonthermal continuum radiation, TNCR)

In §19.2.2 we just described the electrostatic oscillations of low intensity which almost always exist in the magnetosphere. Apparently, there is an analogous *continuous background* between the plasmapause and the magnetopause, consisting of transverse *electromagnetic waves*; these were observed in the range $f \simeq 5$ to 20 kHz with IMP 6 (Gurnett & Shaw, 1973). The spectra of these oscillations are of the noise type. They are cut off at the bottom at frequencies $f \simeq 5$ to 10 kHz, equal to the Langmuir frequencies of the plasma (Fig. 19.20). The upper frequency limit of these oscillations was close to the harmonics sf_H of the electron gyrofrequency. Near the plasmapause these oscillations were sometimes associated with emission at the upper-hybrid frequency f_U. The emission is cut off abruptly on the Earth's dayside near the magnetopause, at distances

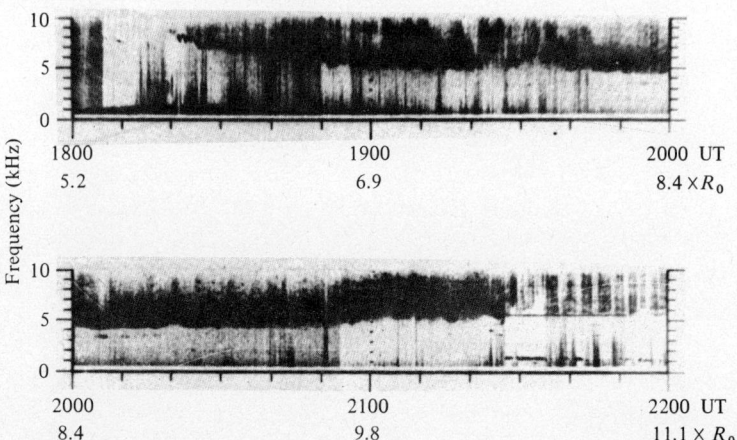

Fig. 19.20. Spectrograms of an emission at $f > f_0$, trapped in the magnetosphere between the plasmapause and the magnetopause (trough region). The emission is cut off at the Langmuir frequency f_0; IMP 6, 6 April 1972 (Gurnett & Shaw, 1973).

19.2. Magnetosphere, interplanetary medium and solar wind

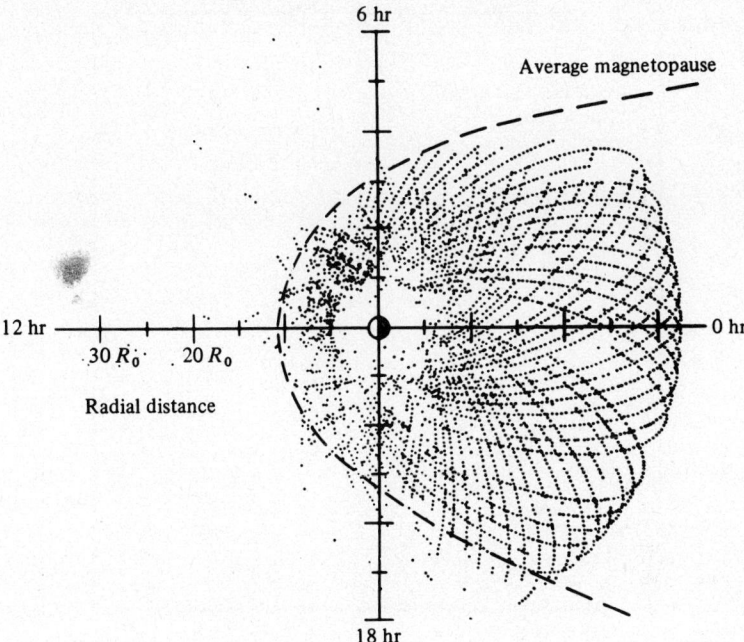

Fig. 19.21. Region of the Earth's vicinity (plasmatrough and magnetotail), in which trapped emission of electromagnetic waves is observed, at frequencies less than the plasma frequency both at the plasmapause and in the solar wind. These measurements were made at $f = 10\,\text{kHz}$ with IMP 6 (Gurnett & Shaw, 1973).

$R \simeq 10 R_0$. On the nightside it was recorded in the magnetotail out to $R \simeq 32 R_0$ (Fig. 19.21). The intensity of this radiation turned out to be stable from case to case. Although it almost always exists, it is very weak. The field strengths are usually $E \simeq 5\,\mu\text{V}\text{m}^{-1}$, and they rarely exceed $10\,\mu\text{V}\text{m}^{-1}$. Mainly at $f = 10\,\text{kHz}$, the energy density is $E^2 \simeq 6 \times 10^{-16}$ to $6 \times 10^{-15}\ \text{V}^2\text{m}^{-2}\text{Hz}^{-1}$ (Fig. 19.22).

This radiation, which is *wide-band* and *trapped* in the plasmatrough between the plasmapause and the magnetopause, is called in the literature *trapped nonthermal continuum radiation* (TNCR). Outside this region, and also in the solar wind, the electron density and thus the Langmuir frequency) is higher than inside it; therefore waves generated in the trough and in the tail cannot escape from these regions, being reflected from the boundaries. Beyond these boundaries the refractive index of the generated waves $n^2 < 0$. The TNCR spectrum is cut off at the top at a frequency

$$\omega_c = (\omega_H/2) + [(\omega_H^2/4) + \omega_0^2)]^{1/2}, \tag{19.3}$$

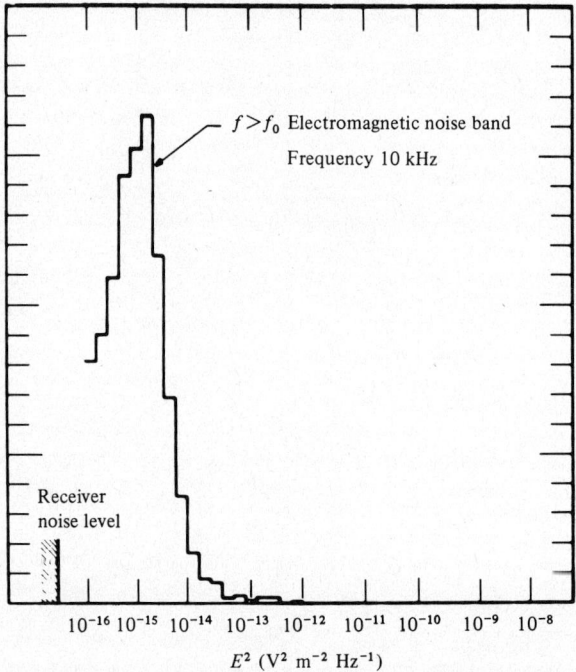

Fig. 19.22. Intensity distribution of electromagnetic waves trapped in the plasmatrough. The histogram shows, on an arbitrary scale, the proportion of the observing time during which their intensity lay in each of the indicated ranges. Data obtained with IMP 6 at $f = 10\,\text{kHz}$ (Gurnett & Shaw, 1973).

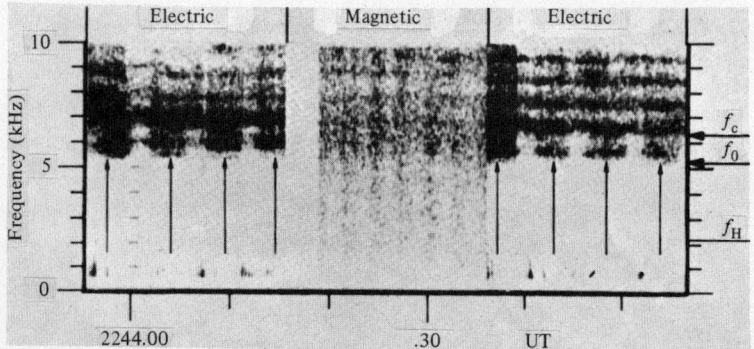

Fig. 19.23. Spectrogram of radiation trapped in the magnetosphere, showing a series of bands with a uniform spacing $\Delta f \simeq 1\,\text{kHz}$; IMP 6, 10 May 1971, $R = 8.3 R_0$ (Gurnett & Shaw, 1973).

19.2. Magnetosphere, interplanetary medium and solar wind

where the refractive index of the fast HF extraordinary wave is zero (see Fig. 4.3, Volume 1).

In some cases the spectrum of the TNCR trapped in the magnetosphere reveals that it is not purely white noise. Fig. 19.23 shows an example, in which a series of bands with a uniform spacing $\Delta f \simeq 1$ kHz was observed. This frequency (see Gurnett & Shaw, 1973), was unrelated to the local value of the electron gyrofrequency f_H for the region in which the measurements were carried out. There were also cases when the trapped-wave spectrum consisted of discrete packets (chorus), the durations of which were a few tenths of a second to a few seconds. Unlike the usual chorus, however, most of these wave packets had frequencies that decreased with time. At present, there are apparently not yet enough data to allow a decision to be made on the most probable generation mechanism for the 'trapped' radiation (see the following section and Gurnett & Shaw, 1973).

Note here, too, that electromagnetic noise of a type different from the above was detected in the magnetosphere with OGO 1, mainly at $L > 5$ on the Earth's nightside (Dunckel et al., 1970). Fig. 19.24 shows some observational data obtained at a geocentric distance from the Earth $R \simeq 7.2R_0$. In the band $f \simeq 0.3$ to 100 kHz a magnetic antenna picked up quite intense oscillations in the form of bursts lasting a few minutes or less. The amplitude of this *broad-band noise* does not exhibit any special features in the vicinity of the characteristic resonance frequencies of the plasma ω_H and ω_0, and it decreases with the frequency approximately as $\omega^{-1.7}$ at $f \simeq 0.3$ to 2.4 kHz and as $\omega^{-1/2}$ at $f \simeq 12$ to 100 kHz. The median

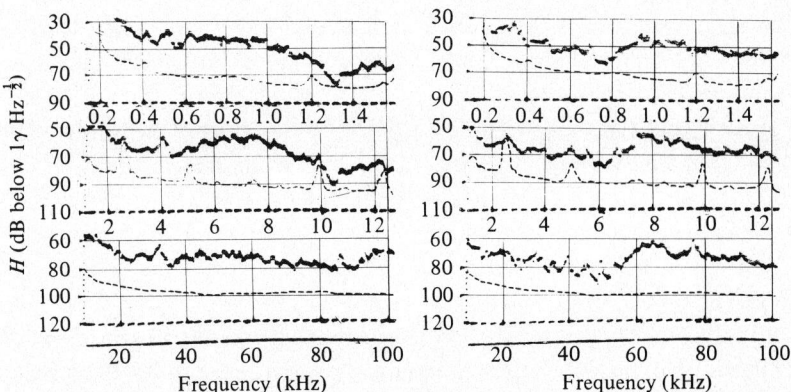

Fig. 19.24. Spectra of emissions of the 'broad-band' noise type. The dashed curve shows the noise level; OGO 1, $L = 9.7$, $\phi = 30°$, $R = 7.2R_0$, 5 May 1965 (Dunckel et al., 1970).

value of the noise intensity varied from $H^2 \simeq 10^{-5}$ to $6 \times 10^{-11} \gamma^2 \, \text{Hz}^{-1}$ as the frequency varied from $f \simeq 0.3$ to 100 kHz. Broad-band noise of this type was apparently generated in the plasma in the vicinity of the observation point (OGO 1), since in part of the frequency range it was not able to escape from these plasma regions. The deep modulation of their intensity is the most important feature of these oscillations. The modulation period was approximately equal to the time of rotation of OGO 1 about its axis. Dunckel *et al.*, 1970, assumed that the generation mechanism of this radiation is similar to the shot effect in an electronic vacuum tube. As far as the author knows, no further results of analogous observations have since appeared in the literature. The properties of this noise and its origin are still unknown.

19.2.4. Continuous spectra of HF radiation from the Earth's magnetosphere travelling beyond its limits (nonthermal continuous radiation, NCR; auroral (or terrestrial) kilometric radiation, AKR, TKR)

Noise spectra of HF waves trapped between the plasmapause and the magnetopause, in the frequency range $f \simeq 5$ to 20 kHz, were considered in the foregoing section. However, these are only a part of the *nonthermal continuum radiation of the magnetosphere* surrounding the Earth. It is generated in different plasma regions over a very wide frequency range $f = 0.5$ to 100 kHz (Gurnett, 1975). The lower boundary of the NCR spectrum is the local Langmuir frequency ω_0. In the distant part of the magnetotail it may correspond to $f_0 \simeq 500$ Hz. At frequencies higher than the Langmuir frequency of the solar wind, this radiation travels beyond the magnetosphere and easily escapes from the Earth. It is generated in the Earth's plasmatrough in a broad region at distances $R \simeq 4R_0$ to $8R_0$, mainly on the dayside from 0400 to 1400 LT, at low and middle latitudes, including the equatorial zone. Fig. 19.25 shows typical spectra of HF waves of this type, observed by IMP 8 in the solar wind and in the magnetotail at various distances from the Earth. The intensity of the radiation decreases rapidly for $f > 20$ kHz; at $f \gtrsim 100$ kHz it is lower than the intensity of the galactic noise. Similar wave spectra were recorded earlier with IMP 6 (Brown, 1973) at $f = 30$ to 110 kHz. Their intensity decreased approximately as $\omega^{-2.8}$.

Fig. 19.26 gives recordings that illustrate the nature of the intensity distribution of the continuum radiation at $f = 56.2$ kHz, as IMP 8 receded from the Earth, starting at the plasmapause, i.e., within the magnetosphere. The location of the plasmapause is marked by the peak at the Langmuir frequency $f_0 = 56.2$ kHz. This value was obtained from an analysis of

19.2. Magnetosphere, interplanetary medium and solar wind

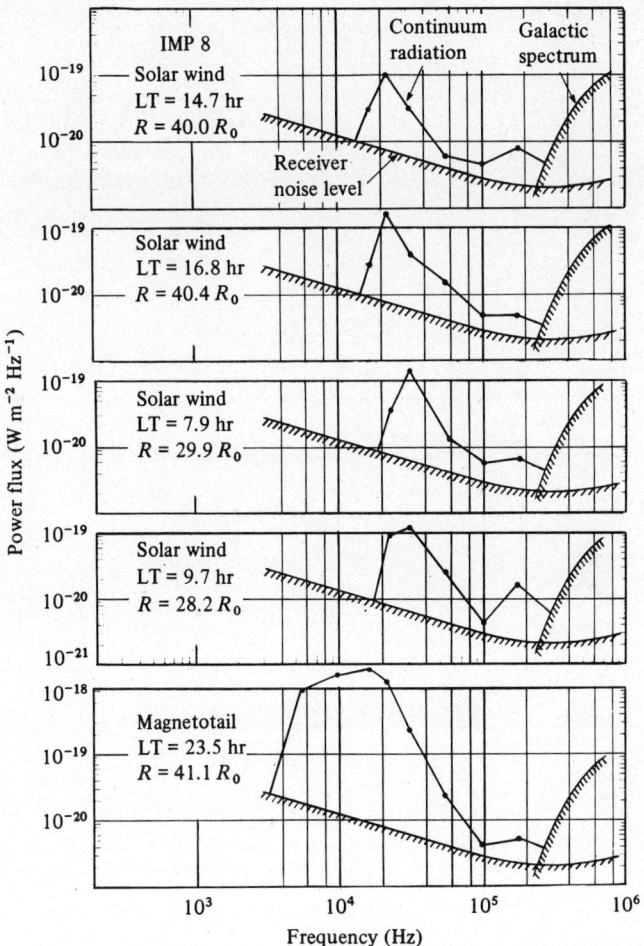

Fig. 19.25. Spectra of transverse waves (nonthermal continuum radiation) generated in the magnetosphere, in the solar wind, and in the magnetotail; ± MP 8, 1974 (Gurnett, 1975).

electrostatic resonance oscillations at the upper-hybrid frequency, excited in this region of the plasma. Further away from the Earth, beyond the plasmapause, at $f > f_0$ NCR was recorded continuously. Analysis of a large set of observational data at $f = 56.2$ kHz indicated that the intensity peak of these waves occurred between $R \simeq 4R_0$ and $8R_0$. Their mean power flux was greater than $S \simeq 5 \times 10^{-20}$ W m^{-2} Hz^{-1} and it varied on the whole from $S \simeq 10^{-17}$ to 10^{-20} W m^{-2} Hz^{-1}. The position of the radiation source deduced from these data agreed well with the results of

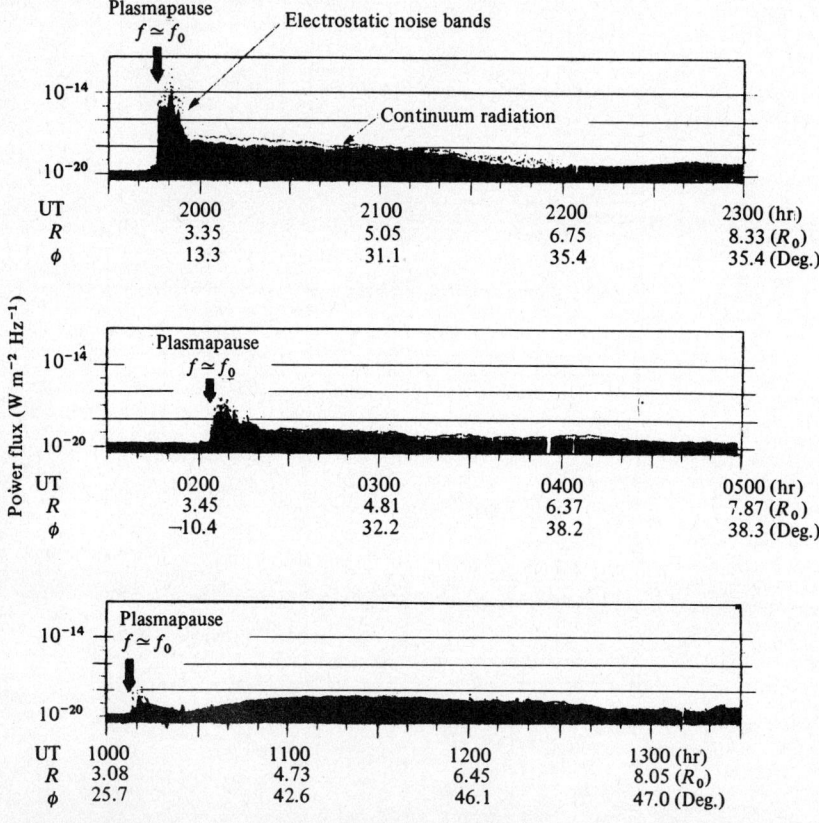

Fig. 19.26. Energy flux of the nonthermal radiation of transverse waves at $f = 56.2$ kHz, as a function of the geocentric distance; IMP 8 (Gurnett, 1975).

locating it from the direction of the incoming waves at distances from the Earth $R \simeq 20R_0$ to $30R_0$. The latter gave an angular source size of $\simeq 40°$.

The mechanism of generation of the nonthermal continuum radiation is assumed to be gyrosynchrotron emission of energetic electrons trapped in the Earth's outer radiation belt (Frankel, 1973). However, analysis of the experimental data revealed a number of facts that do not agree too well with this assumption. For example, studies of this radiation with Hawkeye 1 and IMP 8 during a period of heightened activity following a magnetic storm (Fig. 19.27) indicated that at this time powerful streams of electrons with intensities from $E_E \simeq 1$ to 30 keV were injected into the outer radiation belt (Gurnett & Frank, 1976). An analysis of these observational data showed the most likely cause of the plasma excitation

19.2. Magnetosphere, interplanetary medium and solar wind

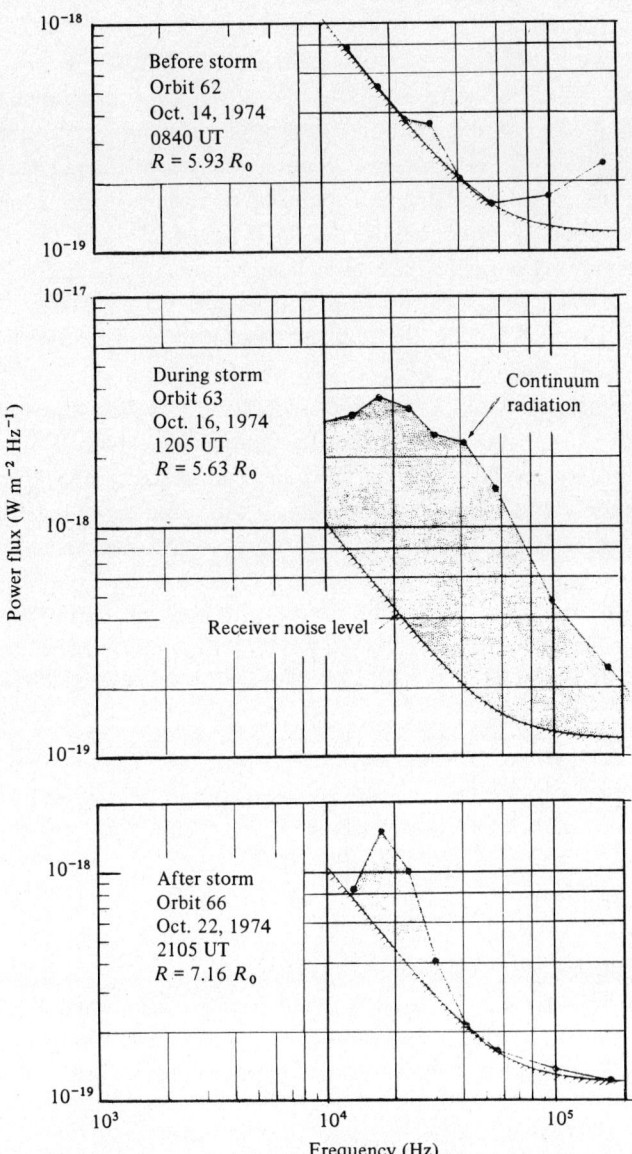

Fig. 19.27. Energy-flux spectra of continuum radiation. Results of observations made in the magnetosphere with Hawkeye 1 during a magnetic storm (Gurnett & Frank, 1976).

of HF waves at this time to be instability arising due to interaction of the plasma with energetic electrons $E_e > 10\,\text{keV}$ and coherent radiation. A serious argument in favour of this is the fact that NCR is often accompanied by the excitation of electrostatic oscillations at $f = (s + \tfrac{1}{2})f_H$ with high values of s. Consequently, there is reason to assume that the nonthermal continuum radiation appeared as a result of transformation and interaction with electrostatic waves generated at the plasmapause.

Fig. 19.28 presents a different type of continuous spectrum of the electromagnetic waves emitted by the Earth's magnetosphere in the range $f \simeq 50$ to $500\,\text{kHz}$, higher than the plasma frequency of the solar wind (Gurnett, 1974). This spectrum was constructed on the basis of IMP 6 data. For comparison, the figure gives spectra of electromagnetic waves trapped in the magnetosphere as well as of the nonthermal continuum radiation that is able to escape. The spectrum is seen to have a broad maximum at $f \simeq 100$ to $300\,\text{kHz}$. The intensity of these waves drops sharply for $f < 100\,\text{kHz}$ and $f > 300\,\text{kHz}$, and it decreases with distance approximately as R^{-2}; it varies rapidly in time, and it may increase in a few minutes. During magnetic storms these waves are observed continually for anywhere from half an hour to several hours, and they disappear completely between storms. Often the amplitude of this radiation is lower than the galactic-noise level. It was apparently first detected with OGO 1 (Dunckel et al., 1970), and then observed with IMP 6 (Brown, 1973). In

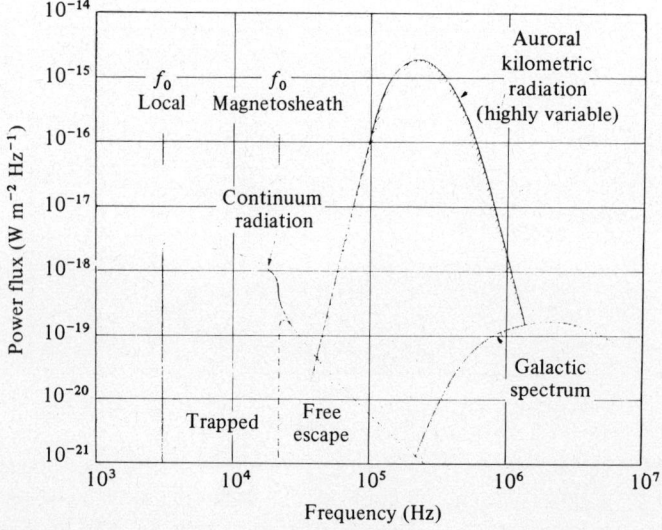

Fig. 19.28. Continuous spectra of HF electrostatic waves generated in the magnetosphere ($R \simeq 30R_0$); IMP 8 (Gurnett et al., 1976).

19.2. Magnetosphere, interplanetary medium and solar wind

these works this radiation was called, respectively, 'high-pass noise' and 'midfrequency noise'. Gurnett, 1974, on the other hand, and a number of authors by whom this radiation of the Earth was studied in detail (Kurth *et al.*, 1975; Voots *et al.*, 1976; Green *et al.*, 1977), called it 'auroral (or terrestrial) kilometric radiation' (AKR or TKR), since its source is located in the polar region and its free-space wavelengths are of the order of a kilometre.

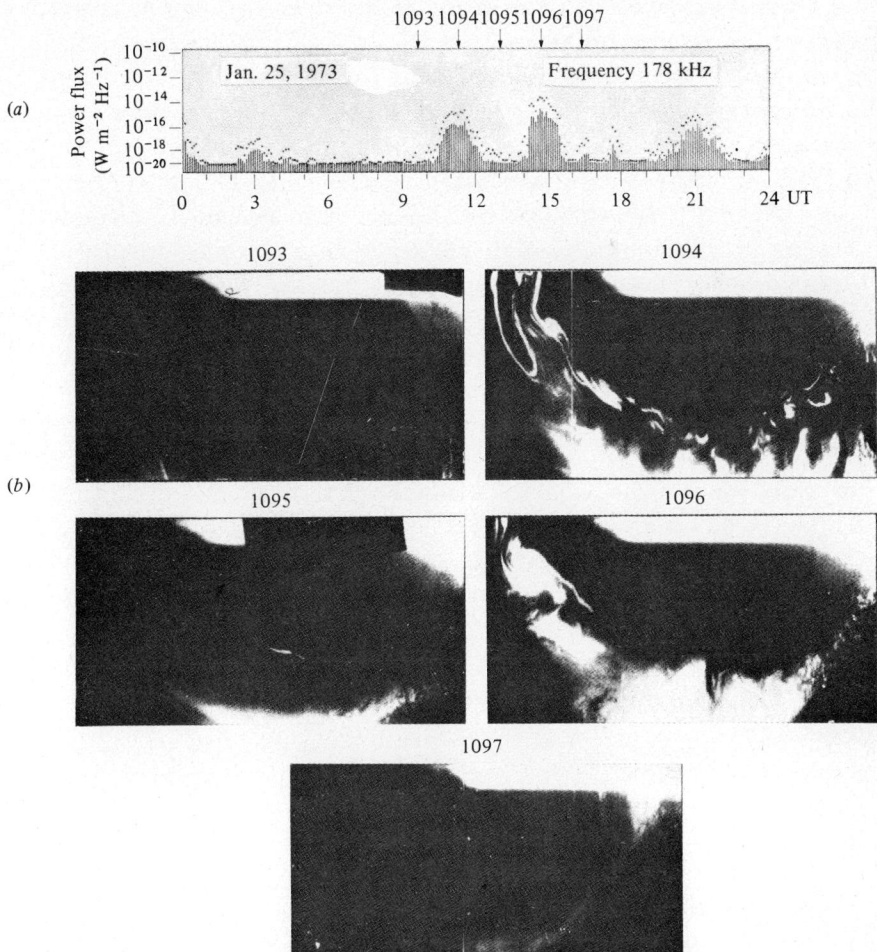

Fig. 19.29. (*a*) Intensity of the continuum radiation of HF electromagnetic waves generated in the polar zone at 178 kHz, recorded with IMP 6, 25 Jan. 1973, orbits 1093–1097 (Kurth *et al.*, 1975). (*b*) Photos of polar aurorae, taken at the same times as the bursts of this radiation.

The appearance of auroral kilometric radiation is intimately associated with the intensities of polar aurorae (see Fig. 19.29, Kurth *et al.*, 1975, and Voots *et al.*, 1976). Direction-finding measurements with IMP 6 and 8 and Hawkeye 1 indicated that the source of the AKR is close to the magnetic force line $\phi \simeq 70°$ at a geocentric distance $R \simeq 2R_0$ in the evening and night sectors of the Earth's polar zone. The distance of the source from the Earth varies in general from $R \simeq 2R_0$ to $3R_0$, and measurements at $R \simeq 32R_0$ give it an angular size $\Omega_T \simeq 12°$.

Hawkeye 1 was also used for a series of measurements along polar force lines at geocentric distances $R \simeq 1.5R_0$ to $2.5R_0$, where the AKR is generated (Gurnett & Green, 1978). In this region of the near-Earth plasma the electron Langmuir frequency ω_0 is usually considerably lower than the electron gyrofrequency ω_H. Thus the lower cutoff of the AKR at $\omega \simeq \omega_H$ which is observed here classifies the emission as the mode of the extraordinary fast wave (see Figs. 4.3 and 4.5, Volume 1), having a clockwise polarization sense. Since here $\omega_H \gg \omega_0$, the cutoff frequency of this mode $\omega_1 > \omega_H$ (see formula (4.25), Volume 1). As in the experiments at great distances, the frequency range of the intense AKR was $f \simeq 100$ to 300 kHz. The maximum power flux at distances $R \simeq 2R_0$ was $W_E \simeq 10^{-11}$ W m^{-2} Hz^{-1}, and it was often $\simeq 10^{-12}$ W m^{-2} Hz^{-1} at $f = 178$ kHz. This radiation was recorded only on the nightside of the Earth.

Observations with Explorer 2, which travelled along a circumlunar orbit, gave us more insight into the location of the AKR (TKR) source and its properties (Alexander & Kaiser, 1976, 1977). Data from numerous experiments with Explorer 2 at $f = 250$ kHz showed that the AKR frequently arrives from the Earth's polar regions from a distance $R = 2R_0$, where $f = f_H$. However, in many cases its source is located at distances $5R_0 < R < 15R_0$. On the dayside the AKR is generated in the polar cusp and the plasmatrough, at geomagnetic latitudes $\phi \simeq 74$ to $80°$, in a sector close to 1200 LT. Its intensity is cut off abruptly at $\phi \simeq 80°$. On the Earth's nightside the source of the AKR is located in the geomagnetic tail at a latitude $\phi \simeq 70°$ at distances $R > 4R_0$ in the 2200 LT sector. During very active periods the AKR is generated on the Earth's dayside in the plasmatrough at distances $R \gtrsim 12R_0$.

The auroral kilometric radiation, in contrast to the nonthermal continuum radiation described earlier in this section, is observed in a limited range of angles, being cut off as the geomagnetic equator is approached. The solid angle of a beam of these waves increases with the frequency.

19.2. Magnetosphere, interplanetary medium and solar wind

For instance, Green et al., 1977, give the following values for the solid angle Ω_r of the AKR cone:

$f = 56.2\,\text{kHz},\ \Omega_r \simeq 1.1\,\text{sr};\ f = 100\,\text{kHz},\ \Omega_r \simeq 1.8\,\text{sr},$
$f = 178\,\text{kHz},\ \Omega_r \simeq 3.5\,\text{sr};\ f = 500\,\text{kHz},\ \Omega_r \simeq 5.3\,\text{sr}.$

Simultaneous observations using different satellites indicated the radiation to be quasi-isotropic – it covers the cone of the beam uniformly. This, together with the direction-finding measurements, was evidence that the source is a 'point source' (i.e. it is small).

The maximum intensity of the emission flux of these waves can be, according to measurements at $f = 178\,\text{kHz}$ at a distance $R \simeq 30 R_0$,

$$E_r^2 = W_r \simeq 10^{-14}\,\text{W}\,\text{m}^{-2}\,\text{Hz}^{-1}. \tag{19.4}$$

Assuming a uniform radiation intensity over Ω_r in the solid angle $\Omega_r = 6.5\,\text{sr}$ and in a frequency band $\Delta f \simeq 300\,\text{kHz}$, we obtain the following value for the *power of the AKR source* (*the source of the Earth's HF radiation*):

$$W \simeq 7 \times 10^8\,\text{W} \tag{19.5}$$

(Gurnett, 1974).

The following generation mechanisms have been proposed (Green et al., 1977): (1) electrostatic electron-beam instability, which is accompanied by transformation into electromagnetic waves at $f \simeq \tfrac{3}{4} f_H$; (2) Doppler-shifted gyroresonance emission, caused by precipitated electrons with $E_e \simeq 50\,\text{keV}$ (so-called inverted-V electrons), moving along magnetic field lines in the polar zone; (3) possible excitation of these waves at twice the upper-hybrid resonance frequency $\omega \simeq 2\omega_U$; (4) a current-driven instability mechanism of generation, which operates because of the activity of the ion-cyclotron instability in the polar current system (Palmadesso et al., 1976). The nature of this auroral kilometric radiation, a phenomenon which is indubitably of great interest, is, however, still unexplained.

20

Energy densities of various types of waves

It will be advisable to present in this chapter some data on the intensities of the wave processes generated in various regions of the ionosphere, interplanetary medium, and solar wind, and to compare these with the thermal energies of the corresponding plasma regions. In spite of the wealth of information available in the literature concerning the wave processes observed in these media, it is often very difficult to derive from these data the quantitative parameters necessary to characterize these processes. For instance, there are no data from simultaneous measurements of such things as the density of charged particles N in the plasma, the magnetic field $\mathbf{H}_0$, and the other parameters needed to ascertain, for instance, the energy fluxes of the observed waves.

The energy fluxes of electromagnetic waves are given by the formula

$$S = (n/\mu_0 c)E^2 \qquad (20.1)$$

or

$$S = (\mu_0 c/n)H^2 \qquad (20.2)$$

where $n = n(N, \omega, H_0 \ldots)$ is the refractive index of the plasma.

Under different conditions and at different frequencies, in the above-described experiments the following values of E and H were observed:

$$E \simeq (10^{-6}-1)\,\mathrm{V\,m^{-1}}, H \simeq (10^{-3}-\ldots 1)\,\gamma. \qquad (20.3)$$

Cases are known in which the value of E has reached tens of volts per metre (!), while H was several gammas (!).

From similar published data it is possible to determine the energy densities of the wave processes:

$$\begin{aligned} W_\mathrm{E} &= \tfrac{1}{2}\varepsilon_0 E^2 n^2 \\ W_\mathrm{H} &= \tfrac{1}{2}\mu_0 H^2. \end{aligned} \qquad (20.4)$$

The corresponding data are given, in CGS units, in Tables 20.1–20.3. In some cases W_E and W_H were evaluated only very approximately, without any claims to great accuracy. In a number of cases they appear to differ from the true values by a factor of $\simeq 1$ to 10. The same is true of the

Table 20.1. *Energy densities of ELF and VLF waves:* $0 \lesssim \omega \lesssim \Omega_H$, $\Omega_H < \omega \simeq \omega_L$

Frequency, f (Hz, kHz)	Plasma region	Energy density, W_E, W_H (erg cm^{-3})	W_E/NkT, W_H/NkT	Notes
		ELF oscillations of the magnetic field		
$\omega \ll, < \Omega_H$ $(10^{-2} < f < 0.3)$ Hz	Magnetosphere, magnetosheath; $R = (5-15)R_0$	$W_H \simeq (10^{-12}-10^{-13})$	$\lesssim 10^{-2}$	Pioneer 5; Coleman, 1964, Electromagnetic (transverse) waves. Magnetic pulsations
$\omega \ll, < \Omega_H(\text{H}^+)$ $(3 \times 10^{-4} < f < 0.5)$ Hz	Interplanetary space	$\bar{W}_H \simeq (10^{-12}-10^{-13})$ $(\bar{H}_\perp^2 \sim \omega^{-3/2})$	$\lesssim 10^{-2}$	Mariner 4; Siscoe et al., 1967. Electromagnetic waves. Magnetic pulsations
$f \simeq (0.5-3)10^{-2}$ Hz	Magnetosphere; $R \simeq (6-10)R_0$	$\bar{W}_H \simeq 10^{-11}$	$\simeq (10^{-1}-10^{-2})$	Explorer 12; Patel, 1965; ATS 1; Cummings et al, 1969, 1972; Dodge; Dwarkin et al, 1971. Magnetic pulsations. PC-3,4,5. Electromagnetic waves
$f \simeq (0.5-2)$ Hz	Magnetosphere; $R \simeq 6R_0$	$W_H \simeq 10^{-11}$	$\simeq (10^{-1}-10^{-2})$	Dodge; Dwarkin et al, 1971; ATS 1; Bossen et al, 1976. Magnetic pulsations PC-1
$f \simeq 7$ Hz $\simeq 0.8 F_H(\text{H}^+)$	Polar cap; $R \simeq 5R_0$	$W_H \simeq 10^{-10}$	$\lesssim 10^{-1}$	ATS 1; McPherron et al., 1972. Quasi-periodic magnetic pulsations
		Ion-cyclotron waves		
$0 < \omega \gtrsim \Omega_H(\text{H}^+)$ $f = 6$ Hz $\simeq 0.4 F_H(\text{H}^+)$	Outer ionosphere, plasmasphere; $R \simeq 3R_0$	$W_E \simeq \ldots 10^{-18}$, $W_H \simeq \ldots 10^{-13}$	$\lesssim 10^{-2}$	Hawkeye 1; Kintner & Gurnett, 1977. Two modes of e.m. waves – transformed electrostatic waves
$\omega \simeq s\Omega_H; s = 1, 2, 3.$ $f \gtrsim (110, 220, 330)$ Hz	Polar region; $R \simeq 2R_0$ $L \simeq 4.7$–6	$W_E \simeq (10^{-13}-10^{-14})$	$\lesssim 10^{-3}$	S3-3; Kintner et al., 1978. Doppler-broadened electrostatic waves

Table 20.1. (contd.)

Frequency, f(Hz, kHz)	Plasma region	Energy density, W_E, W_H(erg cm^{-3})	W_E/NkT, W_H/NkT	Notes
		Ion-acoustic waves		
$\Omega_H < \omega \to \Omega_0$ (0.3 ≲ f ≲ 18)kHz	Outer ionosphere; $z \simeq$ (400–1500) km	$W_E(F_H) \simeq 7 \times 10^{-8}$ ($E_{max} \simeq 40$ mV m^{-1})	≲ 0.5	OGO 2; Guthart et al., 1968. Electrostatic (longitudinal) waves
$\Omega_H \ll \omega < \Omega_0$ (1.7 ≲ f ≲ 14.5)kHz	Outer ionosphere; $z \simeq$ (270–3700) km	$\bar{W}_E \simeq (4-16)10^{-17}$ $W_{E,max} \simeq 10^{-13}$–$10^{-11}$	≪ to ≃ 10^{-8} ≲ 10^{-2}	P 11; Scarf et al, 1965
$\Omega_H < \omega < \Omega_0$ $f \simeq$ (20–300) Hz	Plasmapause; $R \simeq 3.3 R_0$	$W_E \simeq (10^{-17}$–$10^{-18})$	≪ to ≃ 10^{-6}	Explorer 45; Anderson & Gurnett, 1973. Magnetic storm
$\Omega_H < \omega < \Omega_0$ $f \simeq$ (10–100) Hz	Plasmapause; $R > 3 R_0$	$W_E \simeq (10^{-21}$–$10^{-19})$	≪ to ≃ 10^{-8}	Hawkeye 1; Kintner & Gurnett, 1976
$\Omega_H < \omega < \Omega_0$ $f \simeq$ (200–800) Hz	Bow shock; $R \simeq 18 R_0$	$W_E \simeq (10^{-18}$–$10^{-16})$	≪, ≲ 10^{-5}	IMP 6; Rodriguez & Gurnett, 1975, 1976
$\Omega_0 < \omega < \omega_0$	Solar wind;	$W_E \simeq (10^{-19}$–$10^{-17})$	≃ 10^{-6}	Helios 1; Gurnett et al., 1977; Gurnett & Frank 1978a. Doppler-shifted waves
$f \simeq$ (1–10) kHz > $\frac{\Omega_0}{2\pi}$	Interplanetary space, solar wind; $R_\odot \simeq 1.7$ AU $R_\odot \simeq (0.5-1)$ AU	$W_E \simeq 10^{-19}$	≲ 10^{-8}	Voyager; Gurnett et al., 1978b
$\Omega_0 < \omega < \omega_0$ $f \simeq$ (1–3) kHz				
$f = 400$ Hz $< \Omega_0/2\pi =$ 530 Hz	Solar wind; $R \simeq 10^6$ km	$W_E \simeq (10^{-16}$–$10^{-14})$	≪ to ≃ 10^{-3}	Pioneer 8, 9; Scarf et al, 1968b

Note: Most of the quantities in this table, together with the expression for the magnetic energy density, are given in CGS units. The SI expression for the magnetic energy density is $\frac{1}{2}\mu_0 H_0^2$ and the corresponding unit is the joule m^{-3} = 10 erg cm^{-3}.

Table 20.2. *Energy densities of VLF and LF waves:* $\Omega_H < \omega \simeq \omega_L$, $\ll \omega_H$; $\omega_L \ll \omega \lesssim \omega_H$

Frequency, f (Hz, kHz)	Plasma region	Energy density, W_E, W_H (erg cm^{-3})	W_E/NkT, W_H/NkT	Notes
		Broadband VLF turbulence, electromagnetic waves		
$f \simeq (0.3-10)$ kHz	Outer ionosphere; $z \simeq (10^2-10^3)$ km	$W_E \simeq (10^{-17}-10^{-16})$	$\simeq 10^{-8}$	Injun 5; Mosier & Gurnett, 1969a, 1972; Gurnett & Frank 1972a, 1973
$\Omega_H \lesssim \omega \sim \omega_L$, $f \simeq (50-200)$ Hz $f \simeq (100-350)$ Hz	Plasmasphere $2R_0 < R \lesssim 4R_0$	$W_H \simeq (10^{-16}-10^{-14})$ $W_H \simeq (10^{-17}-10^{-14})$	$\ll$ to $\simeq 10^{-3}$	Plasmaspheric hiss; OGO 3; Russell et al., 1969, 1970 OGO 5; Thorne et al., 1973
$0 < \omega < \omega_H$ $f \simeq (2-10^2)$ Hz	Polar cusp; $R \simeq (5-8)R_0$	$W_H \simeq (10^{-14}-10^{-12})$	$\ll$, $\lesssim 10^{-2}$	Hawkeye 1; Gurnett & Frank, 1978b. Broadband magnetic noise. Whistler mode. Uniquely associated with the polar cusp
$\Omega_H \lesssim \omega \to \omega_L$ $f \simeq (1-300)$ Hz	Magnetosheath; $R \simeq (12-17)R_0$	$W_H \simeq (10^{-16}-10^{-12})$	$\ll$ to $\simeq 10^{-2}$	OGO 1; Smith et al., 1967. Broadband magnetic noise
		VLF emission of different type		
$\omega \gtrsim \omega_L, \Omega_0$ $f \lesssim 1$ kHz	Polar zone; $R \simeq 3.2 R_0$	$W_{E\max} \simeq 10^{-14}$ ($E_{\max} \simeq 30$ mV m^{-1})	$\lesssim 10^{-3}$	OGO 5; Scarf et al., 1972. Electrostatic waves
$\Omega_H < \omega < \omega_L$ $f \simeq (10-50)$ Hz	Polar zone; $R \simeq (4-4.5)R_0$	$W_{E\max} \simeq 10^{-14}$ ($E_{\max} \simeq 30$ mV m^{-1})	$\lesssim 10^{-3}$	Hawkeye 1; Gurnett & Frank, 1977. Broadband noise. Electrostatic waves

Table 20.2. (Contd.)

Frequency, f (Hz, kHz)	Plasma region	Energy density, $W_E, W_H (\text{erg cm}^{-3})$	W_E/NkT, W_H/NkT	Notes
Lines $\Delta\omega \simeq \Omega_H$ (H^+, $He\ldots$) $f \simeq (10-10^2)$ Hz	Geomagnetic equator; $R \simeq (2.5-5)R_0$	$W_{H\,max} \simeq 10^{-15}$	$\ll$ to $\simeq 10^{-4}$	Hawkeye 1; Gurnett, 1976. Band of e.m. noise with harmonically spaced lines Ω_{Hi}
$f \simeq (50-200)$ Hz	Outer ionosphere, polar zone; $z < 3000$ km	$W_{E\,max} \simeq 10^{-18}$ $W_H \simeq (10^{-16}-10^{-15})$ $W_E \simeq (10^{-16}-10^{-15})$	$\ll$ to $\simeq 10^{-4}$	'Lion's roar'. Whistler mode. Injun 5, 6; Gurnett & Frank, 1972b
	Magnetosheath; $R \simeq (10-15)R_0$	$W_H \simeq 10^{-14}$ ($f \simeq 10^2$ Hz)	$\ll$ to $\simeq 10^{-3}$	OGO 5; Smith et al., 1969
		LF whistler mode and electrostatic waves		
$0 < \omega < \omega_0$ $f \simeq 1$ Hz–(30–1000)kHz	Polar cusp; $R \simeq (5-8)R_0$	$W_E \simeq (10^{-20}-10^{-16})$	$\ll, \lesssim 10^{-5}$	Hawkeye 1; Gurnett & Frank, 1978b. Broadband electrostatic noise
$\omega_L < \omega < \omega_H$ $f \simeq (20-200)$ Hz ($f = 350$ Hz)	Bow shock; $R \simeq (16-18)R_0$	$W_E(36 \text{ Hz}) \simeq 10^{-18}$ ($E^2 \sim \omega^{-2}, H^2 \sim \omega^{-4}$)	$\lesssim 10^{-8}$	OGO 5; Rodriguez & Gurnett, 1975. Electromagnetic waves
$\omega_L < \omega < \omega_H$ $f \simeq (40-300)$ Hz	Geomagnetic tail; $R \simeq (23-46)R_0$	$W_H \simeq 10^{-16}$	$\lesssim 10^{-5}$	IMP 8; Gurnett et al., 1976. Electromagnetic waves
$(10 < f < 500)$ Hz		$W_E \simeq 10^{-17}$	$\lesssim 10^{-6}$	Electrostatic waves

See the footnote to Table 20.1.

Table 20.3. *Energy densities of HF waves*: $\omega_H, \omega_0 < \omega \to \infty$

Frequency, f(Hz, kHz)	Plasma region	Energy density, W_E(erg cm^{-3})	W_E/NkT	Notes
	Longitudinal (electrostatic) waves connected with the electron gyrofrequency ω_H			
$\omega \simeq s\omega_H \simeq \omega_0$ $f(1-\ldots 10)$ kHz	Plasmatrough; $R \simeq (4-10)R_0$	$W_E \simeq 10^{-22}$	$\lesssim 10^{-10}$	IMP 6; Shaw & Gurnett, 1975
$\omega \gtrsim \omega_H$ $f \simeq (3-5)$ kHz $\simeq (1.1-1.2)f_H$	Polar cusp; $R \simeq (5.5-6.5)R_0$	$W_E(10^{-21}-10^{-19})$	$\ll, \lesssim 10^{-8}$	Hawkeye 1; Gurnett & Frank, 1978b. Doppler-shifted waves
$\omega = (s+\frac{1}{2})\omega_H$ $f \simeq \ldots 1$ kHz	Magnetosphere; $R \simeq (6-7)R_0$	$W_E \simeq (10^{-17}-10^{-15})$	$\ll$ to $\simeq 10^{-4}$	OGO 5; Fredricks & Scarf, 1973
$\omega \simeq \omega_H$ $f \simeq 300$ Hz	Geomagnetic tail; $R \simeq 33 R_0$	$W_E \simeq 10^{-18}$	$\ll$ to $\simeq 10^{-7}$	IMP 8; Gurnett et al., 1976. Doppler-shifted waves
	Longitudinal waves connected with the plasma frequency ω_0			
$\omega \simeq \omega_0 \simeq \omega_H$ $f \simeq 70$ kHz	Boundary of the polar cusp; $R \simeq 3R_0$	$W_E \sim 10^{-14}$ ($E = 27$ mV m^{-1})	$\ll$ to $\simeq 10^{-3}$	OGO 5; Scarf et al., 1972, 1974c. Geomagnetic storm
$\omega = \omega_0, 2\omega_0$ $f \simeq (10-50)$ kHz	Solar wind beyond the bow shock; $R \simeq (10-30)R_0$	$W_E \simeq (10^{-18}-10^{-15})$	$\ll$ to $\simeq 10^{-6}$	OGO 5; Fredricks et al., 1972
$f \simeq (15,30)$ kHz	$R \simeq 43 R_0$	$W_E \simeq 10^{-15}$	$\ll$ to $\simeq 10^{-6}$	IMP 8; Gurnett & Frank, 1975
$\omega \simeq \omega_0$ $f \simeq (20-100)$ kHz	Solar wind near the Sun; $R_\odot \simeq (0.3-1)$ AU	$W_E \simeq (10^{-18}-10^{-17})$	$\ll$ to $\simeq 10^{-8}$	Helios 1; Gurnett & Anderson, 1977

Table 20.3. (Contd.)

Frequency, f(Hz, kHz)	Plasma region	Energy density, W_E(erg cm^{-3})	W_E/NkT	Notes
$\omega \simeq \omega_0$ $f \simeq 20$ kHz	Solar wind; $R_\odot \simeq 1$ AU	$W_E \simeq (10^{-17}-10^{-15})$	$\ll$ to $\simeq 10^{-6}$	Pioneer 8; Scarf et al., 1968b
$\omega \simeq \omega_0$ $f \simeq 31$ kHz	Solar wind; $R_\odot \simeq 0.47$ AU	$W_E \simeq 10^{-19}$	$\lesssim 10^{-10}$	Helios 2; Gurnett et al., 1978c. Waves associated with an interplanetary shock
$\omega \simeq \omega_0$ $f \simeq (18-56)$ kHz $f \simeq 100$ kHz	Interplanetary space, solar wind; $R_\odot \simeq (0.3-1)$ AU	$W_E \simeq (10^{-18}-10^{-15})$ $E \simeq E_0 R_\odot^{-3.5}$	$(10^{-6}-$ $5 \times 10^{-9})$	Associated with type III solar radio bursts. IMP, Helios, Voyager. Gurnett et al., 1978a

Electromagnetic (transverse) waves

Frequency, f(Hz, kHz)	Plasma region	Energy density, W_E(erg cm^{-3})	W_E/NkT	Notes
$f \simeq (0.3-100)$ kHz	Magnetosphere; $R \simeq (7-20)R_0$	$W_H \simeq (10^{-14}-10^{-17})$	$\ll$ to $\simeq 10^{-2}$	OGO 1; Dunckel et al., 1970. Broadband magnetic noise
$\omega < \omega_0$ $f \simeq (5-20)$ kHz	Plasmatrough and geomagnetic tail; $R < 10 R_0$; $R \to 32 R_0$	$W_E \simeq (10^{-22}-10^{-21})$	$\ll$ to $\simeq 10^{-10}$	IMP 6; Gurnett & Shaw, 1973. Trapped continuum radiation
$\omega > \omega_0$ $f \simeq (0.5-100)$ kHz	Outside the magnetosphere. Geomagnetic tail; $R \gtrsim (20-40)R_0$	$W_{E\,max} \simeq (10^{-22}-10^{-20})$	$\ll$ to $\simeq 10^{-9}$	IMP 6, 8; Gurnett, 1975. Escaped continuum radiation. Source: plasma-trough $R \simeq (4-8)R_0$
$\omega > \omega_0$ $f \simeq (50-500)$ kHz	Solar wind; $R \simeq 30 R_0$	$W_{E\,max} \simeq 10^{-17}$ $f \simeq (100-300)$ kHz	$\lesssim 10^{-6}$	IMP 6; Gurnett, 1974. Auroral kilometric radiation. Source: polar region, $R \gtrsim 2R_0$, $\phi \simeq (70-80)°$
$\omega \simeq \omega_0$ $f \simeq 70$ kHz	Solar wind; $R \simeq 10^6$ km	$W_E \simeq (10^{-17}-10^{-13})$	$\ll$ to $\simeq 10^{-6}$	OGO 5; Scarf et al., 1970a, 1971b

See the footnote to Table 20.1.

estimates of NkT, which were based on average values of the parameters of the medium (see Chapter 2, Volume 1). Even so, the information presented in the following tables should create a general picture of the energy levels of the various types of wave processes generated in the vast reaches from the plasma in the immediate vicinity of the Earth out to the plasma regions quite near to the Sun. (In Table 20.3, the parameter $R_\odot$ is the distance from the Sun expressed in astronomical units.)

The ELF and VLF waves in the near-Earth plasma are the most intense, especially ion-acoustic longitudinal electrostatic waves and the so-called PC magnetic pulsations. In general, electrostatic waves are more intense in all the frequency ranges. In the polar zone various kinds of waves have a high intensity, and the relative value of their energy there is $W_E/Nkt \simeq 10^{-2}$ to 10^{-5}.

The relative values of the energy densities of the different kinds of waves vary in general as follows:

$$W_E, W_H \simeq 10^{-2} NkT \text{ to } 10^{-10} NkT. \tag{20.5}$$

The most frequent values are

$$W_E, W_H \simeq 10^{-3} NkT \text{ to } 10^{-6} NkT. \tag{20.6}$$

The value $W_E/Nkt \simeq 0.5$, apparently observed with OGO 2 when recording ion-acoustic waves (see Table 20.1), constitutes an exception (Guthart et al., 1968).

References

Aikyo, K. & Ondoh, T. 1971. Propagation of nonducted VLF waves in the vicinity of the plasmapause, *J. Radio Res. Lab. Jap.* **18**, 153.

Aikyo, K., Ondoh, T. & Nagayama, M. 1972. Nonducted whistlers observed in the plasmasphere, *J. Radio Res. Lab. Jap.* **19**, 151.

Alexander, J.K. & Kaiser, M.L. 1976. Terrestrial kilometric radiation. 1. Spatial structure studies, *J. Geophys. Res.* **81**, 5948.

Alexander, J.F. & Kaiser, M.L. 1977. Terrestrial kilometric radiation. 2. Emission from the magnetospheric cusp and dayside magnetosheath, *J. Geophys. Res.* **82**, 98.

Aliev, Yu.M. & Silin, V.P. 1965. Theory of plasma oscillations in a high-frequency electromagnetic field, *ZhETF* **48**, 901 [In Russian].

Aliev, Yu.M. & Zyunder, D. 1969. Parametric excitation of lower-hybrid and upper-hybrid resonances, *ZhETF* **57**, 1324 [In Russian].

Allen, E.M., Thome, G.D. & Rao, P.B. 1974. HF-phased array observations of heater-induced spread-F, *Radio Science* **9**, 905.

Al'pert, Ya.L. 1948. On the trajectories of rays in a magnetoactive ionized medium – the ionosphere, *Izv. Akad. Nauk. SSSR. Ser. Fiz.* **12**, 241, 267 (see also 1946, *Dokl. Akad. Nauk SSSR* **53**, 703) [In Russian].

Al'pert, Ya.L. 1960. Investigation of the ionosphere and interplanetary space with the help of satellites and space rockets, *Usp. Fiz. Nauk* **71**, 369 [In Russian].

Al'pert, Ya.L. & Pitaevskii, L.P. 1961. Scattering of electromagnetic waves on disturbances caused in a plasma by a rapidly moving body, *Geomagnetizm i Aeronomiya*, **1**, 709 [In Russian] [1963, *AIAA Journal* **1**, 1001].

Al'pert, Ya.L., Gurevich, A.V. & Pitaevskii, L.P. 1963. Effects produced by a satellite rapidly moving in the ionosphere or in interplanetary space, *Usp. Fiz. Nauk* **79**, 23 [In Russian] (1963, *Space Sci. Rev.* **2**, 680).

Al'pert, Ya.L. 1965. On electromagnetic effects in the neighbourhood of a satellite or a vehicle moving in the ionosphere or interplanetary space, *Geomagnetizm i Aeronomiya* **5**, 3 [In Russian] (1965, *Space Sci. Rev.* **4**, 373).

Al'pert, Ya.L., Gurevich, A.V. & Pitaevskii, L.P. 1965. *Space physics with artificial satellites*, Consultants Bureau, New York [1964, Nauka, Moscow [In Russian]].

Al'pert, Ya.L., Bashilov, I.P., Mikhailova, G.A., Kapustina, O.V. & L'vova, G.A. 1975. An experimental installation for investigating the fine structure of electromagnetic processes and some preliminary results, *Kosmicheskie Issledovaniya* **13**, 544 [In Russian].

Al'pert, Ya.L. & Fligel, D.S. 1977. The locating of hydromagnetic whistler sources and determination of their generating proton spectra, *Planet. Space Sci.* **25**, 487.

Al'pert, Ya.L. 1980. 40 years of whistlers, *J. Atmos. Terr. Phys.* **42**, 1.

References

Anderson, R.R. & Gurnett, D.A. 1973. Plasma wave observations near the plasmapause with the S^3-A satellite, *J. Geophys. Res.* **78**, 4756.

Anderson, R.R. & Maeda, K. 1977. VLF emissions associated with enhanced magnetospheric electrons, *J. Geophys. Res.* **82**, 135.

Angerami, J.J. & Carpenter, D.L. 1966. Whistler studies of the plasmapause in the magnetosphere. 2. Electron density and total tube electron content near the knee in magnetospheric ionization, *J. Geophys. Res.* **71**, 711.

Angerami, J.J. 1970. Whistler duct properties deduced from VLF observations made with the OGO 3 satellite near the magnetic equator, *J. Geophys. Res.* **75**, 6115.

Ashour-Abdalla, M. 1972. Amplification of whistler waves in the magnetosphere, *Planet. Space Sci.* **20**, 639.

Ashour-Abdalla, M., Chanteur, G. & Pellat, R. 1975. A contribution to the theory of the electrostatic half-harmonic electron gyrofrequency waves in the magnetosphere, *J. Geophys. Res.* **80**, 2775.

Ashour-Abdalla, M. & Kennel, C.F. 1978. Nonconvective and convective electron cyclotron harmonic instabilities, *J. Geophys. Res.* **83**, 1531.

Astrelin, V.T., Bogashchenko, I.A., Buchelnikova, N.S. & Eidman, Yu.I. 1971. Similarity of flow past bodies of different sizes in a magnetoactive plasma, *Preprint IYaF* 41-71 [In Russian].

Astrelin, V.T., Bogashchenko, I.A., Buchel'nikova, N.S. & Eidman, Yu.I. 1972. Flow past a plate in a magnetoactive plasma, *ZhETF* **42**, 1715 [In Russian].

Barfield, J.N. & Coleman, P.J. 1970. Storm-related wave phenomenon observed at the synchronous equatorial orbit, *J. Geophys. Res.* **75**, 1943.

Barkhausen, H. 1919. Zwei mit Hilfe der neuen Verstärker entdeckte Erscheinungen, *Phys. Zeits.* **20**, 401.

Barkhausen, H., 1930. Whistling tones from the earth, *Proc. IRE* **18**, 1155.

Barrett, P.J. 1964. Electron wakes in a plasma, *Phys. Rev. Lett.* **13**, 742.

Barrington, R.E. & Belrose, J.S. 1963. Preliminary results from the very-low-frequency receiver aboard Canada's Alouette satellite, *Nature* **198**, 651.

Barrington, R.E., Belrose, J.S. & Keeley, D.A. 1963. VLF noise bands observed by the Alouette I satellite, *J. Geophys. Res.* **68**, 6539.

Barrington, R.E., Belrose, J.S. & Nelms, G.L. 1965. Ion composition and temperatures at 1000 km deduced from simultaneous observations of a VLF plasma resonance and topside sounding data from the Alouette I satellite, *J. Geophys. Res.* **70**, 1647.

Barrington, R.E., Belrose, J.S. & Mather, W.E. 1966. A helium whistler observed in the Canadian satellite Alouette II, *Nature* **210**, 80.

Beghin, C. & Siredey, C. 1964. Un procédé d'analyse fine des sifflements atmosphériques, *Ann. Geophys.* **20**, No. 3-4, 30.

Beghin, C. & Debrie R. 1972. Characteristics of the electric field far from and close to a radiating antenna around the lower hybrid resonance in the ionosphere, *J. Plasma Phys.* **8**, 287.

Benioff, H. 1960. Observations of geomagnetic fluctuations in the period range 0.3 to 120 seconds, *J. Geophys. Res.* **65**, 1413.

Bernstein, W., Leinbach, H., Cohen, H., Wilson, P.S., Davis, T.N., Hallinan, T., Baker, B., Martz, J., Ziemke, R. & Huber, W. 1975. Laboratory observations of RF emissions at ω_{pe} and $(n+\frac{1}{2})\omega_{ce}$ in electron beam-plasma and beam-beam interactions, *J. Geophys. Res.* **80**, 4375.

Bitoun, J., Graff, P. & Aubry, M. 1970. Ray tracing in warm magnetoplasma and applications to topside resonances, *Radio Science* **5**, 1314.

Bogashchenko, I.A., Gurevich, A.V., Salimov, R.A. & Eidel'man, Yu.I. 1970. Flow past bodies in a magnetoactive plasma, *Preprint IYaF* 15; 1971, *ZhETF* **59**, 1540 [In Russian].

Booker, H.G. 1975. Electromagnetic and hydromagnetic waves in a cold magnetoplasma, *Phil. Trans. Roy. Soc. Lond.* **A280**, 57.

Bossen, M., McPherron, R.L. & Russell, C.T. 1976. A study of Pc-1 magnetic pulsations at a synchronous orbit, *J. Geophys. Res.* **81**, 6083.

Bourdeau, R.E., Donley, J.L., Serbu, G.P. & Whipple, E.C. 1961. Measurements of sheath currents and equilibrium potential on the Explorer VIII satellite, *J. Astron. Sci.* **8**, 65.

Bourdeau, R.E. 1962. On the interaction between a spacecraft and an ionized medium, *Space Sci. Rev.* **1**, 719.

Bourdeau, R.E. & Donley, J.L. 1964. Explorer 8 satellite measurements in the upper atmosphere, *Proc. Roy. Soc. Lond.* **A281**, 487.

Bowen, P.J., Boyd, R.L., Henderson, C.L. & Willmore, A.P. 1964. Measurement of electron temperature and concentration from a spacecraft, *Proc. Roy. Soc. Lond.* **A281**, 514.

Brice, N.M. 1963. An explanation of triggered VLF emissions, *J. Geophys. Res.* **68**, 4626.

Brice, N.M. 1964. Fundamentals of VLF emission generation mechanisms, *J. Geophys. Res.* **69**, 4515.

Brice, N.M. & Smith, R.L. 1965. Lower hybrid resonance emissions, *J. Geophys. Res.* **70**, 71.

Brown, L.W. 1973. The galactic radio spectrum between 130 kHz and 2600 kHz, *Astrophys. J.* **180**, 359.

Brundin, C.L. 1963. Effects of charged particles on the motion of an earth satellite, *AIAA Journal*, **1**, 2529.

Brundin, C.L. (Editor), 1967. *Rarefied Gas Dynamics (5th Symp.)*, Academic Press, New York.

Bud'ko, N.I. 1966. Asymptotic formulas for the wake of a rapidly moving body in rarefied plasma, *Geomagnetizm i Aeronomiya* **6**, 1008 [In Russian].

Bud'ko, N.I. 1969a. Disturbance of a nonisothermal plasma by a body moving at a supersonic velocity, *ZhETF* **57**, 686 [In Russian].

Bud'ko, N.I. 1969b. Lower-hybrid resonance excitation around a moving body by whistlers, *Geomagnetizm i Aeronomiya* **9**, 430 [In Russian].

Bud'ko, N.I. 1970. Investigations of disturbances in the far zone of a body moving rapidly in a plasma, Candidate's Dissertation *IZMIRAN* [In Russian].

Bullough, K., Denby, M., Gibbons, W., Hughes, A.R., Kaiser, T.R. & Tatnall, A.R. 1975. ELF/VLF emissions observed on Ariel 4, *Proc. Roy. Soc. Lond.* **A343**, 207.

Bullough, K., Tatnall, A.R. & Denby, M. 1976. Man-made ELF/VLF emissions and the radiation belts, *Nature* **260**, 401.

Burtis, W.J. & Helliwell, R.A. 1969. Banded chorus: A new type of VLF radiation observed in the magnetosphere by OGO 1 and OGO 3, *J. Geophys. Res.* **74**, 3002.

Burton, R.K. & Holzer, R.E. 1974. The origin and propagation of chorus in the outer magnetosphere, *J. Geophys. Res.* **79**, 1014.

Call, S.M. 1969. *The interaction of a satellite with the ionosphere*, Rep. 46, Plasma Lab., Columbia Univ. (NASA CR-106555).

References

Calvert, W. & Goe, G.B. 1963. Plasma resonances in the upper ionosphere, *J. Geophys. Res.* **68**, 6113.

Calvert, W. & Van Zandt, T.E. 1966. Fixed frequency observations of plasma resonances in the topside ionosphere, *J. Geophys. Res.* **71**, 1799.

Calvert, W. & McAfee, J.R. 1969. Topside-sounder resonances, *Proc. IEEE* **57**, 1089.

Campbell, W.H. & Stilner, E.C. 1965. Some characteristics at frequencies near 1 cps *Radio Sci. J. Res. NBS* **69D**, 1089.

Carpenter, D.L. 1963. Whistler evidence of a 'knee' in the magnetospheric ionization density profile, *J. Geophys. Res.* **68**, 1675.

Carpenter, D.L., Dunckel, N. & Walkup, J.F. 1964. A new, very low frequency phenomenon: whistlers trapped below the protonosphere, *J. Geophys. Res.* **69**, 5009.

Carpenter, D.L. & Dunckel, N. 1965. A dispersion anomaly in whistlers received on Alouette 1, *J. Geophys. Res.* **70**, 3781.

Carpenter, D.L. 1966. Whistler studies of the plasmapause in the magnetosphere. 1. Temporal variations in the position of the knee, *J. Geophys. Res.* **71**, 693.

Carpenter, D.L. 1968. Ducted whistler mode propagation in the magnetosphere: A half gyrofrequency upper intensity cutoff and some associated wave growth phenomena, *J. Geophys. Res.* **73**, 2919.

Carpenter, D.L., Park, C.G., Taylor, H.A. & Brinton, H.C. 1969a. Multi-experiment detection of the plasmapause from EOGO satellites and Antarctic ground stations, *J. Geophys. Res.* **74**, 1837.

Carpenter, D.L., Stone, K. & Lasch, S. 1969b. A case of artificial triggering of VLF magnetospheric noise during the drift of a whistler duct across magnetic shells, *J. Geophys. Res.* **74**, 1848.

Carpenter, D.L. & Miller, T.L. 1976. Ducted magnetospheric propagation of signals from the Siple, Antarctica, VLF transmitter, *J. Geophys. Res.* **81**, 2692.

Cartwright, D.G. & Kellogg, P.J. 1971. Controlled experiment on wave–particle interactions in the ionosphere, *Nature*, **231**, 11.

Cartwright, D.G. & Kellogg, P.J. 1974. Observations of radiation from an electron beam artificially injected into the ionosphere, *J. Geophys. Res.* **79**, 1439.

Cerisier, J.C. & James, H.G. 1970. Etude à l'aide du satellite FR 1 de la propagation d'ondes très basse fréquence dans l'ionosphere à basse latitude, *Ann. Geophys.* **26**, 829.

Cerisier, J.C. 1973. A theoretical and experimental study of nonducted VLF waves after propagation through the magnetosphere, *J. Atmos. Terr. Phys.* **35**, 77.

Cerisier, J.C. 1974. Ducted and partly ducted propagation of VLF waves through the magnetosphere, *J. Atmos. Terr. Phys.* **36**, 1443.

Chan, K.W., Holzer, R.E. & Smith, E.J. 1974. A relation between ELF hiss amplitude and plasma density in the outer plasmasphere, *J. Geophys. Res.* **79**, 1989.

Chan, K.W. & Holzer, R.E. 1976. ELF hiss associated with plasma density enhancements in the outer magnetosphere, *J. Geophys. Res.* **81**, 2267.

Chopra, K.P. 1961. Interactions of rapidly moving bodies in the terrestrial atmosphere, *Rev. Mod. Phys.* **33**, 153.

Christiansen, P., Gough, P., Martelli, G., Bloch, J.-J., Cornilleau, N., Etcheto, J., Gendrin, R. Jones, D., Beghin, C. & Decreau, P. 1978, Geos 1: Identification of natural magnetospheric emissions, *Nature* **272**, 682.

Clark, D.H., Raitt, W.J. & Willmore, A.P. 1973. A measured anisotropy in the ionosphere electron temperature, *J. Atmos. Terr. Phys.* **35**, 63.

Clayden, W.A. & Hurdle, C.V. 1964. Laboratory simulation of the redistribution of charged particles caused by a neutral gas cloud expanding into the ionosphere. In *Rarefied Gas Dynamics*, Academic Press, New York, **4**, 1717.

Coleman, P.J. 1964. Characteristics of the region of interaction between the interplanetary plasma and the geomagnetic field: Pioneer 5, *J. Geophys. Res.* **69**, 3051.

Coroniti, F.V., Fredricks, R.W., Kennel, C.F. & Scarf, F.L. 1971. Fast time-resolved spectral analysis of VLF banded emissions, *J. Geophys. Res.* **76**, 2366.

Crawford, F.W., Harp, R.S. & Mantei, T.D. 1967. On the interpretation of ionospheric resonance stimulated by Alouette 1, *J. Geophys. Res.* **72**, 57.

Cummings, W.D., O'Sullivan, R.J. & Coleman, P.J. 1969. Standing Alfvén waves in the magnetosphere, *J. Geophys. Res.* **74**, 778; see also 1971, **76**, 926.

Cummings, W.D., Masson, F. & Coleman, P.J. 1972. Some characteristics of low-frequency oscillations observed at ATS 1, *J. Geophys. Res.* **77**, 748; see also 1978, **83**, 697.

D'Angelo, N., Bahnsen, A. & Rosenbauer, H. 1974. Wave and particle measurements at the polar cusp. *J. Geophys. Res.* **79**, 3129.

Davis, A.H. & Harris, I. 1961. Interaction of a charged satellite with the ionosphere. In *Rarefied Gas Dynamics*, Academic Press, New York, **2**, 691.

Dougherty, J.P. & Monaghan, J.J. 1966. Theory of resonances observed in ionograms taken by sounders above the ionosphere, *Proc. Roy. Soc. Lond.* **A289**, 214.

Dowden, R.L. 1972. Trigger delay in whistler precursors, *J. Geophys. Res.* **77**, 695.

Dubovoi, A.P. 1972. Electrical potential and particle concentration of a plasma in the vicinity of a rapidly moving charge, *ZhETF* **63**, 951 [In Russian].

Dubovoi, A.P. & Yaroslavtsev, A.A. 1976. On the tail of a body rapidly moving in a magnetoactive plasma, Preprint *IZMIRAN* No. 21 (164) [In Russian].

Dunckel, N. & Helliwell, R.A. 1969. Whistler mode emissions on the OGO 1 satellite, *J. Geophys. Res.* **74**, 6371.

Dunckel, N., Ficklin, B., Rorden, L. & Helliwell, R.A. 1970. Low-frequency noise observed in the distant magnetosphere with OGO 1, *J. Geophys. Res.* **75**, 1854.

Dungey, J.W. 1963. Loss of Van Allen electrons due to whistlers, *Planet. Space Sci.* **11**, 591.

Dwarkin, M.L., Zmuda, A.J. & Radford, W.D. 1971. Hydromagnetic waves at 6.25 earth radii with periods between 3 and 240 seconds, *J. Geophys. Res.* **76**, 3668.

Dysthe, K.B. 1971. Some studies of triggered whistler emissions, *J. Geophys. Res.* **76**, 6915.

Eckersley, T.L. 1928. Letter to the Editor, *Nature* **122**, 768.

Eckersley, T.L. 1929. An investigation of short waves, *J. Inst. Elec. Engrs.* **67**, 992.

Eckersley, T.L. 1935. Musical atmospherics, *Nature* **135**, 104.

Edgar, B.C. 1976a. The upper and lower frequency cutoffs of magnetospherically reflected whistlers, *J. Geophys. Res.* **81**, 205.

Edgar, B.C. 1976b. The theory of VLF Doppler signatures and their relation to the magnetospheric density structure, *J. Geophys. Res.* **81**, 3327.

Fejer, J.A. & Calvert, W. 1964. Resonance effects of electrostatic oscillations in the ionosphere, *J. Geophys. Res.* **69**, 5049.

Fejer, J.A. 1970. Radio wave probing of the lower ionosphere by cross-modulation techniques, *J. Atmos. Terr. Phys.* **32**, 597.

Forslund, D.W. 1970. Instabilities associated with heat conduction in the solar wind and their consequences, *J. Geophys. Res.* **75**, 17.

Fournier, G. 1971. Positively biased ionospheric probes, *Phys. Rev. Lett.* **34A**, 241.

Fournier, G. & Pigache, D. 1972. Transverse ion temperature in an ionospheric wind tunnel, *J. Appl. Phys.* **43**, 4548.

Frank, L.A. 1971. Plasma in the earth's polar magnetosphere, *J. Geophys. Res.* **76**, 5202.

Frank, L.A. & Ackerson, K.L. 1971. Observations of charged particle precipitation into the auroral zone, *J. Geophys. Res.* **76**, 3612.

Frankel, M.S. 1973. LF radio noise from the earth's magnetosphere, *Radio Science* **8**, 991.

Fredricks, R.W., Crook, G.M., Kennel, C.F., Green, I.M., Scarf, F.L., Coleman, P.J. & Russell, C.T. 1970. OGO 5 observations of electrostatic turbulence in bow shock magnetic structures, *J. Geophys. Res.* **75**, 3751.

Fredricks, R.W. 1971. Plasma instability at $(n+\frac{1}{2}) f_b$ and its relationship to some satellite observations, *J. Geophys. Res.* **76**, 5344.

Fredricks, R.W., Scarf, F.L. & Frank, L.A. 1971. Nonthermal electrons and high-frequency waves in the upstream solar wind, *J. Geophys. Res.* **76**, 6691.

Fredricks, R.W., Scarf, F.L. & Green, I.M. 1972. Distributions of electron plasma oscillations upstream from the earth's bow shock, *J. Geophys. Res.* **77**, 1300.

Fredricks, R.W. & Russell, C.T. 1973. Ion cyclotron waves observed in the polar cusp, *J. Geophys. Res.* **78**, 2917.

Fredricks, R.W. & Scarf, F.L. 1973. Recent studies of magnetospheric electric field emissions above the electron gyrofrequency, *J. Geophys. Res.* **78**, 310.

Fredricks, R.W., Scarf, F.L. & Russell, C.T. 1973. Field-aligned currents, plasma waves, and anomalous resistivity in the disturbed polar cusp. *J. Geophys. Res.* **78**, 2133.

Fredricks, R.W. 1975. Wave–particle interactions in the outer magnetosphere, *Space Sci. Rev.* **17**, 741; Wave–particle interactions and their relevance to substorms, *Space Sci. Rev.* **17**, 449.

Gaponov, A.V. & Miller, M.A. 1958. On the potential for charged particles in a high-frequency electromagnetic field, *ZhETF* **34**, 242 [In Russian].

Gendrin, R. & Stefant, R.J. 1962. Analyse de fréquence des oscillations en perles, *Comp. Rend.* **255**, 752.

Gendrin, R. 1971. Phénomènes TBF d'origine magnétosphérique, *Handbuch der Physik*, **49/3**, 461.

Gendrin, R. 1974. The French–Soviet 'ARAKS' experiment, *Space Sci. Rev.* **15**, 905.

Gendrin, R. 1975. Waves and wave–particle interactions in the magnetosphere, *Space Sci. Rev.* **18**, 145.

Gendrin, R. & Jones, D. (Editors), 1978. Measurements of electric and magnetic wave fields and of cold plasma parameters on board GEOS 1, Preliminary results, *Planet. Space Sci.* **27**, 682.

Gershman, B.N. 1953. Propagation of electromagnetic waves in a magnetoactive plasma (kinetic theory), *ZhETF* **24**, 659 [In Russian].

Gordon, W.E. & LaLonde, L.M. 1961. The design and capabilities of an ionospheric radar probe, *IRE Transactions* **7**, 17.

Grard, R. (Editor), 1973. *Particle and photon interactions with surfaces in space*, North Holland, Amsterdam.

Green, J.L., Gurnett, D.A. & Shawhan, S.D. 1977. The angular distribution of auroral kilometric radiation, *J. Geophys. Res.* **82**, 1825.

Gulel'mi, A.V. & Troitskaya, V.A. 1973. *Geomagnetic pulsations and diagnostics of the magnetosphere*, Nauka, Moscow [In Russian].

Gurevich, A.V. 1960. Disturbances in the ionosphere caused by a moving body,

Transactions of *IZMIRAN* **17** (27), 173 [In Russian] [1962, *Planet. Space Sci.* **9**, 321].

Gurevich, A.V. 1963. The structure of the disturbed zone around a large charged body in a plasma, *Geomagnetizm i Aeronomiya* **3**, 1021 [In Russian].

Gurevich, A.V. 1964. Instability of a disturbed zone in the vicinity of a charged body in a plasma, *Geomagnetizm i Aeronomiya* **4**, 247 [In Russian].

Gurevich, A.V. 1965. Nonlinear effects for powerful radio waves in the ionosphere, *Geomagnetizm i Aeronomiya* **5**, 70; see also 1972, **12**, 24 [In Russian].

Gurevich, A.V. & Pitaevskii, L.P. 1966. Scattering of radio waves in the wake of a body moving in a plasma, *Geomagnetizm i Aeronomiya* **6**, 842 [In Russian].

Gurevich, A.V., Pitaevskii, L.P. & Smirnova, V.V. 1969a. Ionosphere aerodynamics, *Usp. Fiz. Nauk* **99**, 3 [In Russian] [*Space Sci. Rev.* **9**, 805].

Gurevich, A.V., Salimov, R.A. & Buchel'nikova, N.S. 1969b. Investigation of the stability of a plasma, *Teplofiz. Vys. Temp.* **7**, 852 [In Russian].

Gurevich. A.V. 1972. Travelling ionization disturbances in a field of strong electromagnetic waves, *Radiofizika* **15**, No. 1, 11 [In Russian].

Gurevich, A.V., Pariiskaya, L.V. & Pitaevskii, L.P. 1972. Ion acceleration on expansion of a rarefied plasma, *ZhETF* **63**, 516 [In Russian].

Gurevich, A.V. & Shlyuger, I.S. 1975. Investigations of nonlinear phenomena with a powerful radio pulse in the lower ionosphere, *Radiofizika* **18**, 1237 [In Russian].

Gurnett, D.A. & O'Brien, B.J. 1964. High-latitude geophysical studies with satellite Injun 3. Part 5, very-low-frequency electromagnetic radiation, *J. Geophys. Res.* **69**, 65.

Gurnett, D.A., Shawhan, S.D., Brice, N.M. & Smith, R.L., 1965. Ion cyclotron whistlers, *J. Geophys. Res.* **70**, 1665.

Gurnett, D.A. & Brice, N.M. 1966. Ion temperature in the ionosphere obtained from cyclotron damping of proton whistlers, *J. Geophys. Res.* **71**, 3639.

Gurnett, D.A. & Shawhan, S.D. 1966. Determination of hydrogen ion concentration, electron density, and proton gyrofrequency from the dispersion of proton whistlers, *J. Geophys. Res.* **71**, 741.

Gurnett, D.A. & Burns, T.B. 1968. The low frequency cutoff of ELF emission, *J. Geophys. Res.* **73**, 7437.

Gurnett, D.A. & Mosier, S.R. 1969. VLF electric and magnetic fields observed in the auroral zone with the Javelin 8.46 sounding rocket, *J. Geophys. Res.* **74**, 3979.

Gurnett, D.A., Pfeiffer, G.W., Anderson, R.R., Mosier, S.R. & Cauffman, D.P. 1969. Initial observations of VLF electric and magnetic fields with the Injun satellite, *J. Geophys. Res.* **74**, 4631.

Gurnett, D.A. & Rodriguez, P. 1970. Observations of 8-amu/unit charge ion cyclotron whistlers, *J. Geophys. Res.* **75**, 1342.

Gurnett, D.A., Mosier, S.R. & Anderson, R.R. 1971. Color spectrograms of very-low-frequency Poynting flux data, *J. Geophys. Res.* **76**, 3022.

Gurnett, D.A. & Frank, L.A. 1972a. VLF hiss and related plasma observations in the polar magnetosphere, *J. Geophys. Res.* **77**, 172.

Gurnett, D.A. & Frank, L.A. 1972b. ELF noise bands associated with auroral electron precipitation, *J. Geophys. Res.* **77**, 3411.

Gurnett, D.A. & Frank, L.A. 1973. Observed relationships between electric fields and auroral particle precipitation, *J. Geophys. Res.* **78**, 145.

Gurnett, D.A. & Shaw, R.R. 1973. Electromagnetic radiation trapped in the magnetosphere above the plasma frequency, *J. Geophys. Res.* **78**, 8136.

Gurnett, D.A. 1974. The earth as a radio source: Terrestrial kilometric radiation, *J. Geophys. Res.* **79**, 4227.

Gurnett, D.A. 1975. The earth as a radio source: The nonthermal continuum, *J. Geophys. Res.* **80**, 2751.

Gurnett, D.A. & Frank, L.A. 1975. The relationship of electron plasma oscillations to type III radio emissions and low-energy solar electrons, *Solar Phys.* **45**, 477.

Gurnett, D.A. 1976. Plasma wave interactions with energetic ions near the magnetic equator, *J. Geophys. Res.* **81**, 2765.

Gurnett, D.A. & Frank, L.A. 1976. Continuum radiation associated with low-energy electrons in the outer radiation zone, *J. Geophys. Res.* **81**, 3875.

Gurnett, D.A., Frank, L.A. & Lepping, R.P. 1976. Plasma waves in the distant magnetotail, *J. Geophys. Res.* **81**, 6059.

Gurnett, D.A. & Anderson, R.R. 1977. Plasma wave electric fields in the solar wind: Initial results from Helios I, *J. Geophys. Res.* **82**, 632.

Gurnett, D.A. & Frank, L.A. 1977. A region of intense plasma wave turbulence on auroral field lines, *J. Geophys. Res.* **82**, 1031.

Gurnett, D.A., Anderson, R.R., Scarf, F.L. & Kurth, W.S. 1978a. The heliocentric radial variation of plasma oscillations associated with type III radio bursts, *J. Geophys. Res.* **83**, 4147.

Gurnett, D.A. & Frank, L.A. 1978a. Ion-acoustic waves in the solar wind, *J. Geophys. Res.* **83**, 58.

Gurnett, D.A. & Frank, L.A. 1978b. Plasma waves in the polar cusp: Observations from Hawkeye I, *J. Geophys. Res.* **83**, 1447.

Gurnett, D.A. & Green, J.L. 1978. On the polarization and origin of auroral kilometric radiation, *J. Geophys. Res.* **83**, 689.

Gurnett, D.A., Marsch, E., Pilipp, W., Schwenn, R. & Rosenbauer, H. 1978b. *Ion-acoustic waves and related plasma observations in the solar wind*, University of Iowa 78-50, *J. Geophys. Res.* **84**, 2029: 1979.

Gurnett, D.A., Neubauer, F.M. & Schwenn, R. 1978c. *Plasma wave turbulence associated with an interplanetary shock*, University of Iowa 78-9, *J. Geophys. Res.* **84**, 541: 1979.

Guthart, H., Crystal, T.L., Ficklin, B.P., Blair, W.E. & Yung, T.J. 1968. Proton gyrofrequency band emission observed aboard OGO 2, *J. Geophys. Res.* **73**, 3592.

Hagg, E.L., Howens, E.J. & Nelms, G.L. 1969. The interpretation of topside ionograms, *Proc. IEEE* **57**, 949.

Hall, D.F., Kemp, R.F. & Sellen, J.M. 1964. Plasma-vehicle interaction in a plasma stream, *AIAA Journal* **2**, 1032.

Hall, D.F., Kemp, R.F. & Sellen, J.M. 1965. Generation and characteristics of plasma wind-tunnel streams, *AIAA Journal* **3**, 1490.

Hamelin, M. & Beghin, C. 1976. Electromagnetic and electrostatic waves in a multi-component plasma near the lower hybrid frequency, *J. Plasma Phys.* **15**, 115.

Hartz, T.R. & Barrington, R.E. 1969. Nonlinear plasma effects in the Alouette recordings, *Proc. IEEE* **57**, 1108.

Hayakawa, M. 1974. Nonducted two-hop whistlers in the inner magnetosphere deduced from rocket measurements, *Planet. Space Sci.* **22**, 638.

Hayakawa, M., Tanaka, Y. & Ohtsu, J. 1975a. Satellite and ground observations of magnetospheric VLF hiss associated with the severe magnetic storm on 25–7 May 1967, *J. Geophys. Res.* **80**, 86.

Hayakawa, M., Tanaka, Y. & Ohtsu, J. 1975b. The morphologies of low-latitude and auroral VLF "hiss", *J. Atmos. Terr. Phys.* **37**, 517.

Hayakawa, M., Bullough, K. & Kaiser, T.R. 1977. Properties of storm-time magnetospheric VLF emissions as deduced from the Ariel 3 satellite and ground-based observations, *Planet. Space Sci.* **25**, 353.

Helliwell, R.A. 1963. Whistler-triggered periodic VLF emissions, *J. Geophys. Res.* **68**, 5387.

Helliwell, R.A., Katsufrakis, J.P. Trimpi, M. & Brice, N. 1964. Artificially-stimulated VLF radiation from the ionosphere, *J. Geophys. Res.* **69**, 2391 (see also paper presented at URSI Meeting, Washington, D.C., April 1964).

Helliwell, R.A. 1965. *Whistlers and related ionospheric phenomena*, Stanford University Press, Palo Alto, Calif.

Helliwell, R.A. 1967. A theory of discrete VLF emissions from the magnetosphere, *J. Geophys. Res.* **72**, 4773.

Helliwell, R.A. & Crystal, T.L. 1973. A feedback model of cyclotron interaction between whistler mode waves and energetic electrons in the magnetosphere, *J. Geophys. Res.* **78**, 7357.

Helliwell, R.A., Katsufrakis, J.P. & Trimpi, M.L. 1973. Whistler-induced amplitude perturbation in VLF propagation, *J. Geophys. Res.* **78**, 4679.

Helliwell, R.A. 1974. Controlled VLF wave injection experiments in the magnetosphere, *Space Sci. Rev.* **15**, 781.

Helliwell, R.A. & Katsufrakis, J.P. 1974. VLF wave injection into the magnetosphere from Siple Station, Antarctica, *J. Geophys. Res.* **79**, 2511.

Helliwell, R.A., Katsufrakis, J.P., Bell, T.F. & Raghuram, R. 1975. VLF line radiation in the earth's magnetosphere and its association with power system radiation, *J. Geophys. Res.* **80**, 4249.

Henderson, C.L. & Samir, U. 1967. Observations of the disturbed region around an ionospheric spacecraft, *Planet. Space Sci.* **15**, 1499.

Hester, S.D. & Sonin, A.A. 1970a. A laboratory study of the wakes of ionospheric satellites, *AIAA Journal* **8**, 1090.

Hester, S.D. & Sonin, A.A 1970b. Laboratory study of the wakes of small cylinders under ionospheric satellite conditions, *Phys. Fluids* **13**, 641.

Higuchi, Y. & Jacobs, J.A. 1970. Plasma densities in the thermal magnetosphere determined by using hydromagnetic whistlers, *J. Geophys. Res.* **75**, 7105.

Hines, C.O. 1951. Wave packets, the Poynting vector and energy flow, *J. Geophys. Res.* **56**, 63, 197, 207, 535.

Hines, C.O. 1957. Heavy ion effects in audio-frequency radio propagation, *J. Atmos. Terr. Phys.* **11**, 36.

Hoffman, J.H., 1967. Composition measurements of the topside ionosphere, *Science* **155**, 322.

Hoffman, W.C. 1960. Conditions for the presistence of purely longitudinal or purely transverse propagation, *J. Atmos. Terr. Phys.* **18**, 1.

Hultqvist, B. 1966. Plasma waves in the frequency range 0.001–10 cps in the earth's magnetosphere and ionosphere, *Space Sci. Rev.* **5**, 599.

Istomin, Ya.I., Karpman, V.I. & Shklyar, D.R. 1976. Contribution to the theory of triggered emissions, *Geomagnetizm i Aeronomiya* **16**, 67[In Russian].

Jacobs, J.A. & Watanabe, T. 1963. Trapped charged particles as the origin of short-period

geomagnetic pulsations, *Planet. Space Sci.* **11**, 869; 1964, *J. Atmos. Terr. Phys.* **26**, 825.
Jacobs, J.A. 1970. *Geomagnetic micropulsations*, Springer Verlag, Berlin.
James, H.G. 1973. Whistler mode hiss at low and medium frequencies in the dayside cusp ionosphere, *J. Geophys. Res.* **78**, 4578.
James, H.G. 1976. VLF saucers, *J. Geophys. Res.* **81**, 501.
Jastrow, R. & Pearse, C.A. 1957. Atmospheric drag on the satellite, *J. Geophys. Res.* **62**, 413.
Joselyn, J.A. & Lyons, L.R. 1976. Ion-cyclotron wave growth calculated from satellite observations of the proton ring current during storm recovery, *J. Geophys. Res.* **81**, 2275.
Kaiser, T.R., Orr, D. & Smith, A.J. 1977. Very low frequency electromagnetic phenomena: whistlers and micropulsations, *Phil. Trans. Roy. Soc. Lond.* **B279**, 225.
Karpman, V.I., Istomin, Ya.N. & Shklyar, D.R. 1974. Nonlinear theory of a quasi-monochromatic whistler mode packet in a homogeneous plasma, *Plasma Phys.* **16**, 685.
Kasha, M.A., Osborne, F.J.F., Knight, D.J.E., Johnston, T.W. & Shkarofsky, I.P. 1965. Laboratory experiments on satellite sheaths, *Proc. 2nd Symp. on Interactions of Space Vehicles with an Ionized Atmosphere*, University of Miami.
Kasha, M.A. & Johnston, T.W. 1967. Laboratory simulation of a satellite-mounted plasma diagnostic experiment, *J. Geophys. Res.* **72**, 4028.
Kasha, M.A. 1969. *The ionosphere and its interaction with satellites*, Gordon and Breach, New York.
Kellogg, P.J., Cartwright, D.G., Hendrickson, R.A., Monson, S.J. & Winckler, J.R. 1976. The University of Minnesota Electron Echo experiments, *Space Research*, XVI, 589.
Kennel, C.F. & Petschek, H.E. 1966. Limit on stably trapped particle fluxes, *J. Geophys. Res.* **71**, 1.
Kennel, C.F., Scarf, F.L., Fredricks, R.W., McGehee, J.H. & Coroniti, F.V. 1970. VLF electric-field observations in the magnetosphere, *J. Geophys. Res.* **75**, 6136.
Kenney, J.F. & Knaflich, H.B. 1968. Geomagnetic micropulsations, *Boeing Sci. Res. Lab., Geo-Astrophys. Lab. Rev.*, p. 62.
Kenney, J.F., Knaflich, H.B. & Liemohn, H.B. 1968. Magnetospheric parameters determined from structured micropulsations, *J. Geophys. Res.* **73**, 6737.
Kiel, R.E., Gey, F.C. & Gustafson, W.A. 1968. Electrostatic potential fields of an ionospheric satellite, *AIAA Journal* **6**, 690.
Kimura, I., Smith, R.L. & Brice, N.M. 1965. An interpretation of transverse whistlers, *J. Geophys. Res.* **70**, 5961.
Kimura, I. 1966. Effects of ions on whistler-mode ray tracing, *Radio Science* **1**, 269.
Kimura, I. 1967. On observations and theories of the VLF emissions, *Planet. Space. Sci.* **15**, 1427.
Kimura, I. 1968. Triggering of VLF magnetospheric noise by a low-power (~ 100 watts) transmitter, *J. Geophys. Res.* **73**, 445.
Kimura, I. 1974. Interrelation between VLF and ULF emissions, *Space Sci. Rev.* **16**, 389; see also *Rep. Ion. Space. Res. Jap.* **25**, 360.
Kintner, P.M. & Gurnett, D.A. 1976. *Evidence of drift waves at the plasmapause*, University of Iowa 76–24, *J. Geophys. Res.* **83**, 39: 1978.
Kintner, P.M. & Gurnett, D.A. 1977. Observations of ion-cyclotron waves within the plasmasphere by Hawkeye 1, *J. Geophys. Res.* **82**, 2314.
Kintner, P.M., Kelley, M.C. & Moser, F.S. 1978. Electrostatic hydrogen cyclotron waves

near one earth radius altitude in the polar magnetosphere, *Geophys. Res. Lett.* **5**, 139.
Knecht, R.W., Van Zandt, T.E. & Russell, S. 1961. First pulsed radio soundings of the topside of the ionosphere, *J. Geophys. Res.* **66**, 3078 (see also 1962, **67**, 1178).
Knyazyuk, V.S. & Moskalenko, A.M. 1966. Distribution of the concentration and density of a particle flux in the vicinity of a small body moving in a rarefied plasma, *Geomagnetizm i Aeronomiya* **6**, 997 [In Russian].
Kraus, L. & Watson, K. 1958. Plasma motions induced by satellites in the ionosphere, *Phys. Fluids* **1**, 480.
Kraus, J.D., Higgy, R.G., Scherr, D.J. & Crone, W.R. 1960. Observation of ionization induced by artificial earth satellite, *Nature* **185**, 520.
Kraus, J.D. 1965. *Interaction of space vehicles with an ionized atmosphere* (S.F. Singer, Editor), Pergamon Press, New York–London.
Kurth, W.S., Baumback, M.M. & Gurnett, D.A. 1975. Direction-finding measurements of auroral kilometric radiation, *J. Geophys. Res.* **80**, 2764.
Laaspere, T. & Wang, C.Y. 1968. Whistler precursors, *Radio Science* **3**, 213.
Laaspere, T., Morgan, M.G. & Johnson, W.C. 1969. Observations of lower hybrid resonance phenomena on the OGO 2 spacecraft, *J. Geophys. Res.* **74**, 141.
Laaspere, T. & Taylor, H.A. 1970. Comparison of certain VLF noise phenomena with the lower hybrid resonance frequency calculated from simultaneous ion-composition measurements, *J. Geophys. Res.* **75**, 97.
Laaspere, T. & Johnson, W.C. 1973. Additional results from an OGO 6 experiment concerning ionospheric and electromagnetic fields in the range 20 Hz to 540 Hz, *J. Geophys. Res.* **78**, 2926.
Landau, L.D. & Lifshitz, E.M. 1957. *Electrodynamics of continuous media*, Gostekhizdat, Moscow [In Russian] [1960, Pergamon Press, Oxford–New York].
Landau, L.D. & Lifshitz, E.M. 1958. *Mechanics*, Fizmatgiz, Moscow [In Russian] [1969, Pergamon Press, Oxford–New York].
Lasch, S. 1969. Unique features of VLF noise triggered in the magnetosphere by Morse-code dots from NAA, *J. Geophys. Res.* **74**, 1856.
Lederman, S., Bloom, M.H. & Widhonf, V.C. 1969. Laboratory measurements of electron-density distributions in the near wake, *AIAA Journal* **7**, 1421.
Lefeuvre, F. & Bullough, K. 1973. Ariel 3 evidence of zones of VLF emission at medium invariant latitudes which corotate with the earth, *Space Research* **XIII**, 699.
Liemohn, H.B. & Scarf, F.L. 1962. Exopheric electron temperatures from nose whistler attenuation, *J. Geophys. Res.* **67**, 1785, 4163.
Liemohn, H.B. & Scarf, F.L. 1963. Dispersion function for a plasma with a Cauchy equilibrium distribution, *Phys. Fluids* **6**, 388, 490.
Liemohn, H.B. & Scarf, F.L. 1964. Whistler determination of electron energy and density distributions in the magnetosphere, *J. Geophys. Res.* **69**, 883.
Liemohn, H.B. 1965a. Partial electron velocity spectra from cyclotron absorption of whistler power, *J. Geophys. Res.* **70**, 4817.
Liemohn, H.B. 1965b. Radiation from electrons in the magnetoplasma, *Radio Science*, **69D**, 741.
Liemohn, H.B. 1969. *ELF propagation and emission in the magnetosphere*, Boeing Sci Res. Lab. Document DI-82-0890.
Liemohn, H.B. 1971. *Tabulation of rapid-run geomagnetic micropulsation stations*, Boeing Sci. Res. Lab. Document DI-82-1043.

Liu, V.C. 1967. Particles trapped in the potential well behind a mesothermally moving satellite, *Nature* **215**, 127.

Liu, V.C. & Jew, H. 1967. In *Rarefied Gas Dynamics (5th Symp.)*, Academic Press, New York, **2**, 1703.

Liu, V.C. 1969. Ionospheric gas dynamics of satellites and diagnostic probes, *Space Sci. Rev.* **9**, 423.

Lockwood, G.E. 1963. Plasma and cyclotron spike phenomena observed in topside ionograms, *Can. J. Phys.* **41**, 190.

Lockwood, G.E. 1965. Excitation of cyclotron spikes in the ionospheric plasma, *Can. J. Phys.* **43**, 291.

Lucas, C. & Brice, N. 1971. Irregularities in proton density deduced from cyclotron damping of proton whistlers, *J. Geophys. Res.* **76**, 92.

Lyons, L.R., Thorne, R.M. & Kennel, C.F. 1972. Pitch-angle diffusion of radiation-belt electrons within the plasmasphere, *J. Geophys. Res.* **77**, 3455.

Mainstone, J.S. & McNicol, R.W. 1962. *Micropulsation studies at Brisbane, Queensland I. Pearl pulsations and 'screamers'*, Proc. Intern. Conf. on Ionosphere, London, p. 163.

Martin, A.R. 1974. Numerical solutions to the problem of charged particle flow around an ionospheric spacecraft, *Planet. Space Sci.* **22**, 121.

Maslennikov, M.V. & Sigov, Yu.S. 1965. A discrete model for the study of the flow of a rarefied plasma around a body, *Dokl. Akad. Nauk SSSR* **9**, 1063 [In Russian].

Maslennikov, M.V. & Sigov, Yu.S. 1967. Discrete model of medium in a problem on rarefied plasma stream interaction with a charged body. In *Rarefied Gas Dynamics (5th Symp.)*, Academic Press, New York, **2**, 1657.

Maslennikov, M.V. & Sigov, Yu.S. 1969. Rarefied plasma stream interactions with charged bodies of various forms. In *Rarefied Gas Dynamics (6th Symp.)*, Academic Press New York, **2**, 1671.

Matsumoto, H. & Kimura, I. 1971. Linear and nonlinear cyclotron instability and VLF emissions in the magnetosphere, *Planet. Space Sci.* **19**, 567.

Matsumoto, H., Miyatake, S. & Kimura, I. 1971. Fundamental experiments on wave phenomena in space chamber plasma, *Rep. Ion. Space Res. Jap.* **25**, 40.

Matsumoto, H. 1972. *Theoretical studies on whistler-mode wave–particle interactions in the magnetopheric plasma*, Kyoto University, Japan, December.

Matsumoto, H., Miyatake, S. & Kimura, I. 1974. Frequency spectra of VLF plasma waves observed by Japanese ionospheric sounding rocket K-9M-41, *Rep. Ion. Space Res. Jap.* **28**, 89.

McAfee, J.R. 1968. Ray trajectories in an anisotropic plasma near plasma resonance, *J. Geophys. Res.* **73**, 5577.

McAfee, J.R. 1969a. Topside resonances as oblique echoes, *J. Geophys. Res.* **74**, 802.

McAfee, J.R. 1969b. Topside ray trajectories near the upper hybrid resonance, *J. Geophys. Res.* **74**, 6403.

McAfee, J.R. 1970. Topside plasma frequency resonance below the cyclotron frequency, *J. Geophys. Res.* **75**, 4287.

McEwen, D.J. & Barrington, R.E. 1967. Some characteristics of the lower hybrid resonance noise bands observed by the Alouette 1 satellite, *Can. J. Phys.* **45**, 13.

McKeown, P. 1961. A new method for the measure of sputtering in the region near the threshold. In *Rarefied Gas Dynamics.*, Academic Press, New York, **1**, 29.

McPherron, R.L. & Coleman, P.J. 1970. Magnetic fluctuations during magnetospheric substorms, *J. Geophys. Res.* **75**, 3927.

McPherron, R.L. & Coleman, P.J. 1971. Satellite observations of band-limited micropulsations during a magnetospheric substorm, *J. Geophys. Res.* **76**, 3010.

McPherron, R.L., Russell, C.T. & Coleman, P.J. 1972. Fluctuating magnetic fields in the magnetosphere, II. VLF waves, *Space Sci. Rev.* **13**, 411.

Meckel, B.B. 1961. Experimental study of the interaction of a moving body with a plasma. In *Rarefied Gas Dynamics*, Academic Press, New York, **1**, 701.

Medved, D.B. 1969. Measurement of ion wakes and body effects with the Gemini/Agena satellite. In *Rarefied Gas Dynamics (6th Symp.)*, Academic Press, New York, p. 1525.

Melrose, D.B. 1976. An interpretation of Jupiter's decametric radiation and the terrestrial kilometric radiation as direct amplified gyroemission, *Astrophys. J.* **207**, 651.

Miller, M.A. 1960. Candidate's Dissertation, Gorki University [In Russian]; *Proc. Inter. Conf. High-Energy Accel. and Instr.*, Cern, p. 661.

Miyatake, S., Matsumoto, H. & Kimura, I. 1974. Rocket experiments on nonlinear wave-wave interaction in the ionospheric plasma, *Space Research*, XIV, 369.

Monson, S.J., Kellogg, P.J. & Cartwright, D.G. 1976. Whistler-mode plasma waves observed on Electron Echo 2, *J. Geophys. Res.* **81**, 2193.

Monson, S.J. & Kellogg, P.J. 1978. Ground observations of waves at 2.96 MHz generated by an 8-to 40-keV electron beam in the ionosphere, *J. Geophys. Res.* **83**, 121.

Morgan, M.G., Brown, P.E., Johnson, W.C. & Taylor, H.A. 1977. An improved comparison of the LHR frequency observed by a VLF receiver and calculated from ion-density measurements on the same spacecraft, *Radio Science* **12**, 811.

Moiser, S.R. & Gurnett, D.A. 1969a. VLF measurements of the Poynting flux along the geomagnetic field with the Injun 5 satellite, *J. Geophys. Res.* **74**, 5675.

Mosier, S.R. & Gurnett, D.A. 1969b. Ionospheric observation of VLF electrostatic noise related to harmonics of the proton gyrofrequency, *Nature*, **223**, 605.

Mosier, S.R. 1970. *VLF measurements of the Poynting flux along the geomagnetic field with the Injun 5 satellite*, University of Iowa 70–2.

Mosier, S.R. 1971. Poynting flux studies of hiss with the Injun 5 satellite, *J. Geophys. Res.* **76**, 1713.

Mosier, S.R. & Gurnett, D.A. 1972. Observed correlations between auroral and VLF emissions, *J. Geophys. Res.* **77**, 1137.

Mosier, S.R., Kaiser, M.L. & Brown, L.W. 1973. Observation of noise bands associated with the upper hybrid resonance by the IMP 6 radio astronomy experiment, *J. Geophys. Res.* **78**, 1673.

Moskalenko, A.M. 1964a. Particle distribution in a central-symmetrical field in the Presence of a running stream 1, *Geomagnetizm i Aeronomiya* **4**, 260 [In Russian].

Moskalenko, A.M. 1964b. Distribution of particles in a potential field of cylindrical symmetry, *Geomagnetizm i Aeronomiya* **4**, 1026 [In Russian].

Moskalenko, A.M. 1964c. Distribution of particles in a central-symmetrical field, I and II *Geomagnetizm i Aeronomiya*, **4**, 509 [In Russian].

Moskalenko, A.M. 1969. On the particle distribution and the electric field in the vicinity of a body moving slowly in a rarefied plasma, *ZhETF* **57**, 1790; 1970, *Geomagnetizm i Aeronomiya*, **10**, 974 [In Russian].

Muldrew, D.B. 1967. Delayed generation of an electromagnetic pulse in the topside ionosphere, *J. Geophys. Res.* **72**, 3777 (see also 1972, **77**, 1794).

Muldrew, D.B. 1972. Electron resonances observed with topside sounders, *Radio Science* **7**, 779.

Muldrew, D.B. & Estabrooks, M.F. 1972. Computations of dispersion curves in a hot magnetoplasma with application to the upper-hybrid and cyclotron frequencies, *Radio Science* **7**, 579.

Muzzio, J.L. 1968. Ion cutoff whistlers, *J. Geophys. Res.* **73**, 7526.

Nelms, G.L. & Lockwood, G.E. 1966. Early results from the topside sounder in the Alouette II satellite, *Space Res.* **VII**, 604.

Neubauer, F.M., Musman, G. & Dehmel, G. 1977. Fast magnetic fluctuations in the solar wind, *J. Geophys. Res.* **82**, 3201.

Nunn, D.A. 1974. A self-consistent theory of triggered VLF emissions, *Planet. Space Sci.* **22**, 349.

Obayashi, T. 1965. Hydromagnetic whistlers, *J. Geophys. Res.* **70**, 1069.

Ondoh, T., Tanaka, Y., Nishizaki, R. & Nagayama, M. 1974. VLF emissions and whistlers observed during geomagnetic storms, *J. Rad. Res. Lab. Jap.* **21**, 361.

Ondoh, T. & Murukami, T. 1975. Mid-latitude VLF emissions observed in the topside ionosphere, *Rep. Ion. Space Res. Jap.* **29**, 23.

Oran, W.A., Samir, U. & Stone, N.H. 1974. Parametric study of the near-wake structure of spherical and cylindrical bodies in the laboratory, *Planet. Space Sci.* **22**, 379; see also 1972, *Planet. Space Sci.* **20**, 1787.

Oran, W.A., Stone, N.H. & Samir, U. 1975. The effects of body geometry on the structure in the near-wake zone of bodies in a flowing plasma, *J. Geophys. Res.* **80**, 207.

Osborne, F.J. & Kasha, M.A. 1967. The (VxB) interaction of a satellite with its environment, *Can. J. Phys.* **45**, 263.

Oya, H. 1970. Sequence of diffuse plasma resonances observed on Alouette II ionograms, *J. Geophys. Res.* **75**, 4279.

Oya, H. 1972. Turbulence of electrostatic electron-cyclotron harmonic waves observed by OGO 5, *J. Geophys. Res.* **77**, 3483.

Oya, H. 1975. Plasma flow hypothesis in the magnetosphere relating to frequency shift of electrostatic plasma waves, *J. Geophys. Res.* **80**, 2783.

Palmadesso, P., Coffey, T.P., Ossakow, S.L. & Papadopoulos, K. 1976. Generation of terrestrial kilometric radiation by a beam-driven electromagnetic instability, *J. Geophys. Res.* **81**, 1762.

Pan, Y.S. & Vaglio-Laurin, R. 1967. Trail of an ionospheric satellite I, *AIAA Journal* **10**, 1801.

Panchenko, Yu.M. & Pitaevskii, L.P. 1964. The influence of an electric field on disturbances around a moving body in a plasma, *Geomagnetizm i Aeronomiya* **4**, 256 [In Russian].

Panchenko, Yu.M. 1965. Asymptotic formulas of the wake of a body moving in a rarefied plasma. In *Studies of Outer Space*, Nauka, Moscow, p. 254 [In Russian].

Papadopoulos, K., Goldstein, M.L. & Smith, R.A. 1974. Stabilization of electron streams in type-III solar radio burst, *Astrophys. J.* **190**, 175.

Park, C.G. 1977. VLF wave activity during a magnetic storm: a case study of the role of power line radiation, *J. Geophys. Res.* **82**, 3251.

Patel, V.L., 1965. Low-frequency hydromagnetic waves in the magnetosphere: Explorer 12, *Planet. Space Sci.* **13**, 485.

Pitaevskii, L.P. 1961. On disturbances around a body moving rapidly in a plasma,

Geomagnetizm i Aeronomiya **1**, No. 1, 194 [In Russian].
Pitaevskii, L.P. & Kresin, V.Z. 1961. On disturbances caused by a rapidly moving body in a plasma, *ZhETF* **40**, 271 [In Russian].
Pope, J.H. 1964. An explanation of the apparent polarisation of some geomagnetic micropulsations (pearls), *J. Geophys. Res.* **69**, 399.
Proceedings of the IEEE, 1969. Special issue on topside sounding and the ionosphere, **57**, No. 6.
Reeve, C.D. & Rycroft, M.J. 1976. A mechanism for precursors to whistlers, *J. Geophys. Res.* **81**, 5900.
Rodriguez, P. & Gurnett, D.A. 1971. An experimental study of very-low-frequency mode coupling and polarization reversal, *J. Geophys. Res.* **76**, 960.
Rodriguez, P. & Gurnett, D.A. 1975. Electrostatic and electromagnetic turbulence associated with the earth's bow shock, *J. Geophys. Res.* **80**, 19.
Rodriguez, P. & Gurnett, D.A. 1976. Correlation of bow-shock plasma-wave turbulence with solar wind parameters, *J. Geophys. Res.* **81**, 2871.
Rosenberg, T.J., Helliwell, R.A. & Katsufrakis, J.P. 1971. Electron precipitation associated with discrete VLF emissions, *J. Geophys. Res.* **76**, 8445.
Rostoker, N. 1961. Fluctuations of a plasma, I, *Nuclear Fusion* **1**, 101.
Roux, A. & Pellat, R. 1976. In *Magnetospheric Particles and Fields* (Edited by B.M. McCormac), p. 209.
Roux, A. & Pellat, R. 1978. A theory of triggered emissions, *J. Geophys. Res.* **83**, 1433.
Russell, C.T., Holzer, R.E. & Smith, E.J. 1969. OGO 3 observations of ELF noise in the magnetosphere. 1. Spatial extent and frequency of occurrence, *J. Geophys. Res.* **74**, 755.
Russell, C.T., Holzer, R.E. & Smith, E.J. 1970. OGO 3 observations of ELF noise in the magnetosphere. 2. The nature of the equatorial noise, *J. Geophys. Res.* **75**, 755.
Russell, C.T., McPherron, R.L. & Coleman, P.J. 1972. Fluctuating magnetic fields in the magnetosphere. 1. ELF and VLF fluctuations, *Space Sci. Rev.* **12**, 810.
Sagdeev, R.Z. 1964. In *Voprosy Teorii Plazmy*, **4**, p. 20, Atomizdat, Moscow [In Russian] [1966, *Reviews of Plasma Physics*, Consultants Bureau, New York, **4**, 23].
Saito, T. 1962. *Scientific Reports*, Tokyo University, **5**, (14), 81.
Samir, U. & Willmore, A.P. 1965. The distribution of charged particles near a moving spacecraft, *Planet. Space Sci.* **13**, 285.
Samir, U. & Willmore, A.P. 1966. The equilibrium potential of a spacecraft in the ionosphere, *Planet. Space Sci.* **14**, 1131.
Samir, U. & Wrenn, G.L. 1969. The dependence of charge and potential distribution around a spacecraft on ionic composition, *Planet. Space Sci.* **17**, 693.
Samir, U. 1970. A possible explanation of an order of magnitude discrepancy in electron-wake measurements, *J. Geophys. Res.* **75**, 855.
Samir, U. 1972. About the interaction between a satellite and its environmental ionospheric plasma, *Israel J. Technology* **10**, 179.
Samir, U. & Jew, H. 1972. Comparison of theory with experiment for the electron-density distribution in the near wake of an ionospheric satellite, *J. Geophys. Res.* **77**, 6819.
Samir, U. & Wrenn, G.L. 1972. Experimental evidence of an electron temperature enhancement in the wake of an ionospheric satellite, *Planet. Space Sci.* **20**, 899.
Samir, U., Maier, E.J. & Troy, B.E. 1973. The angular distribution of ion flux around an ionospheric satellite, *J. Atmos. Terr. Phys.* **35**, 513.
Samir, U., First, M., Maier, E.J. & Troy, B. 1975. A comparison of the Gurevich *et al.*,

and the Liu–Jew wake models for the ion flux around a satellite, *J. Atmos. Terr. Phys.* **37**, 577.

Sawchuck, W. 1963. Wake of a charged prolate spheroid at angle of attack in a rarefied plasma. In *Rarefied Gas Dynamics*, Academic Press, New York, **2**, 33.

Sayasov, Yu. S. & Zhizhimov, L.A. 1963. Resonance scattering of radio waves on the tails of satellites, *Radiotekhnika i Electronika* **8**, 499 [In Russian].

Scarf, F.L. 1962. Landau damping and the attenuation of whistlers, *Phys. Fluids* **5**, 6.

Scarf, F.L., Crook, G.M. & Fredricks, R.W. 1965. Preliminary report on detection of electrostatic ion waves in the magnetosphere, *J. Geophys. Res.* **70**, 3045.

Scarf, F.L., Fredricks, R.W. & Crook, G.M. 1968a. Detection of electromagnetic waves on OV3-3, *J. Geophys. Res.* **73**, 1723.

Scarf, F.L., Crook, G.M., Green, I.M. & Virobik, P.F. 1968b. Initial results of the Pioneer 8 VLF electric field experiment, *J. Geophys. Res.* **73**, 6665.

Scarf, F.L., Fredricks, R.W., Green, I.M. & Neugebauer, M. 1970a. OGO-5 observations of quasi-trapped electromagnetic waves in the solar wind, *J. Geophys. Res.* **75**, 3735.

Scarf, F.L., Fredricks, R.W., Frank, L.A., Russell, C.T., Coleman, P.J. & Neugebauer, M. 1970b. Direct correlations of large-amplitude waves with suprathermal protons in the upstream solar wind, *J. Geophys. Res.* **75**, 7316.

Scarf, F.L. & Fredricks, R.W. 1971. Space Sci. Lab. California, Doc. 05402-6030-RO-OO.

Scarf, F.L., Fredricks, R.W., Frank, L.A. & Neugebauer, M. 1971a. Nonthermal electrons and high-frequency waves in the upstream solar wind. 1. Observations, *J. Geophys. Res.* **76**, 5162.

Scarf, F.L., Fredricks, R.W., Smith, E.J., Frandsen, A.M. & Serbu, G.P. 1971b. *OGO 5 observations of discrete whistlers and emissions during a large magnetic storm*, Space Sci. Lab. California, Doc. 05402-6031-RO-OO.

Scarf, F.L., Fredricks, R.W. & Green, I.M., 1971c. Space Sci. Lab. California, Doc. 17706-6002-RO-OO.

Scarf, F.L., Fredricks, R.W., Green, I.M. & Russell, C.T. 1972. Plasma waves in the dayside polar cusp. 1. Magnetospheric observations, *J. Geophys. Res.* **77**, 2274.

Scarf, F.L., Fredricks, R.W., Kennel, C.F. & Coroniti, F.V. 1973a. Satellite studies of magnetospheric substorms on August 15, 1968, *J. Geophys. Res.* **78**, 3119.

Scarf, F.L., Fredricks, R.W., Russell, C.T., Kivelson, M., Neugebauer, M. & Chappell, C.R. 1973b. Observation of a current-driven plasma instability at the outer zone-plasma sheet boundary, *J. Geophys. Res.* **78**, 2150.

Scarf, F.L., Fredricks, R.W., Green, I.M. & Crook, G.M. 1974a. Observation of interplanetary plasma waves, spacecraft noise and sheath phenomena on IMP 7, *J. Geophys. Res.* **79**, 73.

Scarf, F.L., Frank, L.A., Ackerson, K.L. & Lepping, R.P. 1974b. Plasma-wave turbulence at distant crossings of the plasma sheet boundries and the neutral sheet, *Geophys. Res. Lett.* **1**, 189.

Scarf, F.L., Fredricks, R.W., Neugebauer, M. & Russell, C.T. 1974c. Plasma waves in the dayside polar cusp. 2. Magnetopause and polar magnetosheath, *J. Geophys. Res.* **79**, 511.

Scarf, F.L. 1978. *Wave–particle interaction phenomena associated with shocks in the solar wind*, D. Reidel, Dordrecht, Holland.

Schmitt, J.T.M. 1972. Laboratoire de Physique des Milieux Ionisies, Ecole Polytechnique, Paris.

Sharp, G.W., Hanson, W.B. & McKibbin, D.D. 1963. *Some ionospheric measurements with satellite-borne ion traps*. Lockheed Missiles Space Company Symposium, COSPAR, 3–28 (see also 1964, *Space Research* IV, 455).

Shaw, R.R. & Gurnett, D.A. 1972. Magnetospheric electron-density measurements from upper hybrid resonance noise observed by IMP 6, Res. Rep. 72–37, Dep. of Phys. and Astron., University of Iowa.

Shaw, R.R & Gurnett, D.A. 1975. Electrostatic noise bands associated with the electron gyrofrequency and plasma frequency in the outer magnetosphere, *J. Geophys. Res.* **80**, 4259.

Shawhan, S.D. 1966. Experimental observations of proton whistlers from Injun 3, VLF data, *J. Geophys. Res.* **71**, 29.

Shawhan, S.D. & Gurnett, D.A. 1966. Fractional concentration of hydrogen ions in the ionosphere from VLF proton whistler measurement, *J. Geophys. Res.* **71**, 46.

Shawhan, S.D. & Gurnett, D.A. 1968. VLF electric and magnetic fields observed with the Javelin 8.45 sounding rocket, *J. Geophys. Res.* **73**, 5649.

Shawhan, S.D. 1977. Magnetospheric plasma waves, University of Iowa 77–23.

Singer, S.F. (Editor), 1965. *Interaction of space vehicles with an ionized atmosphere*, Pergamon Press, New York–London.

Siscoe, G.L., Davis, L., Coleman, P.J., Smith, E.J. & Jones, D.E. 1967. Magnetic fluctuations in the magnetosheath: Mariner 4, *J. Geophys. Res.* **72**, 1 (see also 1968, **73**, 61).

Siscoe, G.L., Scarf, F.L., Green, I.M., Binsack, J.H. & Bridge, H.S. 1971. Very-low-frequency electric fields in the interplanetary medium: Pioneer 8, *J. Geophys. Res.* **76**, 828.

Sitenko, A.G. & Stepanov, K.N. 1956. On the oscillations of an electron plasma in a magnetic field, *ZhETF* **31**, 642 [In Russian].

Skvortsov, V.V. & Nosachev, L.V. 1968. Investigations of the structure of the tail behind a spherical body in a stream of rarefied plasma. *Kosmicheskie Issledovaniya* **6**, 228 [In Russian].

Smirnova, V.V. 1967. Instability of the disturbance zone of a plasma around a body in the presence of emission from its surface, *Geomagnetizm i Aeronomiya* **7**, 33 [In Russian].

Smirnova, V.V. 1969. Amplification of fluctuations of the electric field around a body in a plasma, *ZhETF* **39**, 49 [In Russian].

Smith, E.J., Holzer, R.E., McLeod, M.G. & Russell, C.T. 1967. Magnetic noise in the magnetosheath in the frequency range 3–300 Hz, *J. Geophys. Res.* **72**, 4803.

Smith, E.J., Holzer, R.E. & Russell, C.T. 1969. Magnetic emissions in the magnetosheath at frequencies near 100 Hz, *J. Geophys. Res.* **74**, 3027.

Smith, E.J. & Tsurutani, B.T. 1976. Magnetosheath lion roars, *J. Geophys. Res.* **81**, 2261.

Smith, R.L., Helliwell, R.A. & Yabroff, I.W. 1960. A theory of trapping of whistlers in field-aligned columns of enhanced ionization, *J. Geophys. Res.* **65**, 815.

Smith, R.L. 1961. Propagation characteristics of whistlers trapped in field-aligned columns of enhanced ionization, *J. Geophys.* **66**, 3699.

Smith, R.L. 1964. An explanation of subprotonospheric whistlers, *J. Geophys, Res.* **69**, 5019.

Smith, R.L., Brice, N.M., Katsufrakis, J.P., Gurnett, D.A., Shawhan, S.D., Belrose, J.S. & Barrington, R.E. 1964. An ion gyrofrequency phenomenon observed in satellites, *Nature* **204**, 274.

Smith, R.L. & Angerami, J.J. 1968. Magnetospheric properties deduced from OGO 1 observations of ducted and nonducted whistlers, *J. Geophys. Res.* **73**, 1.

Smith, R.L. 1969. VLF observations of auroral beams as sources of a class of emissions, *Nature* **224**, 351.

Stefant, R.J. 1970. Nonpotential ion-harmonic waves, *J. Geophys. Res.* **75**, 7182.

Stiles, G.S. & Helliwell, R.A. 1975. Frequency-time behavior of artificially stimulated VLF emissions. *J. Geophys. Res.* **80**, 608.

Stone, N.H., Samir, U. & Oran, W.A. 1974. Laboratory simulation of the structure of disturbed zones around bodies in space, *J. Atmos. Terr. Phys.* **36**, 253.

Storey, L.R.O. 1953. An investigation of whistling atmospherics, *Phil. Trans. Roy. Soc. Lond.* **A246**, 113.

Storey, L.R.O. 1957. A method for interpreting the dispersion curves of whistlers, *Can. J. Phys*, **35**, 1107.

Storey, L.R.O. 1958. Protons outside the earth's atmosphere, *Ann. Geophys.* **14**, 144 (see also 1956, A method to detect the presence of ionized hydrogen in the outer atmosphere, *Can. J. Phys.* **34**, 1153).

Storey, L.R.O. 1962. In monograph on radio noise of terrestrial origin, F. Horner (Editor), Amer. Elsevier, New York; see also 1963, Whistler propagation, In *Advances in Upper Atmosphere Research*, Pergamon Press, New York, p. 231.

Sudan, R.N. & Ott, E. 1971. Theory of triggered VLF emissions, *J. Geophys, Res.* **76**, 4463.

Tartaglia, N.A. 1970. *Irregular geomagnetic micropulsations, associated with geomagnetic bays in the auroral zone*, University of Pittsburg.

Tatnall, A.R., Matthews, J.P., Bullough, K. & Kaiser, T.R. 1978. *Power-line harmonic radiation and the electron slot*, Space Phys. Group, Sheffield, Sci. Rep. No. 1.

Taylor, J.C. 1967. Disturbance of a rarefied plasma by a supersonic body on the basis of the Poisson-Vlasov equations–I, II, *Planet. Space Sci.* **15**, 155, 463.

Taylor, W.W. & Gurnett, D.A. 1968. Morphology of VLF emissions observed with the Injun 3 satellite, *J. Geophys. Res.* **73**, 5615.

Taylor, W.W., and Shawhan, S.D., 1974. A test of incoherent Čerenkov radiation for VLF hiss and other magnetospheric emissions, *J.Geophys. Res.* **79**, 105.

Taylor, W.W., Parady, B.K. & Cahill, L.J. 1975. Explorer 45 observations of 1- to 30-Hz magnetic fields near the plasmapause during magnetic storms, *J. Geophys. Res.* **80**, 1271.

Taylor, W.W. & Lyons, L.R. 1976. Simultaneous equatorial observations of (1–30) Hz waves and pitch-angle distributions of ring current ions, *J. Geophys. Res.* **81**, 6193.

Taylor, W.W. & Anderson, R.R. 1977. Explorer 45 wave observations during the large magnetic storm of August 4–5, 1972, *J. Geophys. Res.* **82**, 55.

Tepley, L.R. 1961. Observations of hydromagnetic emissions, *J. Geophys. Res.* **66**, 1651.

Tepley, L.R. & Wentworth, R.C. 1962. Hydromagnetic emissions, X-ray bursts and electron bunches: 1. Experimental results, *J. Geophys. Res.* **67**, 3317.

Thorne, R.M. & Kennel, C.F. 1967. Quasi-trapped VLF propagation in the outer magnetosphere, *J. Geophys. Res.* **72**, 857.

Thorne, R.M. 1968. Unducted whistler evidence for a secondary peak in the energy spectrum near 10 keV, *J. Geophys. Res.* **73**, 4895.

Thorne, R.M., Smith, E.J., Burton, R.K. & Holzer, R.E. 1973. Plasmaspheric hiss, *J. Geophys. Res.* **78**, 1581.

Troitskaya, V.A. 1961. Pulsation of the earth's electromagnetic field with periods of 1–15 sec and their connection with phenomena in the high atmosphere, *J. Geophys. Res.* **66**, 5.

Troitskaya, V.A. and Gulel'mi, A.V. 1969. Hydromagnetic diagnostics of a plasma in the magnetosphere, *Usp. Fiz. Nauk* **97**, 453 [In Russian] [1970, *Ann. Geophys.* **26**, 893].

Troy, B.E., Medved, D.B. & Samir, U. 1970. Some wake observations obtained on the Gemini/Agena two-body system, *J. Astr. Sci.* **18**, 173.

Troy, B.E., Maier, E.J. & Samir, U. 1975. Electron temperatures in the wake of an ionospheric satellite, *J. Geophys. Res.* **80**, 993.

Tsurutani, B.T. & Smith, E.J. 1974. Postmidnight chorus: a substorm phenomenon, *J. Geophys. Res.* **79**, 118.

Vaglio-Laurin, R. & Miller, G. 1971. Ionospheric gas dynamics revisited. In *Rarefied Gas Dynamics* (7th Symp.), Academic Press, New York.

Vas'kov, V.V. 1966. Perturbation of the density of charged particles at large distances from a body moving rapidly in a plasma in the presence of a magnetic field, *ZhETF* **50**, 1124. [In Russian].

Vas'kov, V.V. 1969a. Scattering of radio waves by the wake of a body moving in the ionosphere near the caustic, *Geomagnetizm i Aeronomiya* **9**, 847 [In Russian].

Vas'kov, V.V. 1969b. Scattering of radio waves by the wake of a body moving in the ionosphere, *Kosmicheskie Issledovaniya* **7**, 559 [In Russian].

Vas'kov, V.V. 1969c. Scattering of radio waves by the wake of a rapidly moving body, taking into account the sphericity of the ionosphere, *Geomagnetizm i Aeronomiya* **9**, 426 [In Russian].

Vas'kov, V.V. 1969d. Candidate's Dissertation, *IZMIRAN* [In Russian].

Vas'kov, V.V. & Gurevich, A.V. 1973. Parametric excitation of Langmuir oscillations in the ionosphere in a strong radio-wave field, *Radiofizika* **16**, 188 [In Russian].

Voots, G.R., Gurnett, D.A. & Akasofu, S.I. 1976. Auroral kilometric radiation as an indicator of auroral magnetic disturbances, *J. Geophys. Res.* **81**; University of Iowa 76-31 *J. Geophys. Res.* **82**, 2259: 1977.

Walker, A.D.M. 1976. The theory of whistler propagation, *Rev. Geophys. Space Phys.* **14**, 629.

Walter, F. & Angerami, J.J. 1969. Nonducted mode of propagation between conjugate hemispheres; observation on OGO's 2 and 4 of the 'walking trace' whistler and of Doppler shifts in fixed frequency transmissions, *J. Geophys. Res.* **74**, 6352.

Warren, E.S. & Hagg, E.L. 1968. Observation of electrostatic resonances of the ionospheric plasma, *Nature* **220**, 466.

Whipple, E.C. & Troy, B.E. 1965. Preliminary ion data from the planar ion–electron trap on OGO 1, Amer. Geophys. Union Meeting.

Winckler, J.R. 1974. An investigation of wave–particle interactions and particle dynamics using electron beams injected from sounding rockets, *Space Sci. Rev.* **15**, 751; *J. Geophys. Res.* **79**, 4195.

Young, T.S.T., Callen, J.D. & McCune, J.E. 1973. High-frequency electrostatic waves in the magnetosphere, *J. Geophys. Res.* **78**, 1082.

Author index

Abrams, S. I 279
Ackerson, K.L. II 122, 229, 239
Adachi, S. I 217, 271
Aikyo, K. I 216, 238, 241, 243–5, 271 II 151, 224
Akasofu, S.I. II 242
Akhiezer, A.I. I 28, 37, 43, 61, 94, 101, 105, 109, 111, 118, 135, 207, 210, 271
Akhiezer, I.A. I 271
Alexander, J.K. II 100, 214, 224
Alexeff, I. I 204, 279
Aliev, Yu.M. I 193–5, 199, 271 II 224
Allen, E.M. I 271 II 224
Al'pert, Ya.L. I 22, 30, 37, 40, 46, 68, 71, 73, 76, 77, 79, 141, 143, 216, 217, 219, 221, 225, 227, 231, 233, 246, 249–51, 260, 264–7, 269, 271, 272 II 3, 4, 15, 38, 63, 65, 67–9, 75, 79, 80, 82, 89, 106–8, 160, 190, 224
Al'tshul', L.M. I 133, 135, 272
Anderson, R.R. II 128, 129, 132, 177, 196, 200, 218, 221, 225, 230, 231, 241
Andreev, N.E. I 200–4, 272
Angerami, J.J. I 15, 272 II 151, 152, 160, 161, 225, 241, 242
Appleton, E.V. I 49, 272
Ashour-Abdalla, M. I 78, 272 II 163, 189, 225
Astrelin, V.T. II 6, 26, 225
Aubry, M.P. I 217, 238, 272, 273 II 226

Bahnsen, A. II 228
Bailey, V.A. I 135, 273
Baker, B. II 225
Barfield, J.N. II 225
Barfield, T.A. I 109, 282
Barkhausen, H. II 160, 225
Barrett, P.J. II 6, 28, 225
Barrington, R.E. II 90, 111, 112, 138, 149, 150, 182, 225, 231, 235, 240
Bashilov, I.P. II 224
Baumback, M.M. II 234
Beghin, C. II 190, 225, 227, 231, 241

Belikovich, V.V. I 160, 161, 168, 169, 273
Bell, T.F. I 213, 273 II 232
Belrose, J.S. II 138, 149, 150, 225, 240
Benediktov, E.A. I 273
Benioff, H. II 103, 225
Berezin, Yu.A. I 135, 210, 211, 273
Bernstein, I.B. I 102, 133, 135, 273
Bernstein, W. II 191, 225
Binsack, J.H. I 280 II 240
Biondi, M.A. I 163, 273, 280
Bitoun, J. I 217, 238, 272, 273 II 186, 226
Blair, W.E. II 231
Bloch, J.J. II 227
Blood, D.W. I 169, 172, 281
Bloom, M.H. II 234
Bogashchenko, I.A. II 6, 25, 28, 29, 225, 226
Bohm, D. I 108, 273
Booker, H.G. I 27, 47, 171, 172, 174, 216, 217, 247, 273 II 226
Borisov, N.D. I 167, 273
Bossen, M. II 110, 217, 226
Bourdeau, R.E. II 5, 226
Bowen, P.J. II 5, 23, 34, 226
Bowhill, S.A. I 167, 273
Boyd, R.L. II 226
Brace, L.H. I 22, 274
Bremmer, H. I 216, 274
Brice, N.M. I 36, 277, 281 II 90, 111–13, 138, 163, 226, 230, 232, 233, 235–40
Bridge, H.S. I 280 II 240
Brinton, H.C. I 281 II 227
Brown, L.W. I 23, 274 II 208, 212, 226, 236
Brown, P.E. II 236
Brundin, C.L. II 4, 226
Buchel'nikova, N.S. II 225, 230
Bud'ko, N.I. II 47–52, 55–57, 88, 90, 91, 226
Budden, K.G. I 47, 66, 217, 269, 274
Bullough, K. II 167, 226, 232, 234 241
Burns, T.B. II 117, 118, 230
Burtis, W.J. II 169–172, 226

Burton, R.K. II 171, 172, 226, 241

Cahill, L.J. II 241
Call, S.M. II 226
Callen, J.D. II 242
Calvert, W. II 180, 181, 183–6, 227, 228
Campbell, W.H. II 104, 227
Carlson, H.C. I 146, 147, 160, 181–3, 274, 276
Carpenter, D.L. I 15, 171, 177, 272, 274 II 149–51, 160, 161, 163, 166, 225, 227
Cartwright, D.G. II 97, 227, 233, 236
Cauffman, D.P. II 230
Cerisier, J.C. II 154, 227
Chan, K.W. II 135, 227
Chanteur, G. II 225
Chapman, S. I 39, 274
Chappell, C.R. I 15, 16, 17, 274 II 239
Cherepovskii, V.A. I 273
Chopra, K.P. I 3, 227
Christiansen, P. II 194, 227
Clark, D.H. II 17, 227
Clayden, W.A. II 6, 24, 25, 40, 228
Coffey, T.P. II 237
Cohen, H. II 225
Cohen, R. I 142, 160, 162, 167, 274, 281
Coleman, P.J. I 275 II 108–10, 217, 225, 228, 229, 236, 238–40
Conte, S.D. I 91, 275
Copal Rao, M.S. I 27, 217, 247, 274
Coronilleau, N. II 227
Coroniti, F.V. II 173, 190, 228, 233, 239
Cowling, T.G. I 39, 274
Cragin, B.L. I 167, 274
Crawford, F.W. II 186, 228
Crone, W.R. II 234
Crook, G.M. I 275 II 229, 239
Crystal, T.L. I 213, 277 II 163, 231, 232
Cummings, W.D. II 109, 217, 228

D'Angelo, N. II 122, 228
Daniell, G.J. I 217, 274
Davis, A.H. II 3, 228
Davis, L. II 240
Davis, T.N. II 225
Debrie, R. II 225, 241
Decreau, P. II 227
DeForest, S.E. I 21, 274
Dehmel, G. II 237
Denby, M. II 226
Deshmukh, A.R. I 281
Domrin, V.I. I 194–6, 274
Donley, J.L. II 5, 226
Dougherty, J.P. II 182, 186, 228
Dowden, R.L. II 151, 228
Doyle, P.H. I 93, 109, 275, 279
Dreicer, H. I 120, 141, 275

DuBois, D.F. I 135, 192, 275
Dubovoi, A.P. II 33, 34, 40, 46, 53, 54, 60, 61, 228
Dunckel, N. II 150, 151, 169, 170, 207, 208, 212, 222, 227, 228
Dungey, J.W. II 168, 228
Dwarkin, M.L. II 109, 217, 228
Dyce, R.B. I 47, 273
Dysthe, K.B. II 163, 228

Eckersley, T.L. II 160, 228
Edgar, B.C. I 216, 275 II 151, 228
Eidel'man, V, Ya. I 276
Eidel'man, Yu.I. II 226
Eidman, Yu.I II 225
Erukhimov, L.M. I 273
Estabrooks, M.F. II 186, 237
Etcheto, J. II 227

Fadeeva, V.N. I 91, 275
Fainberg, Ya.B. I 101, 109, 271
Fejer, J.A. I 135, 167, 274, 275 II 186, 228
Fialer, P.A. I 169, 170, 173, 275
Ficklin, B. II 228, 231
First, M. II 238
Fligel, D.S. I 272 II 106–8, 224
Forslund, D.W. II 130, 228
Försterling, K. I 54, 66, 275
Foster, H.G. I 277
Fournier, G. II 6, 228
Frandsen, A.M. I 280 II 239
Frank, L.A. I 18, 19, 20, 275, 277 II 122, 131, 136–8, 142, 144, 145, 156, 157, 195, 199, 210, 211, 218–21, 229–31, 239
Frankel, M.S. II 210, 229
Fredricks, R.W. I 14, 275, 280 II 99, 116, 129, 187–9, 197, 221, 228, 229, 233, 239
Fried, B.D. I 91, 109, 149, 275
Fritz, T.A. I 282
Frolov, V.L. I 273

Gaponov, A.V. I 275 II 229
Gardner, C.S. I 210, 275
Gendrin, R. II 97, 99, 104, 111, 160, 194, 227, 229
Georges, T.M. I 167, 275
Gershman, B.N. I 84, 90, 236, 275 II 147, 229
Getmantsev, G.G. I 160, 273, 276
Gey, F.C. II 233
Gibbons, W. II 226
Ginzburg, V.L. I 28, 39, 100, 276
Goe, G.B. II 181, 185, 227
Goldman, M.V. I 135, 192, 275
Goldstein, M.L. II 237
Gorbunov, L.M. I 203, 204, 276
Gordeev, G.V. I 120, 276

Gordon, W.E. I 146, 147, 160, 181, 183, 274, 276 II 176, 229
Gough, P. II 227
Graff, P. I 272, 273 II 226
Grard, R. II 4, 229
Grebowsky, J.M. I 281
Green, I.M. I 275, 280 II 229, 239, 240
Green, J.L. II 213–15, 229, 231
Greene, J.M. I 133, 273
Gross, E.P. I 103, 108, 273, 276
Gulel'mi, A.V. I 276 II 106, 229, 242
Gurevich, A.V. I 43, 120, 135, 137–41, 147, 150, 153–6, 167, 169, 179, 180, 184, 186–90, 192, 271–3, 276, 281, 282 II 3, 6, 21–5, 27, 28, 32–6, 38, 63–9, 75, 84, 92, 140, 224, 226, 229, 230
Gurnett, D.A. I 15, 37, 277, 280 II 100, 111–24, 128–32, 136–8, 142, 144–6, 150, 154–7, 159, 174–6, 186–9, 191–5, 198–201, 204–15, 217–22, 225, 229–31, 233, 234, 236, 238, 240–2
Gustafson, W.A. II 233
Guthart, H. II 124, 126, 218, 223, 231

Hagg, E.L. II 181–3, 231, 242
Hake, R.D. I 273
Hall, D.F. I 182, 277 II 6, 40, 231
Hallinan, T. II 225
Hamelin, M. II 231, 241
Hanson, W.B. II 240
Harp, R.S. II 228
Harris, I. II 228
Harris, K.K. I 274
Hartree, D.R. I 49, 277
Hartz, T.R. II 182, 231
Haselgrove, C.B. I 216, 238, 277
Haselgrove, J. I 216, 238, 277
Hashimoto, K. I 238, 277
Haslett, J.C. I 163, 277
Hayakawa, M. II 148, 154, 231
Helliwell, R.A. I 25, 26, 213, 273, 277, 279, 280 II 111, 151, 160–72, 226, 228, 232, 238, 240, 241
Henderson, C.L. II 5, 17, 23, 34, 226, 232
Hendrickson, R.A. II 233
Herring, R.N. I 217, 277
Hess, W.N. I 19, 277
Hester, S.D. II 6, 41, 42, 58, 232
Higgy, R.G. II 234
Higuchi, Y. II 105, 232
Hines, C.O. II 150, 232
Hoffman, J.H. II 232
Hoffman, R.A. I 19, 20, 280
Hoffman, W.C. II 151, 232
Holway, L.H. I 278
Holzer, R.E. II 135, 171, 172, 226, 227, 238, 240, 241

Howens, E.J. II 231
Huber, R.W. II 225
Hughes, A.R. II 226
Hultqvist, B. I 277 II 107, 232
Hurdle, C.V. II 6, 24, 25, 40, 228
Huxley, L.G.H. I 135, 277

Ignat'ev, Yu.A. I 273
Issledovaniya Kosmicheskogo Prostranstva I 19, 192, 277
Istomin, Ya.L. I 212, 277 II 163, 232, 233

Jacobs, J.A. II 104, 106, 232, 233
James, H.G. II 144, 147, 154, 227, 233
Jastrow, R. II 3, 233
Jew, H. II 4, 36–8, 45, 235, 238
Johnson, W.C. II 139, 140, 147–9, 168, 169, 234, 236
Johnston, T.W. II 6, 233
Jones, D. I 37, 277 II 194, 227, 229, 240
Jones, D.E. II 240
Joselyn, J.A. II 114, 233

Kaiser, M.L. II 100, 214, 224, 236
Kaiser, T.R. II 167, 226, 232, 233, 241
Kantor, I.J. I 181, 182, 277
Kapustina, O.V. II 224
Karpman, V.I. I 133, 135, 208, 210–12, 272, 273, 277, 278, 282 II 163, 232, 233
Kasha, M.A. II 6, 233, 237
Katsufrakis, J.P. II 163–6, 232, 238, 240
Kaufman, R.N. I 217, 246, 248–50, 256–8, 272, 278
Kawai, M. I 217, 278
Keeley, D.A. II 225
Kelley, M.C. II 233
Kellogg, P.J. II 97, 98, 227, 233, 236
Kemp, R.F. I 277 II 231
Kennel, C.F. II 117, 275, 278 II 150, 168, 171, 172, 188, 189, 225, 228, 229, 233, 235, 239, 241
Kenney, J.F. II 104, 105, 233
Kiel, R.E. II 4, 233
Kimura, I. I 216, 217, 238, 242, 243, 278 II 99, 150–2, 162, 163, 233, 235, 236
Kintner, P.M. II 114, 115, 124, 129, 217, 218, 233
Kirii, A.Yu. I 200, 272
Kitsenko, A.B. I 111, 115, 236, 278
Kivelson, M. II 239
Knaflich, H.B. II 104, 105, 233
Knecht, R.W. II 179, 234
Knight, D.J.E. II 233
Knyazuk, V.S. II 69–72, 234
Komrakov, G.P. I 273
Konradi, A. I 282
Korobkov, Yu.S. I 273

Kotik, Yu.S. I 273, 276
Kovner, M.S. I 111, 115, 278
Kraus, J.D. II 76, 234
Kraus, L. II 3, 55, 234
Kreppel, R. I 177, 178, 279
Kresin, V.Z. II 56, 238
Kruskal, M.D. I 133, 135, 210, 211, 273, 275, 282
Kugelman, P. I 279
Kumagai, H. I 277
Kurth, W.S. II 213, 214, 231, 234

Laaspere, T. II 139, 140, 147–9, 151, 168, 169, 234
Laird, M.J. I 217, 278
LaLonde, L.M. I 181, 276, II 176, 229
Landau, L.D. I 31, 34, 86, 278 II 76, 234
Lapshin, V.I. I 271
Lasch, S. II 147, 162, 227, 234
Lassen, H. I 49, 278
Laviola, M. I 279
Lederman, S. II 6, 234
Lefeuvre, F. II 167, 234
Leinbach, H. II 225
LeLevier, R.E. I 142–5, 159, 278
Lepping, R.P. II 231, 239
Liemohn, H.B. I 117, 278 II 104, 106, 161, 193, 233, 234
Lifshitz, E.M. II 76, 234
Liu, C.H. I 128, 282
Liu, V.C. II 4, 36, 38, 45, 46, 235
Lockwood, G.E. II 181, 182, 235, 237
Lucas, C. II 113, 235
L'vova, G.A. II 224
L'vovich, V.V. I 135, 278
Lyubarskii, T.Ya. I 135, 271
Lyons, L.R. I 19, 20, 282 II 114, 168, 233, 235

Maeda, K. I 216, 278 II 177, 225
Maier, E.J. II 238, 242
Maier, J.R. I 14, 280
Mainstone, J.S. II 104, 235
Mandel'shtam, L.I. I 193, 278
Mantei, T.D. II 228
Mariani, F. I 14, 278
Marsch, E. II 231
Martelli, G. II 227
Martin, A.R. II 4, 45, 235
Martin, D.F. I 273
Martz, J. II 225
Maslennikov, M.V. II 4, 235
Masson, F. II 228
Mather, W.E. II 225
Matsumoto, H. I 217, 278 II 98, 99, 163, 235, 236

Matthews, J.P. II 241
McAfee, J.R. II 180, 185, 186, 227, 235
McCleod, M.G. II 240
McCune, J.E. II 242
McEwen, D.J. II 90, 235
McGehee, J.H. II 233
McIlwain, C.E. I 15, 21, 274, 278
McKeown, P. II 13, 235
McKibbin, D.D. II 240
McNicol, R.W. II 104, 235
McPherron, R.L. II 108–10, 217, 226, 236, 238
Meckel, B.B. II 6, 236
Medved, D.B. II 9, 236, 242
Megill, L.R. I 163, 277
Melrose, D.B. II 100, 236, 242
Meltz, G. I 142–7, 159, 165–7, 191, 278
Mikhailova, G.A. II 224
Mikhailovskii, A.B. I 97, 98, 109, 111, 114, 278
Miller, G. II 4, 242
Miller, M.A. I 275, 278 II 229, 236
Miller, T.L. II 163, 166, 227
Mimura, I. I 277
Minkoff, J. I 169, 171–3, 176–8, 191, 279
Mironenko, L.F. I 276
Mityakov, N.A. I 161, 273, 276
Miura, R.M. I 275
Miyatake, S. II 97, 235, 236
Mlodnosky, R.F. I. 25, 279
Moiseyev, B.S. I 217, 222, 247, 260, 264–7, 269, 279
Monaghan, J.J. II 182, 186, 228
Monson, S.J. II 98, 233, 236
Montgomery, D. I 204, 279
Morgan, M.G. II 140, 234, 236
Moser, F.S. II 233
Mosier, S.R. II 118, 119, 123, 124, 142, 145, 146, 186, 219, 230, 236
Moskalenko, A.M. II 41, 69–74, 234, 236
Muldrew, D.B. II 183, 185, 186, 236, 237
Murukami, T. II 148, 237
Musman, G. II 237
Muzzio, J.L. II 120, 124, 237

Nagayama, M. I 279 II 224, 237
Nelms, G.L. II 181, 182, 225, 231, 237
Ness, N.F. I 14, 278
Neubauer, F.M. II 198, 231, 237
Neufeld, J. I 93, 109, 275, 279 II 239
Neugebauer, M. I 280
Newton, C.C. I 277
Nicolet, M. I 39, 279
Nishikawa, K. I 192, 279
Nishizaki, R. I 279 II 237
Nosachev, L.V. II 6, 39, 240
Nunn, D.A. I 217, 278 II 163, 237

Author index

Obayashi, T. II 106, 237
Oberman, C. I 279
O'Brien, B.J. II 120, 121, 159, 230
Ohtsu, J. II 231
Ondoh, T. I 15, 216, 238, 241, 243–5, 271, 279 II 148, 151, 224, 237
O'Neil, T. I 135, 279
Oran, W.A. II 6, 42, 43, 237, 241
Orr, D. II 233
Osborne, F.J.F. II 6, 233, 237
Ossakow, S.L. II 237
O'Sullivan, R.J. II 228
Ott, E. II 163, 241
Owens, H.D. I 19, 20, 275
Oya, H. II 181, 189, 237

Palmadesso, P. II 215, 237
Pan, Y.S. II 4, 237
Panchenko, Yu.M. II 47, 48, 51, 237
Papadopoulos, K. II 200, 237
Parady, B.K. II 241
Pariiskaya, L.V. II 230
Park, C.G. I 279 II 167, 227, 237
Patel, V.L. II 109, 217, 237
Paul, A.K. I 281
Pearse, C.A. II 3, 233
Pellat, R. II 163, 225, 228, 229, 233, 235, 238, 239, 241
Perkins, W.F. I 167, 182, 183, 279
Petschek, H.E. I 117, 278 II 168, 172, 233
Pfeiffer, G.W. II 230
Phillipp, W. II 231
Pigache, D. II 6, 229
Pitaevskii, L.P. I 214, 271, 272, 279 II 3, 28, 47, 51, 56, 75, 76, 79, 80, 82, 84, 224, 229, 230, 237, 238
Polovin, R.V. I 271
Pope, J.H. II 106, 238
Porter, D. I 279
Proceedings of the IEEE 1969 special issue II 179, 238

Radford, W.D. II 228
Radio Science (journal) 1974 special issue I 142, 279
Raghuram, R. II 232
Raitt, W.J. II 227
Rao, P.B. I 172–4, 271, 279 II 224
Rapoport, V.O. I 111, 273, 276, 279
Ratcliffe, J.A. I 47, 279
Rawer, K. I 47, 49, 279, 281
Razin, V.A. I 273
Reeve, C.D. II 151, 238
Rodriguez, P. II 111, 120, 129, 130, 156, 157, 174, 198, 218, 220, 230, 238
Roederer, J.G. I 19, 247, 279
Rorden, L. II 228

Rosenbauer, H. II 228, 231
Rosenberg, T.J. II 168, 238
Rostoker, N. II 125, 126, 238
Roux, A. II 163, 238
Rukhadze, A.A. I 28, 100, 276, 280
Russell, C.T. I 275, 280 II 103, 111, 116, 135, 154, 219, 226, 229, 236, 238–40
Russell, S. II 234
Rycroft, M.J. II 151, 238
Rytov, S.M. I 101, 216, 280

Sagdeev, R.Z. I 280 II 238
Saito, T. II 103, 238
Salimov, R.A. II 226, 230
Samir, U. II 5, 8, 9, 15–17, 23, 34, 36–8, 232, 237, 238, 241, 242
Sawchuck, W. II 22, 239
Sayasov, Yu.S. II 239
Sazonov, V.O. I 273
Sazonov, Yu.A. I 276
Scarabucci, R.R. I 217, 280
Scarf, F.L. I 14, 180, 275, 280 II 122, 125–7, 129–31, 141, 142, 154, 161, 176, 187–9, 193, 197, 198, 201–4, 218, 219, 221, 222, 228, 229, 231, 233, 234, 239, 240
Scherr, D.J. II 234
Schmitt, J.T.M. II 6, 239
Schwenn, R. II 231
Sellen, J.M. I 277 II 231
Serbu, G.P. I 14, 280, II 226, 239
Shabanskii, V.P. I 19, 280
Sharp, G.W. I 274 II 16, 240
Shavin, P.V. I 273
Shaw, R.R. I 15,280 II 186–9, 191–4, 204–7, 221, 222, 230, 240
Shawhan, S.D. I 277 II 99, 112, 128, 186, 229, 230, 240, 241
Shkarofsky, I.P. II 233
Shklyar, D.R. II 232, 233
Shlyuger, I.S. I 161, 187, 276, 280 II 140, 230
Showen, R.L. I 274
Sigov, Yu.S. II 4, 235
Silin, V.P. I 28, 135, 192–7, 199, 201, 203, 204, 271, 276, 280
Singer, S.F. II 4, 240
Sipler, D.P. I 273, 280
Siredey, C. II 190, 225
Siscoe, G.L. I 14, 280 II 108, 130, 217, 240
Sitenko, A.G. I 84, 90, 93, 94, 271, 280 II 240
Skvortsov, V.V. II 6, 39, 240
Smirnova, V.V. II 92, 230, 240
Smith, A.J. II 233
Smith, C.R. I 281
Smith, E.J. I 280 II 155–7, 171, 173, 174,

219, 227, 238–42
Smith, P.H. I 19, 20, 280
Smith, R.A. II 237
Smith, R.L. I 36, 217, 247, 277, 280, 281
  II 90, 110, 138, 146, 150–3, 160, 220,
  226, 230, 233, 241
Sonin, A.A. II 6, 41, 42, 58, 232
Stefant, R.J. II 104, 114, 229, 241
Stepanov, K.N. I 84, 90, 93, 94, 111, 115,
  236, 271, 278, 280, 281 II 240
Stiles, G.S. II 163, 167, 241
Stilner, E.C. II 104, 227
Stix, T.H. I 28, 91, 96, 216, 281
Stone, K. II 227
Stone, N.H. II 6, 44, 237, 241
Storey, L.R.O. I 262–4, 268–70 II 160, 241
Stott, G.F. I 217, 269, 274
Sturrock, P.A. I 219, 281
Suchy, K. 47, 49, 217, 279, 281
Sudan, R.N. II 163, 241

Tanaka, Y. I 279 II 231, 237
Tartaglia, N.A. II 106, 241
Tatnall, A.R. II 167, 226, 241
Taylor, H.A. I 15, 281 II 139, 140, 227,
  234, 236
Taylor, J.C. II 4, 241
Taylor, W.W. II 114, 116, 129, 186, 195,
  196, 241
Tepley, L.R. II 103, 104, 241
Teplykh, A.I. I 273
Terent'ev, N.M. I 91, 275
Terry, P.D. I 217, 274, 281
Theis, R.F. I 22, 274
Thome, G.D. I 169, 172–4, 271, 279, 281
Thorne, R.M. II 133–5, 150, 154, 171,
  235, 241
Tomchinskii, A.M. I 273
Tomljanovich, N.M. I 278
Trakhtengerts, V.Yu. I 273, 276, 279
Trimpi, M. II 232
Troitskaya, V.A. I 276 II 103, 106, 229,
  242
Troy, B.E. II 5, 9, 15, 16, 238, 242
Tsurutani, B.T. II 157, 171, 173, 174, 240,
  242
Tsytovich, V.N. I 135, 281

Utlaut, W.F. I 142, 160–2, 167, 168, 281

Vaglio-Laurin, R. II 4, 237, 242
Valeo, E.J. I 279
VanZandt, T.E. II 183, 184, 227, 234
Vas'kov, V.V. I 167, 169, 179, 180, 273,
  281, 282 II 28, 48, 52, 56, 59, 75, 77,
  83–5, 242
Violette, E.J. I 160, 167, 281
Virobik, P.F. II 239
Voots, G.R. II 213, 242

Walker, A.D.M. I 217, 282 II 160, 242
Walkup, J.F. II 227
Walsh, W.J. I 281
Walter, F. I 216, 238, 282 II 151, 242
Wang, C.Y. II 151, 234
Warren, E.S. II 181, 242
Watanabe, T. II 104, 232
Watson, G.N. I 3, 249, 282
Watson, K. II 55, 234
Weissman, I. I 279
Wentworth, R.C. II 104, 241
Whipple, E.C. II 16, 226, 242
Whitehead, J.D. I 167, 274
Whittaker, E.T. I 249, 282
Widhonf, V.C. II 234
Williams, D.J. I 19, 20, 282
Willmore, A.P. II 5, 23, 34, 226, 227, 235
Wilson, P.S. II 225
Winckler, J.R. II 97, 233, 242
Wrenn, G.L. II 5, 9, 15–17, 238

Yabroff, I.W. I 216, 238–40, 243, 280, 282,
  II 160, 240
Yaroslavtsev, A.A. II 34, 53, 54, 60, 61,
  228
Yeh, K.C. I 128, 282
Young, T.S.T. II 189, 242
Yung, T.J. II 231

Zabusky, N.J. I 135, 210, 211, 282
Zakharov, V.E. I 135, 282
Zeleznyakov, V.V. I 282
Zhizhimov, L.A. II 239
Ziemke, R. II 225
Zmuda, A.J. II 228
Zuikov, N.A. I 273, 276
Zyunder, D. I 199, 271 II 224

Subject index

absorption coefficients I 81, 168, 180
absorption of particles II 12
accommodation of particles, partial and total II 11, 12
accumulation of particles near body II 63
Aerobee rocket II 149
AKR, see auroral kilometric radiation
Alfvén refractive index I 8, 53, 63
Alfvén velocity I 8, 43, 63, 203
Alfvén wave I 53, 58, 63, 64, 93, 94, 101, 102, 123, 124, 127, 203, 222, 237 II 104
Alouette satellites II 97, 149, 186
Alouette 1 II 90, 91, 110, 138, 151, 179
Alouette 2 I 23 II 112, 147, 148, 179–83, 193
ambipolar diffusion coefficient I 154
amplitude-modulation index I 186
angular dependence of electric field near body II 55, 56
angular dependence of particles near body II 19–25, 30, 33–41, 47–54, 57–61
angular dependence of temperature near body II 16, 17
anisotropic distribution of electron velocities II 178
anisotropic electron distribution II 130
anisotropic instability I 110ff
anisotropic Maxwellian distribution I 110, 117, 126
anisotropic pitch angle distribution II 100
anisotropic temperature distribution I 106, 107 II 100, 106
anisotropic velocity distribution I 106, 110ff, 116ff, 126 II 106
anomalous absorption I 167–9
anomalous Doppler effect I 111
Antarctica, see Siple, Eights station
Apollo model II 43, 44
Appleton–Hartree formula I 48, 49
Appleton–Lassen formula I 49
Ariel 1 satellite II 5, 17, 23, 34, 35
Ariel 3 satellite II 148, 167
Ariel 4 satellite II 167

artificial sporadic layer I 163, 164, 167 II 97
artificially created emissions II 194, 195
artificially stimulated emissions (ASE) II 162ff
ASE, see artificially stimulated emissions, stimulated emission
ATS I satellite II 109, 110
ATS 5 satellite I 21
attenuation coefficient I 70ff, 82ff, 147, 184
attenuation factor I 28, 38, 43, 45, 70, 71
aurorae II 213, 214
auroral activity II 136, 176
auroral hiss (AH) II 137, 142ff, 146, 147
auroral kilometric radiation (AKR) II 208ff, 222
auroral oval II 124, 144
auroral region I 21

backscatter I 163, 172, 191
band structure of chorus II 174
band structure of TNCR II 207
beam deformation I 190
beam instability I 31, 106, 107ff, 118ff II 100, 101, 178, 215
beam instability of longitudinal waves I 111ff
beam instability of transverse (e.m.) waves I 115ff, 123ff
beam of particles I 31, 107ff
beam of electromagnetic waves I 153, 184, 188
Bernstein modes I 102 II 185, 194
Bessel function (imaginary argument) I 102, 196 II 20, 61, 77
Bessel function (real argument) I 194, 261, 262 II 77
bi-Maxwellian function I 117
Boltzmann distribution function I 29, 30, 214
boundary conditions at body's surface I 7 II 10 (ch. 10)
boundary layer near body II 66

boundary problem I 246, 253
bow shock I 11, 14, 24, 206, 213 II 129, 130, 137, 170, 174, 175, 178, 196–9, 218, 220
branching spectra II 162
bremsstrahlung I 111, 115
broad-band emission II 133, 136–8, 142
broad-band noise II 207, 208
broad-band VLF emission II 154ff, 219
broad lines I 182
bunching of particles II 72
bursts of protons II 122
bursts of scattered radiation II 81, 82, 85
bursts of wave amplitude II 126, 128, 132, 133, 136, 142, 156, 175, 201, 207
bursts of X-rays II 168

caustic II 75, 84–6
centre scattering line I 175–8
Cerenkov absorption I 94
Cerenkov attenuation I 33, 34, 47, 94
Cerenkov damping I 89, 90, 92
Cerenkov excitation I 125 II 88, 89
Cerenkov growth rate I 110
Cerenkov radiation, coherent II 147
Cerenkov resonance I 82, 111, 112, 113, 125, 132 II 186
Cerenkov type emission II 144
Cerenkov–Vavilov effect I 33
Chapman layer I 143
characteristic plasma field I 137
characteristic time of nonlinear process I 133
chorus II 100, 110, 116ff, 119, 144, 148, 169ff, 207
chorus bands II 148
circular polarization I 51, 55
circular scanning antenna I 165, 166
classification of waves I 56
cloud-like inhomogeneities I 160ff
coherent radiation II 212
cold beam I 109
cold magnetoplasma I 35 (ch. 4)
cold plasma, cold beam I 113, 115, 120
cold plasma, hot beam I 111, 115, 122, 125
collisional plasma I 70ff, 128, 133
collision cross-section I 138
collision frequency I 36, 134, 138ff, 248 II 83
collision frequency, effective I 106
collision integral I 30, 37, 42 II 10
collisionless attenuation I 64, 169
collisionless plasma I 34, 35, 65ff, 128, 130, 193, 205 II 50
collisionless shock I 206, 213 II 198
collisions I 36, 222, 235 II 63, 75, 83
collisions, effect on wake of body II 27, 28

collisions of like particles I 42
combination waves I 192
complex frequency I 28, 107
complex mass I 37
complex refractive index I 26, 32, 73
complex tensor elements I 149
concentration regions near body II 31
condenser I 193
conductivity tensor I 32, 40
cone in electron density II 88, 89
cone in radiation polar diagram I 262–4
confluent hypergeometric function I 249
continuity equation I 208
continuum radiation II 137, 192, 204ff, 208ff, 222
Coulomb centre II 63, 64
Coulomb field II 33, 63, 64, 69
Coulomb interactions I 161
Coulomb logarithm I 36, 39
coupled equations I 54
coupling between harmonics I 205
critical collision frequency I 51
cross-modulation I 135, 151, 192
crossover frequency I 54, 66, 72, 78, 79, 81, 232–5, 270 II 112, 117
crossover region I 73
current-driven instability II 101, 138, 176, 177, 215
curtain spectrogram II 139
cutoff for guided waves I 247
cutoff for whistlers II 160
cutoff frequency I 50, 70, 255, 259, 268 II 117, 119, 138, 153
cutoff of ELF hiss II 117–9, 121, 122
cyclotron attenuation coefficient I 94 II 113, 160
cyclotron cutoff II 160
cyclotron damping I 89
cyclotron excitation I 125
cyclotron instability I 116
cyclotron resonance I 82, 111, 113, 123, 125
cyclotron resonance, electrons I 63
cyclotron waves I 104
cylinder in plasma stream II 26, 31ff, 42, 43, 47, 51, 52
cylindrical coordinates I 260

D region I 186, 192, 248 II 168
damping of ion whistlers II 113
Debye length I 2, 7, 86, 152, 179, 196, 207, 213 II 3, 6, 35, 42, 88, 126
Debye screening II 29, 33, 65, 66
decay instability I 160, 182, 203
decay line I 182
decay rate I 28
defocusing I 192

Subject index

defocusing of beam I 184
demodulation I 187
dielectric moving body II 13, 15
differential scattering cross-section II 75–7, 81, 84
diffuse spectra of emission II 191–3
diffuse type resonances II 181
diffusion I 152 II 100
diffusion coefficients I 154, 179
dimensionless coordinate system I 260, 261
dipole field (magnetic) I 25 II 152
Dirac delta function I 261
disintegration of soliton I 210
disk in ion stream II 25, 32, 33
dispersion denominator II 88, 89
dispersion equation I 28, 32, 35, 53, 82, 194, 207, 219, 221, 252
dispersion length I 208
dispersive medium I 205, 206, 208
displaced Maxwellian velocity distribution I 107
distribution function (Boltzmann) I 29, 30, 214
Dodge satellite II 109, 217
Doppler broadening of resonance oscillations II 124
Doppler effect I 33, 34, 165, 166, 176
Doppler shift of excited waves II 129, 132
Doppler-shifted cyclotron resonance damping I 82
Doppler-shifted gyroresonance II 100, 124, 172, 196, 215
drift instability II 101, 129, 204
drift motion and instability I 127 II 100
ducts I 51, 235, 236, 243, 251, 260 II 160, 161, 166
dusk bulge (of plasmapause) I 15–17

E layer I 162
E region I 186, 192
Earth-ionosphere wave guide II 168
Earth's bow shock I 11, 14, 24, 206, 213 II 129, 130, 137, 170, 174, 175, 178, 196–9, 218, 220
Earth's magnetic field I 7, 12
Earth's rotational energy II 99
effective collision frequency I 106
effective length of wake, for scattering II 85
effective mass (complex) I 37, 71, 77
effective mass of ion mixture I 69
Eights station, Antarctica II 168
elastic diffuse reflection of particles II 11
electric field, angular distribution near body II 55, 56
electric field, effect on particles near body II 29ff

electron-acoustic waves I 96ff II 198
electron-cyclotron wave I 102ff
electron density I 7, 10, 16, 18
electron density, altitude dependence I 22, 23 II 152, 187
electron–electron collisions I 42, 235
electron flux at body surface II 14
electron gyrofrequency I 25, 26, 27, 56 II 136, 144, 155, 160, 180, 191ff, 214
electron gyroresonance I 49, 140 II 96, 100
electron–ion collisions I 185
electron–neutral collisions I 186
electron oscillations II 98
electron resonances II 91
electron ring current II 148
electron run-away condition I 119
electron sound I 97
electron temperature I 2, 6, 10, 14, 138ff, 146 II 9, 16ff, 124, 194
electron velocity distribution II 189
electron viscosity I 43
electron-whistler I 56, 64, 90, 95, 115, 116, 126, 224, 247, 251 II 104, 110, 120, 159
electrons injected into plasma II 97, 191
electrostatic gyroresonance instability II 196
electrostatic instability II 144, 215
electrostatic noise II 122, 124, 136, 137, 191–3
electrostatic resonance oscillations II 209
electrostatic waves I 98 II 96, 125, 128, 129, 132, 175, 176, 187, 188, 198, 200, 212, 216–221
ELF, *see* extremely low frequency
ELF hiss II 144
ELF pulsations II 108ff
ELF transverse waves II 108
ellipsoidal body II 20ff
elongated irregularities I 1152
energy density of waves II 125, 126, 131, 216 (ch. 20)
energy density, time average I 218
energy flux I 216, 218 II 216
energy sources for waves II 99
energy spectra of electrons II 154, 172
enhancement factor of trapped energy I 257–9
enhancement of dipole field in plasma I 266, 267
enhancement of particle concentration II 31, 72
enhancement of power grid harmonics II 167
enhancement of whistler II 165
equations of motion of charged particles I 40

equator, geomagnetic II 135, 136, 155, 161, 169, 173, 174, 188, 220
erosion by particles II 12
error function II 28, 36, 64
ESRO Ia satellite II 17
Euler–Lagrange system equations I 238
evanescent I 268
evaporation of particles from surface II 10, 12, 13
experiments, bodies in plasma II 5
experiments, near Sun II 133
experiments, waves and oscillations II 96f, 191f
Explorer satellites II 97
Explorer 2 satellite II 214
Explorer 8 satellite II 5
Explorer 12 satellite II 109, 217
Explorer 20 satellite II 179, 184
Explorer 31 satellite II 5, 8, 15–17, 35–7, 92
Explorer 45 satellite I 19, 20 II 114, 115, 128, 129, 176, 195, 196, 218
extraordinary fast wave I 56, 92, 230, 231 II 207, 214
extraordinary slow wave I 56, 92, 227, 255
extraordinary wave I 48, 53, 74, 140, 144, 161, 178, 179 II 110, 115
extremely low frequency (ELF) I 48, 52, 62, 65, 70, 93, 118, 222ff, 230, 233, 248, 270 II 98, 103 (ch. 16), 217, 218

F region I 161 II 75
F-1 layer I 161, 162
F-2 layer I 128
F-2 region I 5, 22
falling tone (faller) II 161, 165, 166, 168, 173, 174
far zone of body in plasma II 4, 8, 9, 22, 23, 29, 31, 34, 41, 47ff, 66, 68, 88
fast ion-acoustic wave II 125
fast magnetoacoustic wave I 63, 64, 66, 90, 93, 95, 102, 123, 224, 225
fast-moving body II 19 (ch. 11), 75 (ch. 13), 87 (ch. 14)
feedback mechanism for ASE II 164
Fermat's principle I 238
field-aligned ducts I 260 II 160, 161, 166, 203
fine structure of HF emissions II 190
five branches of dispersion curves I 56, 57
floating spike II 182, 183
flow instability II 100
flow of plasma II 3 (chs. 9–14), 36, 38
focusing of charged particles II 7, 31, 38, 40, 41, 43, 46, 47, 53, 61, 72, 73, 88
focusing of waves I 153, 167, 184ff, 241 II 84

$f^{(0)}$ F2 I 143, 158, 161, 166, 169, 190, 191
force line, magnetic, curvature I 27
force line, magnetic, length I 26
formations in trough I 17, 18
Fourier–Bessel transform I 261
Fourier component of electron perturbation II 76, 88
Fourier transform I 261
FR 1 satellite II 154
fractional hop whistler II 111, 149
free energy II 100, 101
free particles near body II 63–5
frequency dispersion I 32
frequency modulation II 173
Fresnel zone in wake of body II 83, 86
frictional forces of colliding particles I 43
fringe pattern ionograms II 184, 185
fundamental equations I 28 (ch. 3)

galactic noise I 208, 209, 212
gas flow I 1
Gaussian beam I 154, 189
GBR radio transmitter II 168, 169
Gemini/Agena satellite II 5, 15–17, 92
geomagnetic tail I 24, 25 II 96, 100, 101, 108, 137, 138, 174–6, 187ff, 195, 208, 209, 214, 220–2
geometrical optics I 216, 247
GEOS 1 satellite II 194
GEOS 2 satellite II 122
gravity waves I 128, 164
Gross gaps I 103
group delay time II 106–8, 112
group propagation time II 104
group refractive index I 219 II 152
group velocity I 60, 216 (ch. 8), 218ff, 260ff II 90, 106, 112, 121
group velocity direction I 218, 219, 220, 233ff
group velocity opposite to wave velocity I 236
growing lines I 181
growth rate I 28, 106 (ch. 6), 127–9, 131, 198, 199, 202–4 II 92
guiding of waves I 27, 51, 228, 239, 247 II 103, 109, 110, 160, 166
gyrofrequency, electron I 7, 25 II 180, 195, 214
gyrofrequency, ion I 7, 56, 71ff II 77
gyrofrequency multiples I 89 II 180
gyroresonance I 95, 138 II 96, 164
gyroresonance instability II 100, 106, 148
gyroresonance, odd half-integral II 97, 187ff
gyroresonant interaction II 168, 204
gyrosynchrotron emission II 210

Subject index

half-integral gyroresonances II 97, 187ff, 194, 212
half-integral resonances II 182
half-plane in plasma stream II 26
Hamilton's principle I 238
harmonics of carrier frequency I 184, 205, 208, 209
harmonics of electron gyrofrequency II 180, 184, 193, 195, 204
harmonics of Langmuir frequency II 97, 98, 181, 199, 200
harmonics of power frequency II 166, 167
harmonics of proton gyrofrequency II 123, 124
harmonics of upper-hybrid frequency II 97, 180, 181, 186, 215
Hawkeye 1 satellite II 114, 115, 122,129, 136, 154, 195, 210, 211, 214, 217–21, 231
heat conduction I 143, 154
heat-flux instability II 130
heating of ionosphere I 142ff II 97
heating type instability I 128, 141, 153, 167 II 100
heating type nonlinearity I 135, 136ff II 98
helicon I 101
Helios 1 satellite II 131, 200, 201, 218, 221
Helios 2 satellite II 131, 132, 198, 201, 222
helium ion density I 15
helium ions I 65, 230
helium whistler II 111, 112
HF, see high frequency
high energy electrons and protons I 19
high energy particles I 19
high frequency (HF) I 93 II 98, 178 (ch. 19), 221, 222
high pass noise II 213
hiss II 100, 116ff, 119, 128, 129, 135, 138, 169ff, 220
hole in bulge I 190, 191
hook II 161, 162
horizontal gradients in ionosphere II 151
hot ions I 126
hot plasma, cold beam I 114, 124
hybrid frequencies I 67
hybrid resonance I 68 II 89, 101
hydrodynamic approximation I 127, 128 II 30
hydromagnetic approximation I 37
hydromagnetic whistlers I 22, 24, 93, 251 II 100, 103ff
hysteresis of electron temperature I 140

image decay lines I 182
IMP 6 satellite I 16, 18 II 122, 129–31, 136, 155, 174, 186, 188, 191–3, 198, 201 204–6, 208, 212–4, 218, 221, 222

IMP 7 satellite II 188
IMP 8 satellite II 131, 175, 176, 188, 195, 198, 199, 201, 208–10, 212, 214, 220–2
impedance (wave) of free space II 120
inclination of Earth's magnetic field I 27
incoherent scattering I 175, 179
inelastic reflection of particles II 11
infinities of refractive index I 48, 49, 65
inhomogeneity caused by heating I 152, 153–8, 160ff
inhomogeneity in height I 143
inhomogeneous plasma I 142, 246
inhomogeneous structure, resonance of I 246
injection of electrons into plasma II 97, 191
Injun 3 satellite II 110, 111, 116–8, 121
Injun 5 satellite II 111, 118, 119, 121, 122, 124, 141–50, 156, 157, 219, 220
Injun 6 satellite II 220
instability I 21, 106 (ch. 6), 127
instability near body II 4, 87 (ch. 14)
intensification of emission II 116ff, 162ff
intermediate zone of body in plasma II 4, 7, 21, 31ff, 58
intermixing, turbulence I 22
internal gravity waves I 128, 164,
interplanetary medium II 187ff
interplanetary plasma I 10ff II 130
interplanetary shock II 222
interplanetary space I 5, 7 II 217, 218, 222
intersection frequency I 54
inverted-V electrons II 137, 215
ion-acoustic oscillations I 180
ion-acoustic waves I 98ff, 118–20, 124, 181, 201, 204, 207–9, 212, 236 II 30, 88ff, 96, 123ff, 198, 218, 223
ion composition I 14, 15 (Table 2.3) II 141
ion-cyclotron damping II 112
ion-cyclotron instability II 176, 215
ion-cyclotron resonance I 63
ion-cyclotron wave I 9, 102ff II 100, 104, 110ff, 220
ion-cyclotron whistlers II 110ff
ion flux at body surface II 14
ion gyrofrequency I 56, 71ff
ion gyroresonance I 48
ion–ion collisions I 42, 46
ion–ion hybrid frequency I 69–71, 231–4, 270
ion-Langmuir frequency I 8 II 124, 128
ion-Langmuir waves I 236, 237
ion mass, effective I 14
ion oscillations II 98
ion temperature I 2, 6, 10, 14
ion viscosity I 43
ion whistler I 56, 64, 93, 94, 222 II 110ff

ion whistler O^{++} or He_2^+ II 112
ionic sound I 197
ionization time II 12
ionized clouds I 2, 160, 163
ionogram I 161–3, 166–8
ionogram, fixed frequency II 184, 185
ionogram, topside II 179ff
ionosonde I 161–3
ionosonde, rocket-borne II 179
ionosonde, satellite borne II 179ff
ionosphere I 128, 184, 217, 218 II 36, 75, 77, 81, 82, 84, 117, 125, 149, 159, 168, 169
irregularities I 2
ISIS 1 satellite I 22 II 97, 122, 179
ISIS 2 satellite II 148, 179
isolated packets of waves I 135
isotropic plasma I 88, 96, 98, 107, 130, 139, 148, 184, 199, 200, 204, 207, 214 II 36, 88

Javelin 8 rocket II 123, 124
Javelin 8–45 rocket II 128

K_n and K_r functions I 149, 150
K_p, index I 17 II 135
kinetic attenuation I 94 II 113
kinetic correction to refractive index II 160
kinetic energy density II 125, 126
kinetic equation I 29 II 3, 19, 44, 87
kinetic instability I 106
kinetic refractive index II 101
kinetic theory I 2, 147ff, 236, 247
knee in outer ionosphere II 160
Korteweg–de Vries equation I 210
Kramp function I 47, 90, 91 II 34, 48, 58, 59, 78, 88
K 9M 26 rocket II 154

L value I 15, 16, 20, 24, 117
laboratory measurements, body in plasma II 24, 39, 42
laboratory plasma experiments II 191
Landau damping (Landau attenuation) I 34, 83, 86, 89, 118, 133, 196 II 123, 133
Landau excitation I 82
Langmuir electron oscillations I 9
Langmuir frequency, electrons I 13, 57, 58, 175, 199, 207, 230, 255 II 76, 178, 180, 184, 185, 191ff, 203, 204, 208, 214
Langmuir frequency, ions I 8, 13, 65 II 99, 128, 142
Langmuir ion oscillations I 98
Langmuir oscillations I 114, 118, 192, 197, 212 II 101, 191, 196, 198, 199, 201
Langmuir resonance I 50, 228
Langmuir–Tonks waves I 98,

Langmuir waves I 84–6, 96, 132, 175, 193, 204, 236 II 96, 202
large body I 7 II 46, 47, 52, 57, 66ff
Larmor radius, electrons I 2, 7, 36 II 59
Larmor radius, ions I 2, 7, 36 II 3, 6, 53
Larmor resonance I 224
lateral deviation I 143, 146
LF, see low frequency
lightning II 102, 110, 159, 167, 168
linear polarization I 51, 54, 55, 60, 66, 73
linearization II 3
lion's roar LF waves II 154ff, 220
lobes in scattering function II 78, 86
lobes in wake structure II 30, 50, 51, 55, 75, 89
longitudinal Langmuir waves I 107, 110 II 201
longitudinal VLF waves II 136
longitudinal waves I 85, 198ff II 96, 185
longitudinal waves transformed to transverse II 203
Lorentz force I 29
loss cone II 106
low frequency (LF) I 93 II 98, 141, 142, 148, 159 (ch 18), 219, 220
lower-hybrid frequency I 64, 66, 67, 69, 93, 103, 136, 194, 199, 224, 225, 241, 243 II 96, 134, 135, 138ff
lower-hybrid resonance I 9, 48, 50, 56, 61
luminosity (airglow) I 163
Luxemburg–Gorki effect I 135

Mach cone II 30, 88
magnetic activity I 15, 17
magnetic antenna II 120–2, 124–8, 141, 142, 155, 157, 169, 198, 207
magnetic conjugate point on Earth II 97, 104, 111, 159, 163–5, 169
magnetic field, effect on particles near body II 26ff
magnetic field of Earth I 24 II 99
magnetic pressure I 94, 143
magnetic storm I 19, 20 II 108–10, 114, 116, 128, 129, 142, 148, 173, 174, 189, 195, 203, 211, 212, 218, 221
magnetoacoustic transverse wave I 237
magnetoacoustic wave I 56, 62, 63, 127, 236 II 96
magnetobraking I 33, 82
magnetohydrodynamic waves I 94, 101ff
magnetopause I 5, 10, 11, 24, 213 II 106, 169, 197, 204
magnetosheath I 5, 11, 24 II 122, 155, 156, 158, 197, 212, 217, 219, 220
magnetosphere I 5, 11, 217 II 102, 108, 149, 167, 187ff, 217, 221, 222
magnetospherically reflected (MR)

Subject index

whistlers II 150, 152
magnetotail II 205, 208, 209, *see also* tail, geomagnetic
Mandel'shtam–Brillouin combination scattering I 204
Mariner 4 satellite II 108, 217
mass, complex I 37, 71, 77
mass, effective, of ion mixture I 69
mass spectrometer I 17 II 140
maximum of refractive index I 72, 78
Maxwell–Boltzmann distribution II 3
Maxwellian distribution function I 30, 31 II 92, 99
Maxwell's equations I 29, 261
mean free path I 2 II 19, 75, 91
mean free time (electron) I 130
meteors II 12
micropulsations, magnetic II 101
midfrequency noise II 213
minimum of refractive index I 72, 78
mixing type instabilities II 100
modified Alfvén wave I 53, 58, 63, 64, 93, 124, 127, 228, 237 II 104
modified ion-acoustic oscillations I 196
modifying the ionosphere I 142 II 97
modulation index I 212
momentum conservation in collisions I 38
Morse code dashes II 162, 164, 165, 168
MR whistlers, *see* magnetospherically reflected whistlers
multicomponent plasma I 37, 39, 46, 65, 73 II 5

NAA radio transmitter II 162, 167
narrow-band hiss II 122, 141, 143
narrow-band plasma waves II 198
narrow-band spectra of emissions II 191–3, 195, 200
narrow-band wave packets II 140, 141, 160
NCR, *see* nonthermal continuum radiation
near-Earth plasma I 10ff, 11 (Fig. 2.1), 134
near zone of body in plasma II 5, 7, 26, 30, 31ff, 38, 66, 68
neutral approximation near body II 4, 19ff, 30, 31, 35, 36, 38, 51
neutral particles I 39, 138 II 20, 21
neutral particles near body, experimental measurement II 23, 24
neutral sheet I 24, 25 II 174–6, 189, 195
noise bands I 177 II 186, 195, 206
non-ducted waves II 149ff
non-Maxwellian distribution I 246 II 193
nonisothermal plasma I 200, 204, 207, 237 II 23, 30–2, 38, 39, 46, 50, 88, 91
nonisothermal sound I 98, 197, 207
nonisothermal velocity of sound I 8, 101, 118 II 6

nonisothermality factor I 137
nonlinear differential equations I 143
nonlinear effects I 130 (ch. 7) II 5, 98, 101, 102, 178, 179
nonlinear instabilities I 128
nonlinear Landau damping I 132
nonlinear refractive index I 215
nonlinear waves I 204ff
nonlinearity parameter I 208, 209
nonthermal continuum radiation (TNCR) II 204ff, 208ff
NPG radio transmitter II 162
NSS radio transmitter II 168
nu whistlers (v whistlers) II 150, 153
numerical solution for particle motions II 31

oblique scattering I 180ff
OGO 1 satellite II 89, 152–5, 169, 170, 172, 207, 208, 212, 219, 222
OGO 2 satellite II 120, 124, 140, 141, 218, 223
OGO 3 satellite I 1,20 II 135, 154, 161, 169, 219
OGO 4 satellite I 22 II 120, 140, 141
OGO 5 satellite I 16 II 109, 110, 116, 122, 129, 130, 133, 134, 141, 142, 145, 154, 156, 157, 171, 173, 187–90, 197, 202, 203, 219, 221, 222
OGO 6 satellite II 139, 140, 147–9, 168
Omega (radio navigation system) II 162, 167, 168
opacity, regions of I 70
opaqueness frequency bands I 235
opposite directions of wave and group velocity I 236
orbits of particles II 63
ordered velocities I 40
ordinary fast wave I 56, 230, 231
ordinary ion wave I 62
ordinary wave I 48, 53, 73, 143, 161, 167, 168, 175, 178, II 110, 115
outer ionosphere I 67, 70, 73, 130, 142, 155, 156, 161, 239 II 107, 112, 117, 123 (ch. 17), 160, 179ff, 217–20
OVO 3 satellite II 141
OV3 3 satellite II 126, 127
oxygen ions, atomic I 65, 230 II 36

P 11 satellite II 124–6, 218
packets of waves II 90, 142, 179
parabolic beam I 190
parametric decay instability I 182, 192 II 98, 179
parametric excitation I 198
parametric instability I 128, 160, 182, 193, 198 II 185

Subject index

parametric resonance I 193, 199
parametric type nonlinearity I 135, 192ff
partial reflection I 163
particle distribution near body II 62ff, 73, 92
particle flux I 106, II 137, 138
PC magnetic pulsations II 109, 217, 223
PC-1 II 109, 110, 217
PC-3, PC-4, PC-5 II 109ff
pearl-type micropulsations II 103, 206
pencil beam I 169
pendulum I 193
periodic structure near body in plasma II 27, 28
permittivity of plasma I 134, 189, 205 II 75
permittivity tensor I 28, 35, 37ff
perturbation method I 253 II 76
perturbation of electron density I 157–9 II 75, 88
PH, see plasmaspheric hiss
phase-modulation, phase modulation index I 187
phase perturbation I 185
phase velocity I 28
photoemission II 12, 92
Pioneer 5 satellite II 108, 217
Pioneer 8 satellite II 130, 131, 201, 202, 218, 222
Pioneer 9 satellite II 130, 218
pitch-angle diffusion II 167
pitch-angle distribution, electrons II 172
pitch-angle scattering II 100, 168
plane inhomogeneity I 252
plane-stratified plasma I 247, 251ff
plasma diagnostics II 62
plasma dispersion function, see Kramp function
plasma flow I 29, 31
plasma flow velocity II 16
plasma frequency I 8, 18, 50
plasma instability I 21
plasma instability near body II 4, 87 (ch. 14)
plasma lines I 175–9, 181, 183
plasma oscillations I 2, 8, 21, 33, 60, 97, 106, 182 II 10, 99ff
plasma parameters, near-Earth and interplanetary I 12, 13 II 6
plasma temperature I 134, 135
plasmapause I 5, 10, 11, 14–17, 22, 24, 239, 241 II 100, 106, 114, 128, 129, 133–5, 148, 160, 162, 163, 166, 169, 176, 186, 187, 204, 208, 212, 218
plasmasphere I 5, 11, 15ff, 22, 24, 243 II 133, 135, 136, 153, 154, 162, 163, 166, 167, 191, 193, 217, 219

plasmaspheric hiss (PH) II 133ff, 219
plasmatrough I 18 II 169, 172, 187ff, 196, 204ff, 208ff, 214, 221, 222
plate in plasma stream II 26, 27
point body I 7
point source I 260ff
Poisson equation I 29, 31, 208, 214 II 3, 31, 44, 88
polar cap I 22 II 101, 109, 122, 195, 217, 220
polar cusp II 121, 122, 142, 195, 203, 204, 214, 219, 221
polar HF radiation II 100
polar hiss II 137, 139, 140, 148, 176
polar ionosphere II 136, 139, 147, 156
polar magnetosphere II 138
polar zone observations of hiss II 121, 122, 124, 156–8
polar zone (polar region) I 22 II 137, 140, 143, 148, 213–15, 217, 219, 220, 223
polarization coefficient I 53, 73
polarization of wave I 53ff II 135, 214
polarization, random II 135
polarization reversal I 54, 55, 60 II 117
postmidnight chorus II 173, 174
potential barrier I 213
potential distribution near body II 44–6, 63, 68, 73
potential of body in plasma II 6, 10, 11, 13ff, 62
potential of body in plasma, experimental measurement II 15, 16
potential well I 133
power grid harmonics II 166
power spectrum of hydromagnetic whistler II 105, 106
Poynting vector I 192, 216, 247ff, 252ff II 119
Poynting vector, time average I 218
precipitation of electrons II 122, 168, 172, 215
pressure and nonlinearities I 135, 205
pressure, gas kinetic I 94, 143
pressure, magnetic I 94, 143
pressure on electrons I 213ff
probe measurements of plasma density II 203
proton bursts II 122
proton density I 15, 16, 18
proton flux I 20, 21 II 176
proton gyrofrequency II 108, 110, 117, 123, 124, 155
proton gyrofrequency harmonics II 123, 124
proton gyroresonance II 110, 123
proton plasma I 268 II 36
proton ring current II 129

Subject index

proton whistler II 110–2
protonosphere II 149–51
protons I 65, 230
protons, suprathermal streams II 130
pulsations of magnetic field II 108ff, 217, 223
pulse transmission II 178–80
pulse type emission II 132, 144

quasi-electrostatic waves II 137
quasi-equilibrium plasma I 207
quasi-isotropic propagation I 228
quasi-longitudinal propagation I 51, 222, 224, 248ff, 253
quasi-monochromatic packet I 216
quasi-monochromatic waves II 102, 162
quasi-neutral approximation II 45, 76
quasi-neutral plasma II 35
quasi-periodic chorus II 174
quasi-periodic ELF waves II 109
quasi-periodic structure of wake II 91
quasi-periodic transverse wave II 110
quasi-perpendicular propagation I 51
quasi-plane waves I 216
quasi-static electric fields II 124
quasi-stationary body II 62 (ch. 12)
quasi-transverse propagation I 46, 51, 230, 255, 256

radiation belts I 19 II 100, 135, 167, 210
radio signals as trigger II 140
radio wave scattering II 75 (ch. 13)
radiotelegraph signals II 162, 168
rarefaction in wake II 7, 8, 29, 30, 31, 36, 39, 72, 73, 88
rarefaction regions near body II 31
ray paths II 135
ray tracing I 238
recombination coefficient I 154
reflection of ELF waves II 118, 119, 121
reflection of particles I 10, 11 II 68, 69
reflection of whistlers II 120, 150
reflectrix I 192
refractive index I 28, 35 (ch. 4), 70ff, 82ff, 147 II 115, 216
refractive indexes, equal I 51, 54, 55
relaxation time for scattered intensity I 178
relaxation time for temperature I 137, 154
remote resonance II 183
resonance I 32, 87
resonance absorption I 169
resonance attenuation coefficient I 94, 95
resonance branches I 194ff
resonance cone I 262–4, 267, 269 II 146
resonance excitation I 178 II 90
resonance frequencies I 68, 71, 103

resonance oscillations of plasma II 96
resonances (spikes) in topside sounders II 179ff
resonant electrons I 133
resonant interaction of waves and wake II 90
reversal of sense of polarization I 66, 81
reversal points of whistler ray I 243
reversed Storey cone I 269, 270
ring current in magnetosphere II 114, 128, 129, 148, 155
ring electron distribution II 189
rising tone (riser) II 161, 162, 165, 166, 168, 173
Roberval station, Quebec II 163, 166
rockets II 97, 123
rough surface of body II 81
run-away electrons I 141

S 3–3 satellite II 124–6, 217
satellite I 134 II 83, 89, 123
saturation of whistler enhancement II 165
saucer-shaped emissions (SSE) II 142ff
scattering cross-section I 172–4, 178 II 75–9, 84
scattering function II 77–9
scattering of electrons II 91
scattering of particles II 11
scattering of radio waves II 10, 75 (ch. 13)
second-harmonic gyroresonance I 85, 88, 180
self-action I 151, 167, 184ff
self-action factor I 185
self-focusing instability I 167
self-focusing of beam I 184, 188, 190
self-modulation I 151, 184
shadow in particles near body II 73
shadow zone for waves II 84, 85
sheath near body surface II 66–8
shock, collisionless I 206, 213 II 198
shock front II 101, 142, *see also* Earth's bow shock
shock wave I 206, 213 II 30, 50
shot effect II 208
side scattering I 175
Siple station, Antarctica II 163–7
sky maps of scattering I 164–6
slow electron-acoustic wave II 159
slow ELF waves II 127, 128
slow magnetoacoustic wave I 102
slowly moving body II 69ff
small body I 7 II 41, 47, 55–60, 62ff
smoothly-varying plasma I 238
solar-flare II 201, 203
solar wind I 5, 7, 9–11, 14, 213 II 96, 99, 130–2, 187ff, 205, 218, 221, 222
solitions I 135, 206, 209–12

sound velocity I 8, 210
sound waves I 197
source size (AKR) II 215
space probes I 134
spatial dispersion I 32, 61, 63, 82ff, 103, 222, 236
spatial inertia I 32
spatial inhomogeneity of electrons I 131, 132
spatial nonuniformity I 107, 205
spectra of ELF waves II 114, 115
spectra of scattered waves I 175ff, 180, 181
spectrogram (proton flux) I 21
spectrogram (swept frequency radar) I 170, 171
spectrogram of waves II 91, 104, 105, 111, 112, 116–21, 123, 124, 128, 132, 138–56, 159, 161, 162, 165–77, 188–90, 195, 196, 206, 208, 209
spectrograms in colour II 119, 144–6, 150, 151
spectrum analyser, high time resolution II 173
specular reflection of particles II 11, 20
specular reflection of waves II 78
specular scattering I 175, 182
spherical body II 20ff, 25, 28, 36, 39–56, 68–72, 77, 81–4
spherical coordinates I 238, 266
spherical waves II 83, 84, 86
spikes II 180ff
sporadic layer I 161–4 II 97
square plates in plasma II 45, 46
SSE, see saucer-shaped emissions
stabilization time for temperature I 137, 160
standing waves I 184
stationary body, large II 66ff
stationary body, small II 62ff
steady-state process I 207
steepening of wave front I 206, 211
stimulated emission (ASE) II 161ff
Storey cone I 262–4, 269
streams of electrons II 197, 210
streams of particles II 6
striction type nonlinearities I 136, 215
strong pump waves I 128, 135, 192
strongly ionized trapping layer I 255
structure of Earth's magnetic field I 24
structure of plasma near Earth I 11
subprotonic whistlers II 150
superposition, principle of I 134
supersonic plasma flow II 6, 19
supersonic speed I 209
suprathermal electron beams II 147
suprathermal proton stream II 130
surface materials II 13

tail, geomagnetic I 24, 25 II 96, 100, 101, 108, 137, 138, 174–6, 187ff, 195, 208, 209, 214, 220–2
tail of velocity distribution I 86 II 100
temperature distribution near body II 16, 17, 92
temperature instability II 106
temperature of plasma I 134, 135
temporal decay rate I 196
temporal inertia I 32
temporal variation of electron temperature I 146ff
tensor elements (permittivity) I 81, 83ff, 214, 215, 221, 253, 261
terrestrial kilometric radiation (TKR) II 208ff
thermal energy in plasma regions II 216ff
thermal motion I 83ff, 236
thermal velocity of electrons, ions I 6, 39 II 160
third branch of refractive index I 84
three (ion) component plasma I 221ff, 230ff, 241, 269
threshold field I 200–3
TKR, see terrestrial kilometric radiation
TNCR trapped nonthermal continuum radiation see nonthermal continuum radiation, continuum radiation
topside sounding II 179ff
total internal reflection I 253–5, 258–60
trains of signals II 103, 104, 106
trajectories I 216 (ch. 8), 238
transition frequencies I 268, 269
transport of particles I 152
transverse electromagnetic waves I 109 II 203ff
transverse propagation II 151, 152
transverse wave I 193, 202ff II 96, 120, 200
transverse whistlers II 150, 174–6
trapped particles II 63
trapped whistlers I 243
trapping boundary II 144
trapping cone I 223, 224, 234
trapping of particles by waves II 102
trapping of waves I 62, 251ff II 96, 149ff, 203ff, 212
trapping of waves by magnetic field I 243ff II 157, 160
triggered emission II 140, 162ff
trough region of magnetosphere, plasmatrough I 17, 18 II 204ff
turbulence I 22
turning of Poynting vector I 247ff
two component plasma, electrons and one ion I 56ff
two (ion) component plasma I 76ff
two-stream instability II 101, 130, 198, 199

ULF (ultra low frequency) II 99
ultrashort radio waves I 163
unguided rays, waves I 241ff II 154
units I 3 II 216, 218
universal function for ion density perturbation II 49–53, 55, 57, 58
unstable plasma I 133 II 99
upper hybrid frequency I 179, 194, 198, 228 II 96, 180, 181, 193, 203, 204, 209, 215
upper hybrid resonance I 49, 50, 56, 61, 103, 180, 236 II 180, 181, 186

V, inverted-V electrons II 137, 215
V-shaped emissions II 142ff
Vanguard 3 satellite II 159
velocity (thermal) of electrons, ions I 6
very low frequency (VLF) I 48, 52, 62, 65, 70, 93, 118 II 87, 98, 123 (ch. 17)
very low frequency resonance oscillations I 62
very small irregularities I 169, 179
virtual height I 161
virtual range II 180
viscosity I 43
VLF, *see* very low frequency
Voyager 1 and 2 satellites II 201, 218, 222

wake of body in plasma II 7–9, 16, 17, 19, 20, 26, 27, 30, 32, 33, 38, 42, 43, 46, 47ff, 75 (ch. 13), 87 (ch. 14)

walking-trace whistlers II 151
warm plasma I 82 (ch. 5)
wave guidance I 27, 51
wave normal, measurement of direction II 172
wave packet II 90, 156, 164, 174
wave–particle interaction I 33 II 102, 135, 164, 178, 179
wave shape I 209
wave surface I 221
wave–wave interaction II 102, 178, 189, 200
weakly ionized trapping layer I 252
whistler I 56, 63, 92, 93, 239ff II 150
whistler mode I 63, 66, 93, 224, 225, 239, 247, 263, 266 II 97, 136, 144, 157, 159 (ch. 19), 219, 220
whistler precursors II 151
whistler ray trajectories I 239–46 II 135, 150, 151, 153
whistling atmospheric I 22, 24 II 90, 91, 103, 111, 119, 138–40, 150, 159ff

X-ray bursts II 168

Z mode I 268, 269
zeros of refractive index I 50, 70
zones around body in plasma II 4ff, 21ff, 31ff, 47ff, 57
zones of altitude I 6, 12, 13